Low-Cost, Energy-Efficient Shelter

for the Owner and Builder

Low-Cost, Energy-Efficient Shelter

for the Owner and Builder

Edited by
Eugene Eccli

Illustrations by
Erick Ingraham

Rodale Press, Inc., Emmaus, Pennsylvania 18049

Book Design by Weidner Associates, Inc.

Library of Congress Cataloging in Publication Data

Main entry under title:

Low-cost, energy-efficient shelter for the owner and
    builder.

    Bibliography: p. 374
    Includes index.
    1. House construction—Amateurs' manuals.
I. Eccli, Eugene.
TH4815.L68   690.8'1   75-38886
ISBN 0-87857-116-7   hardcover
ISBN 0-87857-114-0   paperback

12 14 16 18 20 19 17 15 13 11 hardcover
12 14 16 18 20 19 17 15 13   paperback

CONTENTS

CREDITS

Low-Cost, Energy-Efficient Shelter represents the work of many people. The book itself as an idea and a text was conceived, organized, and edited by Eugene Eccli. Several people (other than the authors) were instrumental in helping edit the text and often contributed ideas. Our special thanks in this regard goes to Sandy Eccli and Stan Ockers, as well as to Carol Stoner and Steve Smyser of Rodale Press.

In addition to this help from family, friends, and publisher, all the authors have read the manuscript and contributed suggestions, criticisms, and ideas throughout the book's development. We all felt that the 90 years of combined architectural, engineering, and scientific experience of this group would ensure that the most realistic and practical solutions to problems were conveyed. Listed below are the names of authors who contributed information to chapters other than their own. Credits are also given for photographs taken by someone other than the author of the chapter, and for information taken from other texts:

Ch. 3: Ron Alward, Alex Wade. Special thanks to the Foundation for Cooperative Housing for use of Table 3-3.

Ch. 5: Chris Ahrens, Eugene Eccli

Ch. 8: Special thanks to the U.S. Department of Agriculture for use of figures 1, 2, and 3 from their "Low-Cost Homes for Rural America: Construction Manual," USDA Handbook #364.

Ch. 9: Ron Alward, Eugene Eccli

Ch. 10: Eugene Eccli, David L. Hartman, William K. Langdon. Special thanks to *R and D* magazine in Barington, Illinois, for use of the photograph that appears in this chapter.

Ch. 11: David L. Hartman, Jerome Kerner

Ch. 12: David L. Hartman, Jerome Kerner, Alex Wade

Ch. 13: Eugene Eccli, Alex Wade. Special thanks to Fred S. Dubin

Ch. 15: Steve Ridenour, Fred S. Dubin

Ch. 16: David L. Hartman, William K. Langdon, Steve Ridenour, William Shurcliff

Ch. 18: Eugene Eccli, David L. Hartman. Special thanks to Professor Jack Duffie of the University of Wisconsin at Madison for use of the four photos that appear in this chapter.

Ch. 19: Special thanks to the U.S. Department of Agriculture for plans (other than the site plan) and specifications on house FS-FPL-4. The graphics and descriptions of houses FS-FPL-2, FS-SE-1, FS-SE-3, and FS-SE-5 were taken from USDA literature. Special thanks, also, to Nathanial Lieberman for his photos of the model house in this chapter.

Glossary: Many of the definitions of terms were taken from the U.S. Department of Agriculture's Handbook #364—"Low-Cost Homes for Rural America: Construction Manual."

INTRODUCTION

Almost all the costs for housing are higher today than they've ever been before. And we all know the reasons for today's high costs: inflation. There are many opinions about why inflation exists; but regardless of what is behind it, it presents us with several problems. Prices are already high. They will probably not go down and may even go up, and we have to pay our bills *now*.

For housing, these price increases have been really dramatic. The costs for land, for material, for the labor to build the house, and for the energy we use to maintain it have all doubled in only a few years. Families who want to build a new home are the worst off, as they feel the effects of all these problems at once. Those who already own their homes are somewhat better off; but here, because of the energy situation, many are hurting from the increases in their utility bills.

Learning how to repair, renovate, remodel, and build to meet your housing needs as efficiently as possible can be a really satisfying—and money-saving— experience. You'll discover that there are many opportunities to have housing that is efficient, beautiful, and yet relatively inexpensive. If you do much of the work yourself, you can design your own home to meet your exact needs, and you can save all or part of the money you'd normally spend for labor and later, maintenance as well.

Low-Cost, Energy-Efficient Shelter brings together the information and design principles you need in order to build sensibly and at a price you can afford. The information in the pages that follow can help you do anything from simple winterizing projects that lower your heating bills to a full-scale design and construction program where you completely build or renovate a house.

Several approaches have been emphasized throughout the book. First, we know that the more interest you take in helping yourself, the lower your costs will be. To help you get involved, in the first section of the book we've included information about the kinds of housing that are available and that suit different family needs, as well as an explanation of how you can undertake the financial and other administrative work that goes along with building a home. Whether you buy your home, design a house and then pay a contractor to build it, or do all the work yourself, the more you know about the housing options that are available, the easier and cheaper the whole process will be. For example, one creative way of meeting your housing needs might be a cooperative venture with others. By pooling your skills and resources, opportunities open up for buying or building efficient cluster housing that you might otherwise not be able to afford by yourself. Working with others can be fun, too. This idea and others can expand your horizons so that you'll have a *choice* of how you want to live, instead of being forced to accept whatever housing is immediately available at a price you may or may not be able to afford.

Section II of the book deals with what we feel are the biggest problems of housing today: namely, the high first costs of land and construction. To solve these problems requires that you understand how your home can be designed and built efficiently, including the use of moderately priced land. Enormous savings can be made once you learn how to lay out space and use materials effectively, and these designs are really elegant as well. You'll be able to choose how much space you want for each activity in your home, and then build with reasonably priced materials that enhance its appearance. With these techniques, you can design a home which is comfortable and flexible enough so that you can add more space when you want to, yet is just the size you need for right now.

Once you understand how to build efficiently, with practical beauty, Section III of the book helps you to make sure that the operating expenses and maintenance-upkeep for your house will be as small as possible. There are several areas of major concern here. The first is the high cost of home heating and cooling. Learning how to insulate, winterize, and ventilate your home properly can mean really substantial savings each year. The other large operating expense is the cost of running appliances and equipment, like refrigerators and lights. You can save money here, too, if you learn how to purchase the best equipment and operate it efficiently. *Low-Cost, Energy-Efficient Shelter* explains how you can make these savings. This section also shows you how to satisfy your esthetic needs by taking advantage of a view to the outside and the natural light provided by your windows.

Section IV deals with how to design for your comfort needs by making use of the warmth of the winter sun and by creating a dynamic, healthy, and sensitive relationship with the natural environment around your home. It's even possible to bring the beauty of nature indoors with a greenhouse that can provide warmth and humidity in winter and can help with the food bills during most of the year. Furthermore, the sun can furnish the heat for most of your hot water needs. *Low-Cost, Energy-Efficient Shelter* explains how you can integrate and harmonize the sun into the design of your home.

Each chapter of the book goes into a certain aspect of housing design. As you read, take a pencil and paper and begin to work out plans for *your* home. Doing these designs as you go along will help you learn how to use this information, and you'll be surprised at how quickly you can pick up these techniques. When you come to the end of the book, try to work up layouts and designs for your house in the manner explained in the last chapter. In this way you can compare your solutions and ideas with the designs we've put together, and you should be ready to take up the challenge of meeting your housing needs efficiently, elegantly, and at low cost.

Have fun!

—E.E.

Section I

FINDING A HOUSING SOLUTION

chapter 1

THE OWNER-BUILDER
IN RETROSPECT

by JEROME KERNER

Since the dawn of humanity on this planet, shelter has been one of our primary concerns. We are one of the few mammals that have difficulty withstanding body temperature variations of more than 10°F. Thus, within our shelter systems we must create an environment that will protect that delicate balance.

The unsettled climate of our planet during the last ice age was not conducive to supporting life in all areas. Where the ice and glaciers receded, and in the warmer areas of the globe, we find the beginnings of human development. It was here that our ancestors found shelter in limestone caves, which gave them both warmth and protection. Because there were also animals in the limited region where the glaciers had melted, the need to travel long distances, foraging or hunting, was minimized; and our ancestors could spend their time developing skills, tools, and a culture that would prepare them for the changes ahead. Then, as the ice continually receded and animals and food sources became increasingly dispersed, people had to move on and begin the task of making shelter where there was none before.

It was not long afterward that the human race became completely mobile and could inhabit the diverse climate zones from the tropics to the arctic regions. Each of these climates helped to shape the lifestyle, economy, and physical environment of these early settlers. Some of the shelter solutions these people found let them be comfortable even in extremely cold conditions. The intuitive use and understanding of both the climate and the available materials by these folk were amazing! There was also much esthetic variation in the homes our primitive ancestors built, with differences indicating cultural influences such as religious symbolism, traditions, and status.

3

The dwellings produced by early societies may at first glance and by our technological standards appear elementary. An Eskimo igloo, for example, looks simple until you examine it in detail. On closer analysis, it can be seen that it is the product of resourcefulness, intelligence, and abilities no different from our own. The biggest difference between primitive societies and ours is that, although there is specialization in primitive societies (such as medicine man, hunter, and craftsman), each person still understands the skills necessary for survival, including how to build a house.

Of the primitive peoples around the globe, each group had a shelter model which had been tested over long periods of time and which met the cultural, physical, and maintenance requirements of that group. Generally speaking, within the group, individuality was not a quality of this type of housing, as the model which had been well proven was pretty closely followed by all. Differences were created by decorative or symbolic embellishment, giving identity to the house and at the same time allowing the owner an important outlet for creative ideas.

As certain societies developed professions, such as merchant, miner, toolmaker, herder, etc., there also emerged a building tradesman, but the owner was still involved to a great extent in the design process, as the basic primitive model of the "ideal dwelling" continued to exist. This sharing of responsibility between the owner and the builder ensured that these "ideal dwellings" were perpetuated. So the traditional building forms became laws or moral standards which were honored by collective assent.

This development of shelter-community-society based on tradition lasted many millennia and created solutions that were both innovative and considerate of the environment and the overall ecological system (which included the impact on one's neighbor). On the other hand, with the industrial age came the breakdown of strong family and community settlements and the development of a new urban society whose housing was built by speculative contractors who no longer followed the established moral laws of treating man and his environment considerately.

The unique quality of buildings within the preindustrial context was the character created by the combination of local conditions such as climate, available materials, and cultural norms. With few exceptions, these early builders developed techniques and designs which were compatible with the local environment. Their houses would respond to the daily and seasonal changes in the weather in a way that gave comfort to the occupant, yet used little energy to maintain a steady state between the man-made environment in the house and the natural environment outside.

Let's examine two shelter solutions in North America that used natural climate controls and natural materials found at the site. The Pueblo Indians found a hot climate and limited building materials in their southwestern environment. Here the challenge was to insulate against the harsh daytime sun. Massive adobe or clay brick walls were used to absorb the day's heat. They were thick enough that most of this heat would never get inside the house, but they would allow the house to cool off again at night. The site planning of the Pueblos also showed a keen understanding of the effects of the house's orientation to the sun. See Figure 1-1. Houses were grouped in units in order to minimize exterior wall surface area. All houses opened on a courtyard to the south, with small windows which permitted the low winter sun to enter but which excluded the hot summer sun.

When the European colonists came to the New World, they found examples of Indian dwellings in eastern North America that could have sufficed as their own homes, but the cultural and religious differences they had with the Indians made it essential for them to develop their own shelter or adapt a familiar European type to this climate.

The English colonists brought with them techniques of construction that had been used by the northern Europeans for centuries: heavy wood framing with fill such as bricks placed between the frames. It didn't take long for them to become aware of the differences in climate between England and North America! Here the winters are more severe, so an important modification took place: clapboard siding was developed in order to cover the bricks and create a tight exterior wall. Other features of the house worth noting were the stone foundations, the compact house design (which was usually added to as the family grew), and the small windows and low ceilings necessitated by difficulties in heating and the high cost of glass importation. Roofs were sloped in a "lean-to" design, with the long slope usually to the north. See Figure 1-2.

As can be seen from the examples just given, shelter solutions evolved by earlier people allowed them to adapt to very harsh climates where resources for building and energy were scarce. It was the strong desire to survive which initially called forth innovative solutions, but it was the passing down of a building tradition within each group that allowed for the perfecting of design, leading to comfortable and beautiful houses. Traditions that embodied laws of a sociocultural, environmental, and practical nature usually led to a very basic solution, where novelty or "change for the sake of change" was unknown.

The successful shelter solutions of the Pueblo Indians and the English colonists can also be used to point out the "efficient and low cost" concept in housing. What can we learn from them that is applicable today? First, we note that each climate makes specific demands on shelter. Second, by observing the environment in that climate it will be apparent that certain key elements such as available sunlight, wind,

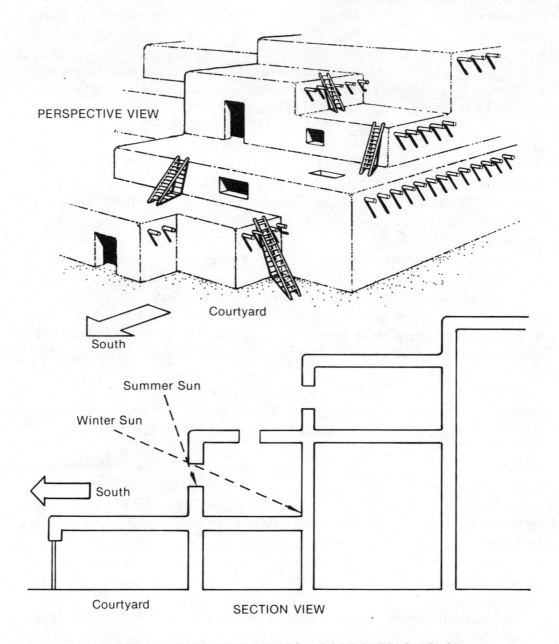

PERSPECTIVE VIEW

Courtyard

South

Summer Sun

Winter Sun

South

Courtyard   SECTION VIEW

FIGURE 1-1: TAOS, NEW MEXICO, ADOBE PUEBLO HOUSE

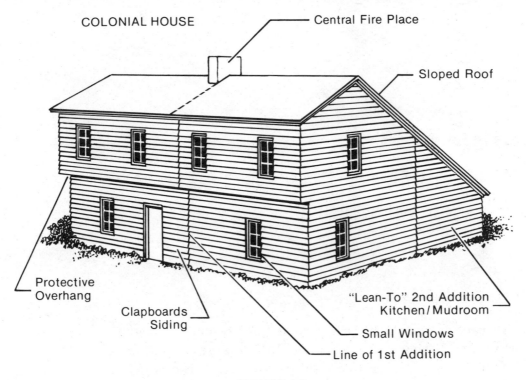

COLONIAL HOUSE — Central Fire Place — Sloped Roof — Protective Overhang — Clapboards Siding — Line of 1st Addition — Small Windows — "Lean-To" 2nd Addition Kitchen/Mudroom

FIGURE 1-2

plants and tree-forms, and building materials on the site will suggest opportunities, and that by using these to respond to the specific environmental demands we can create comfort *efficiently and at low cost.* Last, there is the aspect of preindustrial building where the efficiencies can be traced to the owner-builder concept and the social patterns both within the family unit and in the broader context of the extended family. More specifically, the owner-builder let his house evolve and change as materials became available and as the family grew or diminished in size. There was a sharing of skills among owner-builders in a community and in many instances a sharing of homes too, where both individual and common living spaces were built into the structure, or a row of houses were butted together to form a village street. Aside from the obvious savings with this approach (i.e., common building walls), reduced exposure to hot or cold exteriors, and central waste handling and energy production, there was also the development of the cooperative spirit so lacking in our own culture.

In the following chapters we will explore the solutions open to those who wish to take steps toward evolving their own shelter, and we will examine the practical details that must be considered to achieve efficient, low-cost, beautiful housing.

chapter 2

HOUSING OPTIONS TODAY

by JEROME KERNER

BUILD SMALL AND SAVE

All persons who have thought about the possibility of buying, building, or in some way modifying their housing should be aware by now of some very basic facts in our economy. First, the cost of new construction has risen, some reports showing the average 1976 cost at about $45,000. Mortgage rates have also increased from 5.5 percent 10 years ago to 9 percent at present. From an economic standpoint, therefore, it is imperative to become fully involved in your housing decisions, because you can both cut your initial costs and become involved in a lasting investment. This is especially important in the design stage.

As the square footage of a house very much affects its cost, both from a first cost and from a maintenance cost standpoint, it seems sensible to try to satisfy all your needs within a minimal area. Today, there is a general trend toward smaller homes because of fewer children and the rarity of more than two generations living under the same roof. Ideally, our houses should expand and contract according to our needs. Young couples and older persons whose children have left home require less space than families with growing children. Where children are not noisemakers, spaces can also be left open so that one living area flows into another, and this can help save space too. See Chapter 6 for details.

In the rest of this chapter we will explore various types of housing solutions with degrees of owner involvement. Each could be adapted to a particular stage of a family's space needs, because an important design consideration is expandability. The ability to add rooms as a family grows, so that the addition follows some logical scheme, is an asset to any house plan. The house location on the site should be considered from this aspect as well. Building codes should be consulted to verify the minimum set-back required, and the initial house must be constructed far enough back from this line to allow for future additions. There are other considerations. For example, the original building should be oriented so that any future addition will not shade windows requiring sun, and consideration should be given to the possibility of completing the initial shell and leaving future loft, attic, and basement space unfinished. You can then expand by completing these at a later date.

This last option may not be as easily financed as the completed house, since most lending institutions are very conservative and would rather lend on a house that is finished. Don't forget that their concern lies in being able to resell the house if you cannot meet the payments. Moreover, the smaller initial house option may be discouraged in some areas because of regressive building or zoning codes. In these areas attempts should be made to organize community awareness in order to wage a campaign for repeal of this kind of law.

FOUR BASIC APPROACHES TO MEETING YOUR HOUSING NEEDS

There are several approaches to housing solutions; in their basic appeal to each individual, they are very much like choosing clothes. One type of person will always buy "off-the-rack" clothes because they are less expensive and more convenient. Another type of person will have clothes made to specification. His or her taste is well defined and he or she appreciates quality workmanship. Though he or she may pay more initially, there could be savings in the long run because of the durability of good materials. Still another person will *make* his or her own clothes. These people see the creative experience as being just as important as the end product. Though a lot of patience is required in this last approach, it pays big dividends.

I think these personality types would approach choosing a home in similar frames of mind. That is, some would choose a prebuilt tract house or modular/mobile home. Others would get involved in the designing of their house, then let a contractor build it; still others would both design the home and build it themselves, using recycled materials where possible, sometimes including the renovation of an older house to meet their needs. All these ways are good solutions, *if you know what the options are and then make up your own mind regarding what's best for you.* So let's look at these solutions and get a feeling for what's involved in each approach. You can then evaluate the critical commitment you'll be making with each method and choose the solution that offers the most benefits for you and your family.

The Prebuilt House

This kind of house comes in two varieties today: (a) the tract house built by speculative builders and (b) the industrialized designs delivered in prefinished modules. Neither of these construction types offers many options to the buyer. The layout of the rooms, interior and exterior finishes, and site development are usually fixed. The prebuilt house is often referred to as a "turnkey" project, because the job is completed by the contractor or seller and all that is required of the owner-purchaser is to "turn the key" and move in. Although this route offers the fewest options and possibly quite a few compromises with your personal preferences, initial costs may be less. On the other hand, if there is a lower price, there is a reason. Built into this mass housing system are shortcuts that save money for the builder but may "up" your operating expenses substantially. Only recently have there been campaigns by a few innovative mass builders to offer options like more insulation in the roof and insulating glass in the windows. These are good, but they're only a start. Far more needs to be done.

The problems with prebuilt housing are inherent in the decision-making phases. The site is not analyzed from the point of view of solar orientation, climate adaptation, or privacy, but rather from the standpoint of meeting the minimum code requirements and increasing the density as much as possible to save on land costs. Roads are laid out to maximize this density, and houses are oriented to face the roads. Thus, the layout of windows and outdoor living areas, together with an orientation that makes the best use of the sun, is completely ignored. If you have a modular or mobile home delivered to your lot and placed on your own foundation, you may be able to get around some of these problems, but not always.

If a tract house is your solution, try to look for a builder-developer in your area who *has* considered the climate and the options the particular site offers for using the sun. If one is not available, then look for the specific house in a development which offers a favorable solar orientation and the least wind exposure in winter. You can also look for a tract house that has the most compact layout of space and the greatest potential for expansion later. These could offer both flexibility and immediate economic advantages to a young family, as both the first costs and operating costs would be low. Finally, you should consider the quality of construction. The yardstick is initial costs versus long-term costs. If the builders in your area believe there is no market for smaller and more efficient houses, then perhaps you should start a campaign to educate them. Banks, building materials suppliers, and friends who want the same thing can all be potential influences on area builders. By speaking out, perhaps you can help them make design decisions that benefit all.

The Owner-involved but Contractor-built Home

This second category is made up of those who choose a predesigned plan intended for a specific building site. For a few dollars you can buy a book of house plans at your local building materials supplier. Or, better still, procure one of the outstanding house plans discussed in Chapter 19.

Having a specific site is of paramount importance in this method, because although you may want to use a "proven" house plan that can be visualized at the outset, you still need to adapt the plan to your site. You can, of course, choose a site to fit one of the predetermined plans rather than vice versa; but, either way, it is *your* decision that is required here. A simple checklist of things to look for in a potential site might include: the available sun; possible windbreaks to protect against winter winds; soil type for sanitary waste disposal and drainage; availability of water and how much you desire; a consideration of what the local zoning regulations will permit (and you *can* approach the building inspector early and find out how things stand); scenic views and privacy available; and, finally, room for expansion and the adequacy of space for outdoor living, gardening, animals, etc. More information on site criteria will be found in Chapter 5.

Once you have selected a site for your needs, you then have several choices. You can choose the kind of predesigned plan that includes specifications as to materials, and then have contractors make bids on it. Or, you can use the plan and hire a contractor to build just the shell, while you finish the interior. This means that usually the structural shell, heating, plumbing, and electric wiring are completed by the contractor, while the interior walls, ceilings, and floor finishes are left as work for you to do. For more information on the working arrangements you make with the contractor and on using sheathing materials, see Chapters 4 and 8.

It's also possible to work with companies that supply not only a predesigned plan but precut materials as well. This option of precut and predesign services exists on both a local and national level. On a local level, a firm will offer a number of plan options as well as construction choices (i.e., complete "turnkey," the shell only, or the shell and mechanical systems). In the case of a national firm, there may be local franchises, or you may have to hire your own labor to erect the precut materials package.

Some possible advantages in the precut package system are: a complete supply of materials, which will allow you or the contractor to move much more quickly; usually superior materials, such as seasoned wood or factory fabricated trusses; and, in some cases, such as cedar homes, you may be able to save by buying directly from the dealer on the West Coast. However, if this is your choice, be sure to include any long-distance freight charges. With the prices of fuel skyrocketing, what appear to be savings may not be after transport costs are added in.

The Owner-designed and Owner-built Home

This third category is the most demanding, but also the most rewarding! Now, with all the housing solutions mentioned thus far, it's up to you to figure out your needs and decide on how you want to split up your investment. You may save initially but incur larger operating expenses later, or you could spend a little more initially but save on operating costs over the life of the house.

In addition to doing these things, the owner-builder also takes the time to draw up the plans, and he either constructs the house personally or contracts to have the design built. It's extra work, but the savings can be worth it. If your house budget is between $20,000 and $30,000 and you do the drawings yourself, you can save up to $2,500; and if you act as general contractor, an additional $4,500 is possible. Even more money can be saved—running into many thousands of dollars—if you do the labor yourself. This is not to say that the same money paid to a professional architect or builder isn't earned; but if your budget is limited, these fees saved could mean the difference between building or not. In some areas, codes will require an architect's seal on the plans. What is important here is meeting the structural, plumbing, and possibly electrical codes. If this is the case, it might be possible to have an architect make these specific drawings and do a general review of your own drawings. The fee in this case would be around $500.

In every area there are building materials suppliers that offer services to home builders. Their advice, though, usually will not go beyond specifying materials that are sold by that supplier. Most supply houses carry a wide range of qualities, so it's important to state clearly the budget range you have in mind. To get an idea of how the budget will break down, you can use the following figures (pertaining to a contractor-built house of 1,000 square feet, costing approximately $25,000):

General construction materials	42%
Labor	26%
Electric wiring, materials, and labor	6%
Plumbing, materials, and labor	
(includes a hot-water baseboard heating system)	10%
Contractor's overhead and profit	16%

Once the plan has been drawn and the materials selected, it is possible to get several bids based on common specifications which you can give out to contractors.

Watch for savings that are possible by using discontinued or slightly damaged materials. These are often items left over from the order of a large local builder, and the materials supplier is glad to get rid of them. This is one reason why it can pay to take on the responsibility of purchasing materials yourself, and there are other opportunities for saving money on materials.

In most rural communities you can usually find a small sawmill operated by a local person. It offers local woods precut or milled to your specifications at about a 30 percent savings over the lumberyard prices. Woods will range from oak to hemlock and in age from recently cut to aged. For further information, see Chapter 5.

The ubiquitous old shed, barn, or other outbuilding no longer in use is getting rarer, but the thorough investigator will be rewarded with seasoned wood of interesting color and texture. Sometimes the price is only the labor required to take an entire structure down, which could mean no more than a week's work for two people. Traveling the back roads, leaving signs at the local general store, and placing ads in the local classified shopper are some ways of ferreting out your supply of wood.

Another way to save is by checking those out-of-the-way places in your area, the salvage and junk dealers. They can often be veritable treasure houses of good usable materials, and it's all yours if you do a little investigating. It might be that you find an old window or door with such character and charm that it's worth redesigning a wall so you can incorporate it. Your house begins to evolve from local supplies and spontaneous discoveries, continually held together by your overall conceptual plan.

The Renovated House

Related to the use of recycled materials is the fourth housing solution you have: renovating rather than building a new house. I'll give just a brief word here, as this subject and a checklist on house rehabilitation are included elsewhere in this book. The renovation of an older house can be economical, and it's environmentally and socially a good idea, too. Consider the obvious savings of materials; the improvement of existing neighborhoods that are seriously affected by neglected old houses; and the ability to move in right away and rebuild while you live there, perhaps paying as you go and saving some of the high interest rates you'd have with a new home.

If renovation is your choice, the following investigations should be made:

1. If the house is small, does the present zoning allow expansion or is the house classified a "non-conforming use" that is OK as is, but no additions permitted? Consult the local zoning map and code book, or visit the building inspector for this information.
2. Are there highway or other building projects proposed nearby which might be objectionable or even the reason for the sale? Check this out with the building inspector also, or with the highway department.
3. Determine the extent of necessary repairs. A low selling price may forewarn of extensive repairs, structurally or mechanically. For details of what to look for and how to make the house livable while you do repairs, see Chapter 7.

chapter 3

FINANCING YOUR SHELTER NEEDS

by CHRIS AHRENS

BANK LOANS

Now we come to the important question of money. As everyone knows, the supply of money available for home mortgages in the U.S. and Canada is disastrously low. The result has been a high cost of borrowing that has just about excluded most prospective homeowners from even entering the market! In this case you are at a distinct advantage by building or renovating a small and efficient house, as you will need a smaller loan and a smaller downpayment than someone building a conventional house.

Because of the difficulties most families are having in affording conventional housing, banks have been flooded with applications for trailer loans. Since you will be building or renovating a permanent structure on your own piece of property for approximately the same amount of money as a trailer loan, you are in a very favorable position. Banks would much rather lend for a house that appreciates than for a trailer that depreciates. The most likely snag for you is that some banks may question the resale value of a small house. The federal government may come to your rescue, however, as they have prepared sets of plans for small and efficient houses which are similar in size to some of our plans. We will discuss these in more detail later in the book. The lending officer at the bank may be more encouraged to lend for a small house if you show him that the government encourages small and efficient designs. Another possible difficulty is that the very smallest of our renovation designs (the one in Chapter 7 for a basic living unit) may not meet the bank's minimum standards. If you wish to build this design, you may want to consider taking a trailer loan which carries higher interest and has a shorter payback period than a conventional loan.

14

Banks these days seem to be much more concerned about your ability to repay them than the specifics of the house design. In my discussions with banks, "energy conservation" has been a key phrase. Obviously, if you go broke trying to heat your new house, the bank will have to repossess it. It may take a bit of extra shopping to find a bank that will go along with you, but the long-term savings will be well worth the effort. Finally, remember that interest costs usually exceed the cost of the house during the course of a mortgage, so try to build as economically as you can.

OTHER LOAN OPTIONS

There are many instances when you can't get everything you want through a bank loan. In some cases, the bank won't go along with a small loan on an efficient house, say, if you want to do your own renovation; or it could be that there are loan options and tax breaks through the government which could make your mortgage payments with the bank easier to handle. This section will explain these other opportunities. Though most of the government programs mentioned in this chapter apply specifically to the U.S., Canadian and Mexican readers are encouraged to check with their local government officials for similar projects.

The Credit Union

Many people who want to do part of the construction work themselves, or buy an older home and then renovate it, can run into loan problems. Simply stated, the bank may not trust you to complete the work, or they may not feel a renovation is a good investment. If this is a problem, try your local credit union. They may be far more receptive to the idea of people trying to help themselves.

It is possible for a single family to use their credit union if they want to go the self-help route in construction. Table 3-1 outlines the process. The borrowing is done in $2,000 increments; and as each part of the house is completed and that sum is used up, you can get another $2,000 on the basis of the work already put into the house. With self-help you might expect to have done $4,000 worth of results (including materials, overhead, insurance, etc.) before the credit union recognizes your additional assets for the increased loan. Look into your local credit union or one at your place of work. It could be a way of financing the kind of house you have always wanted.

Government Loans and Tax Benefits

On both a state and federal level, many new forms of legislation are being passed to help people with the housing problems that exist today. We've tried to put

Table 3-1: *Financing Your Home through*
Your Credit Union
A Progressive Construction Program

The Member's Role	*The Credit Union's Role*
Lists assets (money, land, car, etc.)	Credit comm. OK's loan (title search the land, prepare first note and deed of trust for entire loan)
Now you can build	Deposit in member's credit acct's payable
Take loan	
Buy materials	Send bills to your credit union
Begin construction	
When funds are low	Notify borrower Execute another note (on land and work completed)

Borrower pays interest only on money loaned. Government inspector
checks all bills and work statements.

together information about these options, but because these laws are changing rapidly (and often to your benefit), it will pay you to check further. Bankers, contractors, clergy, and others in your area concerned about these matters may be able to help.

First, look into the possibilities of working your loan through the Veterans Administration or Housing and Urban Development (HUD). These agencies don't actually make loans—they only guarantee them. However, their help might make things easier with your local bank, so it's worth a check. Another point is that the paperwork for these loans is enormous, so try to get the local bank to take care of most of it. One area you might want to check out is HUD's program #223 F. It's a bailout program for people caught in the money squeeze and about to lose their mortgages.

Other benefits are opening up too. A number of States already provide tax credits or deductions if you build energy-efficient houses and more have pending legislation about to be passed. A number of demonstration programs are being funded through H.U.D., E.R.D.A., and the National Science Foundation, seeking ways to make homes less wasteful of precious nonrenewable fuels. If you already own an old dilapidated house two programs of home weatherization are now operating mainly through community action agencies, providing caulking and insulation materials, etc. The income guidelines for these programs are shown in Table 3-2. If you are low-income, elderly or disabled you can secure a grant of about $300 for materials to make your house weather-tight. If you can do the work yourself, all the better, but if not able, most programs will supply a work crew of trainees to do the job for you. Having once made your house tight you are ready to consider further use of alternative energy devices for heating and hot water. These programs can be coupled with Farmers Home Loans for home repair and rehabilitation. On page 20 we list how you can arrive at the sliding interest rate scale for FmHA loans in areas under 10,000 population.

Also check on the state level, as there may be local tax benefits there as well. For example, Arizona and Illinois have already enacted legislation favorable to those who install energy-saving equipment.

One other option, depending upon your income, is the Farmers' Home Loan Agency (Fm.H.A.). During the last 10 years if you have been turned down by a bank for a home mortgage and live in a rurally designated area, you have been able to turn to the Farmers' Home Administration, use one of their approved plans, and have a modest house in which to live. However, there are several limitations when taking your loan through Fm.H.A. First, your income must fit into their guidelines. (See Table 3-2.) Just as serious, though, is the restriction that any loan given by Fm.H.A. must be for a home that meets the minimum property standards (M.P.S.).

Unfortunately, these standards were drawn up at a time when the costs of housing were far lower than they are today. Thus, considerations about lowering the operating expenses were generally ignored, as was the need most people have today to start off with a small home and add later. Hence, most of the designs we've suggested in this book for smaller homes won't meet these standards; further, if you build with one of the M.P.S. designs, your operating expenses will be greater than they need to be.

Table 3-2: *Annual Income Guidelines for*
All States Except Alaska and Hawaii (1976)

Family Size	Nonfarm Family	Farm Family
1	$2,800	$2,400
2	3,700	3,160
3	4,600	3,920
4	5,500	4,680
5	6,400	5,440
6	7,300	6,200

NOTE: For more than a six-person family, add $900 for each addition
to a nonfarm family and $760 per addition to farm families.

All this will change, of course, as time goes on. However, even with the higher costs for a larger home than you need right now, Fm.H.A. may still be a good bet. The reason is that the interest rate on their loans is so small, sometimes as low as 1 to 2 percent. Thus, it's very possible that a $15,000 home with a 2 percent mortgage will cost less over a 20-year period than a $10,000 house at 8 percent interest from a bank. If you want to go this route, just make sure to check and see what your total monthly payments will be for mortgages that run over the same period. Then decide what's best for you.

Table 3-2 gives the income guidelines for all states except Alaska and Hawaii. If your income is around these figures, you are eligible for Farmers' Home interest credit programs, which is a way of keeping your monthly housing costs within 20 to 25 percent of your income. Government agencies such as the Fm.H.A. figure your income on an adjusted basis. If your adjusted income is over $12,900 Fm.H.A. will not serve you (as of 1976). But if your adjusted earnings are within the guidelines shown in Table 3-2, you would be eligible for interest credit down to 1 percent with up to 33 years to pay.

A typical example to determine *adjusted income* by Fm.H.A. standards would be as follows:

A family of four has an income of $5,000. The adjusted income for that family would be:

Gross income	$5,000.00
Less 5% (loan, principal, taxes, etc.)	250.00
	$4,750.00

With three dependents (e.g., family of four with head of household claiming three dependents):

3 X $300 each	$ 900.00
Adjusted income	$3,850.00

Interest subsidy by the U.S. Government can reduce the interest to 1 percent, allowing 20 percent of the adjusted income for a shelter allowance, or:

$3,850 X 20% = $770 possible per year, or $64 per month.

Under the adjusted income plan of the Fm.H.A., this family of four, with an income of $5,000 and no major outstanding debts, could afford a house costing about $10,000. Table 3-3 outlines the procedure for getting a Fm.H.A. loan.

While we are on the subject of money, a word should be said about rehabilita-

---

Table 3-3: *Checkpoints on Getting Government Farmers' Home Housing Loans*

OK
Check

_____ 1. Farmers' Home Administration rural loan program: Fm.H.A. office interview credit and employment reviews. Fill out agency forms.

_____ 2. Discuss plans, make rough estimate and materials list, detail budget, timing.

_____ 3. Courthouse trip for ownership check. Get copy of any official papers.

_____ 4. Visit county health department for water and waste permits.

_____ 5. Take picture of property for before and after conditions record of work.

_____ 6. Take all this information to loan office for review.

_____ 7. Loan agency reviews application. (If there is any question, find out specifically why. You may need a co-signer or sponsor.)

_____ 8. Line up possible contractors, helpers, equipment. Get an approximate due date for loan approval.

_____ 9. Loan closed, title search OK, check received.

NOTE: The step-by-step procedure suggested here assumes the loan moves smoothly with no complications of land ownership, hidden indebtedness, or family problems. Getting government loans can be a time-consuming and frustrating job, but hang in there!

tion of existing structures using government insured and direct loans. Under the Farmers' Home Program, it is possible to secure very favorable loans at low interest rates, depending upon your income and the fact that a bank is not interested in helping you rehabilitate a house.

In many respects it is more practical to rehabilitate than build new in this period of tight money, especially if you can do some of the work yourself. In rehabilitating, you will also have fewer problems with codes and inspectors and be able to adapt the house more to your own needs (even if you end up tearing down more than half of the structure!). Community Action programs of the Community Services Administration can help those with low incomes. They will work with you to get loans and may be able to provide free labor in their work-training programs if you cannot do the work yourself. Two loan programs are worth mentioning:

Fm.H.A. 504 Home Repair Loan (for areas under 10,000 population). Loans available up to $5,000 at 1 percent interest rate with repayment periods of:

> 10 years for loans up to $1,500
> 15 years for loans up to $2,500
> 20 years for loans up to $5,000

Fm.H.A. Home Rehabilitation Loan up to $7,000. Has a sliding interest rate scale:

Families with less than $3,000 adjusted income . . . . 1% interest
Families with $3,000–5,000 adjusted income . . . . 2% interest
Families with $5,001–7,000 adjusted income . . . . 3% interest
The payback period for these loans is 25 years.

In both loan plans, the aim is to make the house safe, comfortable, and sanitary with no frills or wasted spending, which fits what we are talking about all through this book. If your income fits these guidelines, see the Farmers' Home agent in your rural area.

Cooperative Housing

What do people think when they hear the word *cooperative?* In New York City, a cooperative usually means a tall high-rise building owned and controlled by the people who live in it. In other cities it might mean a supermarket, grain elevator, department store, or oil and gas station. One of the greatest contributions the cooperative concept has made is the easing of group relations and tensions. In the cooperative housing project, both white- and blue-collar workers, like engineers, painters, and bricklayers, live amicably together. Many know something about construction and as part-owners become deeply involved in the necessary improvement program continually going on with the buildings. Thus, the social aspects of cooperative housing are of vital importance. They must be taken into account, not only in the operational stage by management, but from the very beginning, in the planning stage. A cooperative is a blending of professionals and volunteers working together to provide the community life that is possible when something is owned by all who use it.

A housing cooperative may be of new construction or the rehabilitation of existing buildings. The latter is usually a conversion cooperative where a rental project is turned over to the people who live in it. They become members of a nonprofit corporation that has title to or leases the premises. Conversions of this kind can be most helpful where the area has been experiencing social, management, and operational difficulties due to an inner city location and/or absentee ownership.

What are the advantages? Through cooperative housing, a family can achieve many of the advantages of both individual home ownership and renting, while avoiding the disadvantages of either. In a cooperative, for example, a family pays only what it actually costs to own and operate its dwelling. There is no landlord to deal with! Yet in a co-op, a family is as free to move as if it were simply renting a home with a lease. The co-op usually accepts the responsibility of resale and can work from a list of people waiting for space in the cooperative.

Moreover, the cooperative member has the same tax advantage as an individual homeowner. People can deduct from their taxable income their share of the mortgage interest and real estate taxes the co-op pays. The family that rents, by comparison, can deduct nothing. The question of liability is also very important. In this respect the cooperative is similar to the rental situation. The co-op, not the member, is responsible for paying the mortgage and being responsible to the public, just as the landlord is. Further, a cooperative handles all major repairs, insurance, and maintenance of grounds and other facilities, unlike the private homeowner who must do all these things himself.

So it can be seen that the cooperative approach has many special advantages and can exercise a large measure of community control. Members write reasonable rules designed to keep their neighborhood a pleasant and safe place in which to live. An individual homeowner, on the other hand, has no jurisdiction beyond his or her own property, except by appealing to a court or the local government; and a renter has no control at all. This is summarized in Table 3-4, showing the comparative advantages and disadvantages of the cooperative housing approach.

How is a housing cooperative organized? "Cooperation" means working together. In a housing co-op, families work together to own and control their neighborhood environment. They buy membership shares in the cooperative corporation that holds legal title to the entire development. They sign an occupancy agreement that

Table 3-4: *Co-ops, Private Ownership, Renting*

	Cooperative Plan	*Private Home Ownership*	*Apartment Rental*
MONTHLY COST	Nonprofit co-op buying cuts maintenance-repair costs. Your monthly charges are lower since they cannot exceed actual costs plus reserve funds (to protect your interests).	Individual buying means higher cost.	Includes owner profit. Your rent can be—and frequently is—raised.
OWNERSHIP:	You are an owner of your apartment home together with your fellow members.	You own and are completely responsible for your home.	You have no ownership rights or control: you are only a tenant.
LIABILITY:	You have no personal liability; you sign no mortgages or deeds.	You are personally liable for the mortgage and bond.	
INITIAL COST:	Advantageous long-term financing.	Higher interest rates, larger payments, shorter term.	
RESALE:	Your cooperative usually handles re-sales directly. Any charge is only nominal.	You must re-sell your house yourself, incurring as much as 10% for brokerage fees.	
MANAGEMENT:	Each cooperative member has one vote in the selection of directors. Management is handled by a professional manager.	You alone are completely responsible for all decisions . . . and mistakes.	You have no voice whatever in either management or policy.
MAINTENANCE:	Your cooperative takes care of all repairs and maintenance.	All insurance, repair, and maintenance are your sole responsibility.	You have no control over maintenance standards.

	Cooperative Plan	*Private Home Ownership*	*Apartment Rental*
TAX BENEFIT:	You are entitled to home owners income tax deduction. About 65 percent of your monthly payments are deductible.	You are entitled to deductions as a home-owner.	You have no income tax saving of any kind.
CAPITAL GAINS:	In the event of a sale of a home and subsequent purchase of shares in a cooperative corporation, the tax on capital gains and its deferment pro visions may apply.	Same.	Loss of capital gains advantages if you sell your home and move into a rental apartment.
COMMUNITY CONTROL:	As a cooperative member, you participate in establishing policy as well as rules and regulations governing your community.	You have no jurisdiction, except through a court of law (at your expense).	You have no jurisdiction, except through law or governmental controls.
ELIGIBILITY:	Anyone acceptable to FHA as to credit and the cooperative membership is eligible.	Anyone acceptable to lending institutions.	Usually controlled by lease or prepayment of a number of months' rent.
DURATION:	Your occupancy is permanent as long as you are a cooperative member in good standing.	Permanent.	Your tenancy is for the duration of your lease only.

establishes each family's right to occupy a specific dwelling and spells out the rules of occupancy established by the cooperative itself. All cooperatives have slightly different rules and regulations, depending upon the will of the voting membership. An initial downpayment is required to become a member and you pay those charges necessary for paying off the interest, principal, maintenance, and overhead. Keep maintenance to a minimum and you keep your monthly costs low as well.

Generally, a housing cooperative grows out of a sponsoring organization, which may be a trade union, veteran's organization, church federation, not-for-profit foundation, credit union, or another cooperative. Membership in the housing co-op cannot be restrictive, in accordance with the basic principles indicated in Table 3-5. Table 3-6 suggests some of the more specific requirements usually written into a cooperative contract when dealing with cooperatives only for housing.

Mention should also be made of the term *condominium,* a form of ownership that has become popular in middle- and upper-class urban developments. Here the family has direct ownership of the housing unit or units but shares cooperatively with others the surrounding property owned by the corporation, as well as its use. This would include walks, roads, swimming pools, etc.

Help with organization, literature, and technical assistance can be secured from the Foundation for Cooperative Housing, 1001 15th Street NW, Washington, DC 20007. This foundation will be able to supply you with the latest information about federal legislation and loan programs available in the U.S. with HUD and the Farmers' Home Administration. Their experience in this country is mostly in urban areas where large enough groups of families can be assembled to make a cooperative economically feasible. In other countries, particularly in South America, more rural cooperative developments are possible. The foundation also has an international program.

Table 3-5: *Basic Cooperative Principles*

1. Open to all!
2. No speculation or profit motive
3. Neutral in politics and religion
4. Limited interest
5. One member, one vote
6. Continual education
7. Continual expansion

Table 3-6: *Basic Requirements for Housing Cooperatives*

1. Use of home only for residence
2. Initial payment required of each member
3. Cooperatively planned and directed by committees
4. Acceptance of community-developed rules and regulations
5. Agreement in maintenance standards
6. Cannot "sub-let" without approval
7. Participation in community government
8. Development of a savings plan
9. Maintain monthly payments
10. Right to buy related needs cooperatively
11. Activity in related community activity with children, adult classes, and recreation programs

If you are interested in a cooperative approach to housing, several options open up. Not only is urban housing a possibility, but rural cooperatives can be formed as well. Through the co-op, you can also choose whether you want to build your own home or have the cooperative development corporation do this work for you. In either case, you'll be able to become actively involved in designing the home and the community you'll be living in, and there are real money-saving opportunities with the co-op approach as well.

The Cooperative Credit Union

It is not uncommon in other countries for groups to form their own credit union as a money source for a modest house. I have known of credit union groups in Central America that saved for 8 years before they had enough for the first family to build their house. There was quite a celebration that day because not only were they able to build one house, but they were also able to mortgage that house for the funds necessary to build the next one, and the next one, etc., until all had a new home. Living in a more affluent country, we do not have to wait 8 years, but the principle is a sound one. Although banks will not give you a house mortgage on a set of plans, they will lend money on a house already built. So it means that a small group of families save together until they have enough for the first house, choose the most needy family that should have it, secure a mortgage for the funds needed for the next one, and so on.

The Not-for-Profit Housing Development Corporation

For a larger project, even a whole new community, it's possible to work cooperatively with other families through a nonprofit housing development corporation.

A typical set of bylaws and incorporation papers can be obtained from the Foundation for Cooperative Housing mentioned earlier. The corporation would be able to meet the criteria for building housing today by acting as a large-scale operation in the best interests of many smaller families. Some of the ways the corporation would cut costs include:

1. A land bank or trust, taking land out of the speculative market, holding it in perpetuity, and leasing parcels back to homeowners in the corporation for housing, business, farming, recreation, etc.
2. Sharing the cost of land development, including roads, water and utility supplies, and cluster development where desirable.
3. Having the more costly heavy equipment work done cooperatively to make best use of time and equipment for many units.
4. Using a revolving loan fund with no interest or a low interest rate for the most

needy. Using your credit union or a federal agency like Farmers' Home Loan Agency (Fm.H.A.) where this is possible. Also, taking advantage of grants from private sectors as various experimental and research directed construction is attempted. Above all, being flexible in the kinds of financing used, for each family's situation is different.

5. Eliminating the builder's profit and using basic home designs which incorporate the ideas presented in this book.

6. Using your own counseling staff on a neighbor-to-neighbor basis, thus eliminating the need for real estate fees, selling costs, and collection fees.

Because housing is a complicated business, too much emphasis cannot be placed upon the need to keep an up-to-date counseling person as part of the corporation plan. For new families, a typical program would work with each family in terms of their housing need.

What you'll need is a team consisting of a housing coordinator who helps to present the concept to interested families and determine their qualifications and interest. The family would be introduced to other families in the group contracting or building their own houses. If the family wishes to go ahead, the housing coordinator keeps in close touch with them as they proceed through the planning stages (financial and technical), until they can become fully participating members of the corporation. Other members of the corporation team include a construction foreman, an administrative assistant who keeps account of the money flow, and a project watchman for the storage and protection of building materials and the equipment used for construction.

At this point the family will have decided with the corporation on a particular

---

Table 3-7: *Preconstruction Work of the Housing Coordinator*

Interested family visits corporation	Housing coordinator goes over sites, plans, and financial considerations with family.
Family presents application	Counseling committee with housing coordinator decide upon family, type of house desired, or would they be better off with a conventional housing approach? Tentative approval made.
Family works with housing coordinator	Questions of money management, house design, how to pay for the house, environmental care and maintenance, home management, and referrals to other helpful agencies and people.
Family approved for membership in corporation	Land designated, house plans approved, financing cleared, training begun for self-help approach.

approach to raising the money for construction—whether it will come from the corporation's revolving fund, a Farmers' Home Loan, a credit union, or personal resources. It will be up to the housing coordinator to follow up with the family to see that progress is made and timetables adhered to. Although this procedure may seem complicated, remember that you will be working with people who have the same aims and aspirations.

REPAYMENT PLANNING OF LOANS

When a developer such as our nonprofit corporation borrows money to build a large-scale project, the corporation borrows what is needed as the project develops, paying the construction firm as work is completed. In the case of low-income housing programs, the "construction firm" may be a private contractor-builder, a self-help group, or a nonprofit building corporation. A typical repayment schedule for construction would be as follows:

1st Payment—Rough Enclosure—30 percent of the loan

The building should be framed and sheathed, the roof completed, and window and exterior door frames set. Underflooring should be laid, partition work should be in progress, and plumbing, heating, and electric wiring started. Also, a guaranteed survey should have been submitted showing outside lot lines, building outlines, and projections to roof. (Approximately 200 man-hours per house.)

2nd Payment—Full Enclosure—15 percent of the loan

Exterior finish of brick veneer, siding, shingles, or stucco must be completed. All exterior wood work primed. Chimney must be built and all plumbing and heating, roughing, and electric wiring well along. (Approximately 300 man-hours per house.)

3rd Payment—Interior Roughwork—15 percent of the loan

Interior partitions and wall work completed. Exterior doors and window sash should be installed and some interior trim completed. (Approximately 300 man-hours per house.)

4th Payment—Interior Finishing—15 percent of the loan

All interior work completed except finished floor. All standing trim applied, including door and window casings and jambs. Water and sewer connections completed to street or septic tank. Bathtub set and tile work started. Basement floor laid if required. (Approximately 200 man-hours per house.)

5th Payment—Fixtures—15 percent of the loan

Tile work, plumbing, heating, kitchen cabinets, gutters, and leaders should be in place with a finished floor laid; utility connections should be completed. (Approximately 200 man-hours per house.)

6th Payment—Final Payment—10 percent of the loan

Final painting, decorating, linoleum laid, and floors finished. Sidewalks, walks, driveway, steps, and terraces—including grading, topsoil, shrubbery, and planted lawn. Building entirely completed as called for in plans and specifications, and ready for occupancy. Underwriters and occupancy certificates should be obtained. (Approximately 300 man-hours per house.)

It should be noted that interest charges on advances are made from the date of payment only and on the amount advanced, not on the full amount of the loan from the date of closing.

A number of old established groups have demonstrated the value of working together to produce good housing and community life. Two examples are:

Koinonia Partners in Americus, Georgia, is a religiously centered community which has been together since 1942. In 1968, they established the Fund for Humanity, which was a way to develop a capital fund to invest in low-cost housing, a community center, and numerous small industries in which their neighbors could find jobs and creative work. Their Partnership Housing has developed two housing communities in which loans were made at no interest from the fund.

Bryn Gweled Homesteads is a community of homesteads near Southampton, Bucks County, Pennsylvania, where some 50 families now live on 240 acres. It was organized back in 1940 by a dozen families and has grown steadily through the years. In this case each member family of the corporation buys "certificates" in their corporation to pay off indebtedness resulting from the cost of land, roads, utilities, and the community center.

Recent projects in the planning and construction phase include:

New Town Research Foundation, West Millington, New Jersey

Shannon Community, Charlottesville, Virginia

Experimental Cities, Pacific Palisades, California

In these examples and others you will find people like yourself living in our expensive, competitive system, yet finding less expensive options and other values for their personal needs.

As the cost of materials, insurance, and the rental of construction equipment rises, can we save enough with self-help to make it worth the bother? Although the cost of housing has just about doubled in the last few years, savings are possible with self-help, as can be illustrated in Table 3-8 by the example of a house estimated at $21,000.

There are now about 50 self-help programs working in 70 counties around the country. Thirty operate through Fm.H.A., 10 via other agencies, and 10 through the Department of Labor. These programs serve 5,000 families, with the major group in California. Moreover, many individual self-help builders are doing their thing and are not included in these figures which were reported during the National Conference of the Rural Housing Alliance in January 1975.

Table 3-8: *Self-help Savings*

Loan	Self-help	Contractor
Loan required	$16,000	$21,000
At 9% interest rate	$127 monthly	$166 monthly
At 1% interest subsidy	$47 monthly	$62 monthly
Cost in 33 years at full payment	$50,282	$65,736
Cost in 33 years at minimum interest payment	$18,612	$24,552

SUMMARIZING HOW TO REDUCE HOUSING COSTS BY GROUP EFFORT

The banks and federal insuring and loan agencies will be quick to tell you as a prospective home builder the risks of doing anything for yourself. At the same time, the cost of building as "the system" would have you do it will put you in debt for the rest of your life. What we have suggested in this section are a number of alternatives for financing the housing you need by getting together with others who have the same need; you then by-pass the system as much as possible, or use it where it is in your own best interests, as a group.

Groups such as the Rural Housing Alliance, 1346 Connecticut Avenue NW, Washington, DC 20036 and the Housing Assistance Council, 1601 Connecticut Avenue NW, Washington, DC 20009, to name a few, have the expertise to help local not-for-profit groups get together and produce their own housing corporations. The planning would go like this:

1. A group of people with a similar need find an existing organization such as a church, Community Action Agency, or nonprofit housing corporation interested in sponsoring a housing or community program. If no such organization exists, start your own.
2. Form a local housing committee to research each family's needs, resources, available land, houses that might be rehabilitated, etc., and put it together in a form that can be explained to others.
3. Seek the help of one of the agencies mentioned earlier or a local technical agency if the community has one. (Community Action Agencies often have had experience with formal housing programs or know where to go for information.)
4. Begin a program of training the group in all the factors related to a housing program, including loan programs, local housing restrictions, financing possibilities, and legal pitfalls.
5. At this point you will experience the dropping out of some families, but those that stick it through will be better prepared for the long haul.

6. Attempt to meet some housing need within the community that is small in size, yet which will test the ability of the group to work together.
7. Incorporate under the guidance of the technical assistance group working with you.
8. Begin your own project.

chapter 4

SOME PROBLEMS TO OVERCOME— CODES, APPEALS, PUBLIC ACCEPTANCE, AND CONTRACTOR RESISTANCE

by ALEX WADE

Building or renovating a house these days is a major challenge. This is one of the reasons professional contractors charge so much for their services. Remember, though, that money carefully spent on a house of your own is an "inflation-proof" investment. Also, if you build a small and efficient house, your costs in time, energy, and materials will be far less. Building small and/or doing it yourself may cause you to run into a few problems, though. Even if most homeowners now understand that building or renovating a house that saves space, materials, and money is wise, many building inspectors, bankers, and contractors are still living in the "more is better" past. This chapter will tell you what to look out for and how to sidestep many of these problems if they do arise.

CODES

The most likely difficulties are those concerning various building codes and the officials charged with their enforcement. Fortunately, building codes are generally used only by large and medium-sized cities and some of the more populous states. Also, code officials are much more likely to be sympathetic to a homeowner-builder than to a contractor building for profit.

Before we go into the problems, you should understand just what it is that a building code is supposed to accomplish. A building code is intended to set mini- mum construction standards so as to protect the health and safety of building

31

occupants. Codes control such diverse things as property set-back limits, areas of rooms and windows, sizes and spacing of structural members, and design requirements for stairways and exits. Separate codes are usually written to cover plumbing and electrical wiring. The more recent and progressive codes are written on a performance basis, which means that they state *the desired ends* rather than the exact means of achieving them. An example would be a code requiring a floor construction capable of carrying a load of 40 pounds per square foot, as opposed to one which lists specifically the various materials and spacings of structural members that are acceptable. The first will allow new or different types of construction, while the latter is merely a rubber stamp for the status quo.

The ideal solution would be to have one national code which applied everywhere. The electrical industry has pretty well succeeded in getting the National Electrical Code adopted uniformly throughout the country. The National Plumbing Code is also widely used, but is by no means universal. Three different major attempts have been made to create a national code which covers building construction. Unfortunately, the results were just three more codes. Since writing a good code is an extremely expensive and complicated task, few smaller states or cities write their own. They merely adopt one of the so-called "national codes" with or without their own modification. The three "national codes" are the National Building Code, the Southern Building Code, and BOCA (Building Officials' Code of America), normally pronounced as one word, "Boca." The latter code is the most progressive and fortunately the most widely adopted.

How do you tell which code, if any, you should follow? Normally, the local government unit—city, town, village, or county—administers whatever code is used. In many instances, even though there may be a state code, it is up to the local community to set up enforcement of it through a building inspector. If the community doesn't have an inspector, there is no way it can enforce the code. Even if this is the case, to avoid possible trouble with banks and insurance companies, it would still pay to follow the code as closely as possible.

Whether the code will be a help or a hindrance depends on where you are located. For example, California and Massachusetts have both recently adopted state codes which were written in the "bigger is better" era. Both of these codes make life difficult for those who wish to build small houses or do construction themselves.* Remember, though, that these states require the local community to enforce the code. Careful selection of your homesite may enable you to avoid quite a few

*Governor Brown of California has recently intervened in this situation
by promising to try to get the repressive California code amended.

hassles. On the other hand, many new sensible codes are "in the works" because of the energy crisis. The federal government is considering minimum insulation standards on a national basis. How it plans to enforce these standards is still unknown. In any case, it will pay to insulate as heavily as possible, so don't be afraid of any restrictions such as these.

For those would-be builders who do have to face the reality of codes, we offer some advice on how to deal with them. First, obtain a copy of the code which has been adopted for your area. Also, find out who is responsible for the enforcement of the code. Most towns have one or more building inspectors who issue building permits and then follow up with field inspections to see that the construction conforms to the code. After obtaining the code, *read it thoroughly*. It will pay to get a copy of the code *before* committing yourself to the purchase of a piece of property. Refer carefully to the checklist at the end of this chapter to see if there are any provisions which you can't meet. It is possible for someone to sell you a piece of property on which you can't legally build! Also, study carefully the beginning sections of the code book concerning jurisdiction, enforcement, and appeals. Make sure that you know who is above the building inspector in case he is unreasonable. If you plan some things which are not spelled out by the code or which are different from what the building inspector is used to, make sure that they are labeled (not necessarily too visibly) on the plans which are submitted for approval. Also, remember that some things can be added to your house after the building inspector has made his final inspection. Once the inspector has approved plans in writing, it is much more difficult for him to change his mind. Building inspectors frequently like to show their authority by finding mistakes. One or two obvious "plants" of a very minor nature may help you get by with many things you really want.

Do your homework carefully before approaching the inspector. Try to locate someone who has dealt with him to find out what he is like. It might also pay to have a friend test the waters by going in and asking questions concerning a mythical small house in another part of town. Friends of mine have used another technique very successfully. Since the husband has a full-time job, his wife has done a great deal of the work on both their new homes. Upon finding out that they were faced with a notoriously difficult building inspector, the wife spent several weeks boning up on construction details (for post and beam construction) and confronted the building inspector herself. Dealing with a woman who was knowledgeable on construction matters was such a novel experience that the building inspector was too flustered to be nasty.

Now, let's examine a few specific instances in which you might encounter trouble. In Chapter 6 we make recommendations for small houses and better structural efficiency using post and beam framing and a variety of foundation systems. This may be the most likely area of conflict with your building inspector. It is likely that your code will permit this kind of construction, but that the inspector will be hesitant to approve because he is unfamiliar with these systems. Post and beam is easy for the inexperienced builder to understand, so if you study carefully, you will have the upper hand insofar as knowledge of the system is concerned. An extremely valuable reference which anyone contemplating building a house should obtain is available from the U.S. government. It is entitled *Low Cost Homes for Rural America—Construction Manual,* USDA Pamphlet #364, and is available from the U.S. Government Printing Office for $1.00. This manual covers post and beam framing as well as treated timber foundations. It also takes you step-by-step through the process of building a house. If you take the time to study this book thoroughly, you will be able to talk very knowledgeably to the building inspector. We list this manual again and give additional post and beam references at the end of Chapter 6. Be sure to take your references with you when you visit the inspector. Many of our proposals are strongly reinforced by the government manual mentioned above. You may be able to convince a dubious inspector by showing him that the U.S. government recommends many of these techniques. Finally, remember that you are dealing with a bureaucrat who is near the bottom of the ladder. He needs to feel important. Flatter him, ask his advice. Also bear in mind that it is your tax dollars that are paying his salary, and he is there to serve *you.*

Suppose that you have patiently talked to the inspector and he still turns you down—what do you do next? There are three obvious approaches: first, modify what you are proposing in order to satisfy the inspector; second, sell the property and try again elsewhere; and finally, appeal his decision. Most codes spell out appeal procedures in the introductory sections, so study them carefully. Make sure your request is reasonable, as the inspector will not want to set any bad precedents and neither will the town fathers or the appeal board. If you are in a small or medium-sized town, it may work much better to seek out the people involved and sit down for an informal discussion. Again, remember that the building inspector is a bureaucrat and may worry over his job. If you can get approval from a higher-up without stirring up hard feelings, you will have gotten the inspector off the hook and solved your problem. Whatever you do, *don't hire a lawyer!* Lawyers are notoriously slow and expensive, and very few know anything about the construction codes. Lawyers are basically hired guns, and remember that the town has more money and can hire a bigger gun. Also, bringing in a lawyer tends to escalate the level of hard feelings. Try always to be reasonable and cooperative, as frequently the mood is contagious.

Table 4-1: *Building Code Checklist*				
Item	Code Requirement		Can You Meet It?	
	YES	NO	YES	NO
1. Minimum area of lot (lot size)				
2. Number of families permitted				
3. Minimum or maximum house area				
4. Building height and set-back restrictions				
5. Foundation type (Is masonry required?)				
6. Is post and beam framing permitted?				
7. Is grade stamped lumber required for framing (i.e., new and kiln dried)?				
8. Minimum room sizes				
9. Minimum door sizes and number of exits				
10. Window sizes and ventilation requirements				
11. Forced ventilation for kitchen and baths				
12. Fire retardant enclosure for exitways				
13. Fire alarm system				
14. Riser and tread sizes for stairs				
15. Minimum insulation requirements				
16. Chimney clearances (Is prefab allowed?)				
17. Can owner finish a shell by contractor?				
18. Can attic or some rooms be left unfinished?				

The above list is by no means comprehensive. It is intended to call to your attention the most likely code restrictions that might adversely affect your construction of a house or remodeling of an existing building.

BUILDING CODE CHECKLIST

Before purchasing a piece of property, it would be wise to see if the area in question is governed by a building code. Our checklist (Table 4-1) covers major items which may be covered by any building code. Codes vary widely, so only a few of the items may be covered by a specific code. Check very carefully, though, as the codes have an important bearing on the property price and building cost, and may be a prime factor in your decision to purchase.

PUBLIC ACCEPTANCE

Another possible problem associated with building a small energy-conserving house is that of public acceptance. There is the notorious case of the brash young architect who built a stark, windowless home in the middle of a suburb of tract houses. The owners of the tract houses were furious, went to court, and eventually had the box torn down. Since the energy crisis has been so well publicized, you are in a much better position than the young architect. However, if you plan a windowless north wall or a series of solar collectors facing the street, you might still be in for trouble. Since energy conservation is important to almost everyone these days, you can gain interest and acceptance by explaining to the neighbors what you plan to do before you build.

Another method is to try to start a trend. If you pick an area in which yours is one of the first houses, you will help set the tone for other houses to come. A good example of this would be the use of a pole house to make economical use of a steep slope.

HOW TO FIND A CONTRACTOR

Now that we have all the other problems out of the way, how do you get a contractor to build for you? The simplest answer is to build your home yourself. Many people may not have the time or inclination to do the whole job themselves, though. Even if you can't do all the work, you can easily find a contractor or carpenter to do part of it. Thus, moving up one step from doing the whole house yourself is to hire a contractor to build just the exterior shell of the house. Since this is the work that contractors usually do best and most efficiently, you can save a lot. You also get the most difficult (for a beginner) work out of the way. Moreover, the cost of constructing the shell of a house is simple to estimate, so you can also get some meaningful competitive bids from different contractors. Make sure that you furnish them with a set of construction documents (plans and specifications) which spell out *exactly* what work is to be included. See Chapter 19 for a sample set of construction documents. To finish off the shell, you have your choice of hiring separate tradesmen for the different jobs or doing all or part of the work yourself. Depending upon just how much of the work you do yourself, you can save anywhere from 10 to 40 percent over the cost of hiring a contractor to do the whole job outright.

Assembling, scheduling, and arranging for the work of the various tradesmen is known as general contracting. Over and above his direct costs for building your house, the contractor usually charges about 15 percent for this portion of his services. See the checklist in Table 4-2 for trades which have to be called in. The two trades which require little special skill but lots of hand labor (and can therefore save you the most by doing them yourself) are plumbing and electrical work. (Detailed handbooks for both trades are usually available at your lumberyard.) Both the National Electrical and National Plumbing Codes allow the homeowner to perform his or her own work.

If all this still sounds too involved to you, consider hiring a contractor. But how do you find an innovative and cost-conscious contractor to assume responsibility for the whole project? If you are planning to use one of our cost-saving structural systems, my advice is to find a young carpenter/contractor who will do the job on an hourly basis. Years of experience have shown that there is no craftsman in existence who gets more set in his ways than an older carpenter. More progressive parts of the West Coast have used post and beam framing for years, so you may have an easier time in that part of the country. Next, make absolutely sure that you look at actual samples of work performed by your contractor to see that he is a competent craftsman. Also, make sure that he can read a set of plans. Many otherwise competent craftsmen can't read plans. Once you have a good feeling about being able to work with the contractor, be certain that you provide him with a good set of plans and details to work from. (See Chapter 19 for additional information on plans and specifications.)

Now, how big a work crew do you need? Well, two skilled carpenters can readily frame up one of our small post and beam houses in a few days. I would strongly advise against a large crew. They eat up money in a hurry and too many people just get in each other's way in constructing a small house. Also, finances need to be watched with extreme care as small contractors are notorious for their poor business abilities. If you do hire a contractor on an hourly basis, I would recommend that you purchase materials directly and settle up accounts with the contractor weekly. It may also be necessary to furnish the contractor with a small "petty cash" fund for miscellaneous expenses. Also, make absolutely sure that you have not set up any accounts where he can charge materials for other projects.

Table 4-2: *Checklist for Subcontracting the Construction of Your House*

What follows is a complete step-by-step checklist for all the major areas where it may be necessary to subcontract the various construction operations for a typical single-family house. Review the items carefully and then decide how much of this work you wish to take on and how much to delegate to a contractor.

Item	*Completed*
1. Financial arrangements (loan or other)	
2. Building permit	
3. Drill well or obtain city water connection permit	
4. Septic system permit (from health department) or sewage connection permit	
5. Installation of septic system—generally required to be done by professional; may also be installed last	
6. Temporary electric service (by electrician) or portable generator	
7. Job phone (optional)	
8. Excavation contractor (may be done by hand on pole houses)	
9. Masonry contractor (if you use a masonry foundation, which we do not recommend)	
10. Concrete finisher or rent troweling machine (if you use slab construction)	
11. Materials for shell of house (furnished by contractor or owner?)	
12. Carpenter/contractor to build shell (should include windows, exterior doors and roof)	
13. Installation of rough wiring	
14. Installation of rough plumbing	
15. Inspection of above work if required by code; notify inspector(s)	
16. Install insulation (easy to do-it-yourself)	
17. Install finish siding	
18. Install sheetrock	
19. Finish electric wiring and plumbing	
20. Inspection for above if necessary; notify inspector(s)	
21. Finish carpentry—shelves, hang doors, install trim, finish floors, etc.	
22. Spackle sheetrock and paint interior	
23. Backfill and finish grading (may not be necessary for pole house)	
24. Get occupancy permit from building inspector, if necessary	

Just how do you find this ideal carpenter/contractor? If you follow our advice and use roughsawn lumber (see Chapter 6), the people at the roughsawn mill can probably steer you to someone who is receptive to an innovative project. Another method is to place a newspaper ad in the nearest large or medium-sized city, or even the local or county paper. Contractors frequently place ads themselves these days and thus will look at the want ad section. If possible, place your ad in the contractors' section even if it means faking the ad a bit. Another method is to hunt around the countryside for interesting and unusual houses and then ask the owners who the contractors were.

For a really different approach, write to various schools of architecture to see if one or more upperclassmen will take your house as a summer project. The University of Illinois and the University of Wisconsin both do research on economical houses, so students from these schools may be more likely to be interested in doing practical and economical houses. These young architectural students will require much supervision, so be ready to spend the time coordinating their design and construction work. It's worth the effort, though, as this approach could give you the best house for the least money. Also remember that there are a great many unemployed architects these days. A young architect just out of school with no job may be eager to take on the designing and building of your house just for wages.

Finally, we've compiled a list of people and small companies that might be helpful if you are looking for an energy-conserving contractor in your area; see the "List of Design Groups" at the end of this book.

Section II

LOWERING THE FIRST COSTS OF A HOME

chapter 5

SITE PLANNING AND USE OF ON-SITE RESOURCES

by JEROME KERNER

It should be clear by now that the goal of low-cost, energy-conserving housing can be met only with maximum owner-designer awareness of the key areas in the design and building process, and finally with the synthesis of human need, building, and site. I cannot overemphasize the need to integrate your house with your site. Elsewhere in this book you will find good ideas on building lower-cost homes through more efficient use of space, and energy conservation through such things as window design and location. This information can greatly assist you in beginning to crystallize your ideas and plan for the eventuality of building, even for those without an imminent building program or land ready to go. But keep one thing in mind: that these ideas and suggested designs are generalized and no house plan should be final *until it is placed on a specific site plan.* Usually, nowadays, land is as prohibitively high in cost as new construction is. This chapter will concentrate on using what is normally less costly land, namely, sloped property. Combining the cost-saving construction and energy systems explained throughout this book into a total site-house design will reduce your initial and operating expenses sufficiently that you can afford a larger piece of property. Hence, that garden, the view, or the outdoor play area for children you've always wanted comes into reach.

Why the strong emphasis on the importance of the site in our building designs? Some of the more obvious factors that make the site significant in our design process are (1) *The topography* should guide us in determining the space potential of a house. That is, a sloped site may offer the possibility of a fully habitable lower level opening to grade at little more cost than the basic pier or crawl space foundation would have cost. (2) *The land physiognomy*, physical and ecological features combined with external influences such as available sun, wind, views, and other developments nearby, must be a major influence on your house design. From the standpoint of layout, these factors will determine which rooms get the maximum direct sunlight, have the most private views, are the most protected from unwanted winds, etc. Of course, if we are dealing with a specific piece of land and all relevant information is available, then it can be synthesized into the design process and become as much a part of the design program as the size of the house and other factors which you yourself control. (3) *Layout and required utilities* must be considered. Where is water coming from? Where do wastes go? What is the best road approach? (4) *Other site functions* such as play areas, gardens, and animal shelters ought to be planned from the start, including how these relate to house functions. (5) You should also look for *free materials on the site*. What characteristics of the site offer materials for you to use and may affect your choice of materials for building?

Each of these factors will be elaborated on to develop an understanding of the indivisible nature of site planning and building design. As you review each of the points, try to consider examples of buildings which you have seen that have, in one way or another, failed to consider the positive or negative aspects of the site, thereby creating problems.

TOPOGRAPHY

The slope of the land can have an effect on the spatial design, energy-conserving construction, and character of a home. Sloped land is very adaptable and potentially beneficial to house construction. Since sloped land is usually less suitable for other functions, it can, generally speaking, be less expensive because it is less sought after by speculators, developers, and commercial farmers. Yet the very problems sloped land presents to these people may become the advantages for your building.

Wherever you build your house, and particularly on sloped land, check the type of soil and the water conditions.

Soil Type

Soil types vary from site to site and sometimes within the same site. Thus, it is important to survey the soil in the area of the foundation and determine the soil strength or bearing capacity. The lower the strength of the soil, the broader the "foundation base" must be. The following is a list of the maximum allowable loads or bearing capacities for different soil conditions:

Hard, sound rock	up to 40 tons/square foot
Soft rock	" 8 "
Coarse sand	" 4 "
Hard, dry clay	" 3 "
Fine clay sand	" 2 "
Soft clay	" 1 "

Soils firm enough for standard foundations are those composed of stone, gravel, consolidated dry or wet sand, and compacted clay not subjected to appreciable moisture changes (see Figure 5-1a). Soils composed of vegetable material, dumped earth or loose fill, or weak clays are not suitable for foundations (particularly on sloped land) unless special measures are taken (see Figure 5-1b).

If soils of several types are encountered in the same foundation area, then a design that will equalize the settlement patterns must be used. As an example, if the load from the building on each individual footing were 10,000 pounds, then the area for a footing in a coarse sandy part of the soil would need a base area of 1.25 square feet, while a footing in a fine clay and sand area would require 2.5 square feet, and a footing in a soft clay area would require 5 square feet.

Figures 5-1c through 5-1f describe various foundation solutions. Note that the use of piers will minimize the amount of excavation. However, in this case it is advisable to close off the crawl space to prevent winds from infiltrating under the floor. The best way to do this is with a skirt that stretches all the way around the foundation or by using an insulated barrier. With both the skirt and the insulation, the material runs from the floor all the way down to the ground.

Soil evaluations can be done by the owner-designer by consulting with the local soil conservation, home extension, or university service. Careful excavation and sampling in the manner recommended by these services will yield very informative results.

Water Conditions

Water conditions also help determine the house's relation to the site. If there are underground springs or a very high water table (make sure to verify this during the spring, or during a period of heavy rains), then a basement or other underground space might be ill-advised. If you must go underground in these instances, be prepared to construct your foundation walls and slab like a bathtub in reverse, in order to keep water out. There are several methods for doing this. One is the use of cement-based hydraulic compounds ("Thoroseal" standard drywall products, for example); or you could use a synthetic rubber elastomeric membrane (such as that made by Karnak Chemical Corporation). Asphalt composition vapor barriers, in

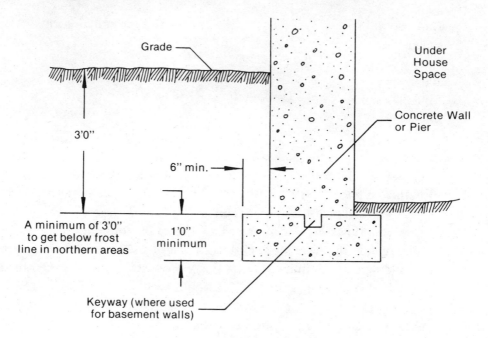

FIGURE 5-1a: FIRM SOILS TAKE A STANDARD FOUNDATION

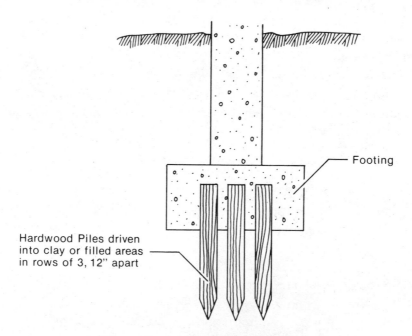

FIGURE 5-1b: WEAK CLAY OR FILL—USE PILES

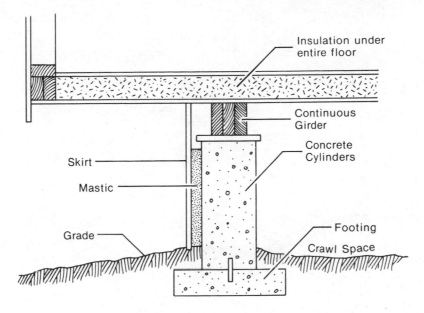

FIGURE 5-1c: PIERS

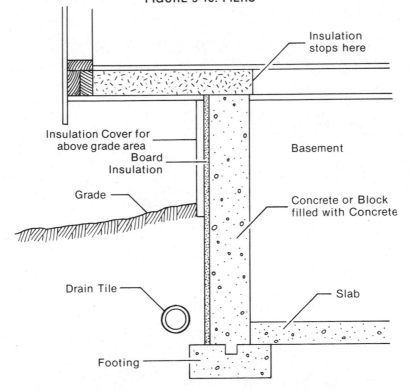

FIGURE 5-1d: BASEMENT

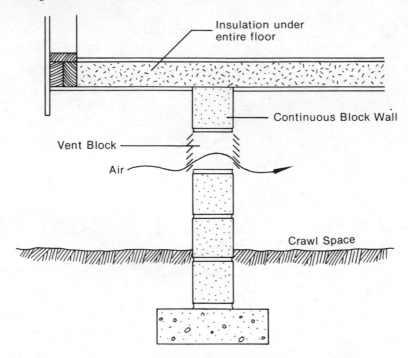

FIGURE 5-1e: CRAWL SPACE WITH CONTINUOUS FOUNDATION

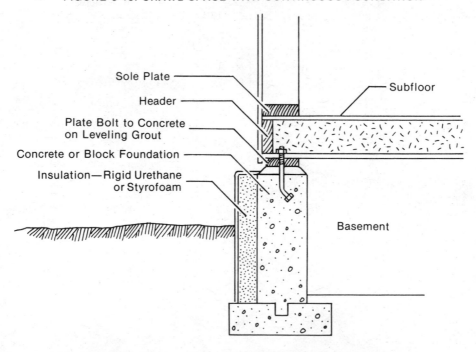

FIGURE 5-1f: TYPICAL BASEMENT OR CRAWL FOUNDATION WALL

conjunction with nonsetting catalytic asphalt sealants, are also good. All these systems are effective but costly in both material and labor. A very careful application (by a professional) is essential to avoid seams or weak joints. The synthetic rubber elastomers, though the most costly, will also be the most flexible and allow for some movement from the settlement or shrinkage of mortar in the foundation without cracking. See Figure 5-2.

The best method of protecting against water pressure build-up at the foundation wall is to provide proper drainage of water away from the walls. If there is land draining toward your home, construct a swale (sloped away from the house to a ditch) which can be provided with a gravel bed or porous pipe, directing water around and away from the foundation. In addition, all basement foundations should include a footing drain, as shown in Figure 5-3. Several types of drainage pipe are available. The most expensive and effective is a porous material of Portland cement, sand, and crushed aggregate made by the Walker Porouswall Pipe Co., Little Ferry, New Jersey. Corrugated plastic tubing is another type. This material offers the advantage of long coils (up to 250 feet long) and rapid installation by companies like the Advanced Drainage Systems, Inc., Columbus, Ohio. The third is a fiber and asphalt perforated drain called "Orangeburg." This is easy to cut and adapts to numerous plastic fittings.

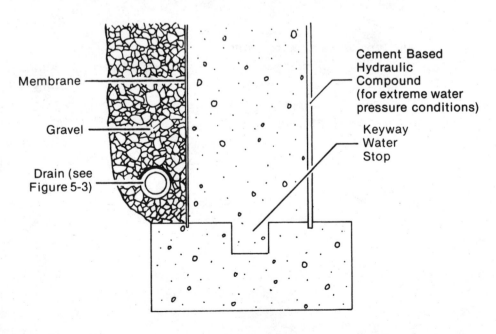

FIGURE 5-2

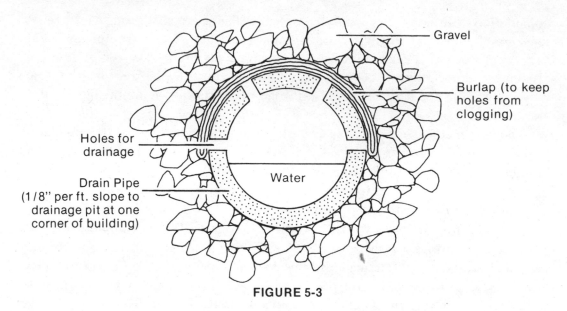

Gravel

Burlap (to keep holes from clogging)

Holes for drainage

Drain Pipe (1/8" per ft. slope to drainage pit at one corner of building)

Water

FIGURE 5-3

Installation of the drain pipe is very important. Holes should be left clear and protected from backfill that may clog them up. In extreme cases, where hardpan or dense clay soils might prevent water from draining, columns of gravel placed in burlap bags could be set along the foundation wall to the drain pipes (see Figure 5-2).

After the soil type and site water conditions are observed, a decision can be made on the advisability of building on piers, using a crawl space, or adding a full basement. If the drop in the slope from one side of the house to the other is 8 feet or more, several money-saving options open up. Both for a house with a crawl space and for a house with piers (for a home 24 × 36 feet), the extra cost of building a full basement on such sloped land would be minimal, about $2,100. This amounts to space at about $2.40 per square foot for the basement (the shell only), which is pretty cheap. Even if this space is initially left unfinished, it could be reproduced at a later date only at a much greater cost.

THE PHYSIOGNOMY OF THE LAND

When you make the decision to build, you will be making a statement that will have a lasting impact on the environment and change the nature of the landscape. Spend the time to walk over your land and study its features before you decide on a site plan.

The process of site planning is a very old one and has meant many different things to different peoples. Some cultures, such as the Chinese, have believed in a spiritual approach to land selection; that is, celestial factors and earth electromagnetic currents have guided them in site utilization. Their artisans tried to combine a sensitivity to human needs with a reverence for the environment in their site planning. Today, we can try to combine a conservation ethic with a site-planning process that emphasizes human comfort. The amount of care and planning of the building site will be reflected in the success of the energy-saving design of the house, specifically in the way the sun, wind, view, and privacy are handled. A checklist for your site should contain the following questions:

1. What is the sun's path over your site in relation to nearby ridges, tree lines, or structures during both winter and summer?
2. What are the prevailing wind directions for winter storms, as well as for summer day-night breezes?
3. What is the orientation of the view you wish to let in?
4. What is the orientation of developed areas around the site that might require screening at or away from the house?

It can be seen from the checklist that most of the items involve window areas in their evaluation. In Chapter 9 you will find a thorough treatment of the problems posed by the conflicts that often arise, such as a southern (solar) exposure facing a busy street, or the best view opening to the northwest, where prevailing winter winds come from. Needless to say, in this section on site physiognomy, we are primarily concerned with focusing on good site features to look for (or build in) to overcome these problems.

Sun and Shade

Try to orient your house directly south so you can take advantage of the winter sun. Ideally, you should experience your site for a whole year to determine the sun's path over it. If this is not possible, make a model duplicating the sun's travel and nearby obstructions. Or cross sections, such as those shown in Figure 5-4, could be drawn to examine the solar effects on your home. The sketches show both winter and summer solar angles for 40° north latitude. In this latitude the primary concern is collecting the winter sun, with a less important concern being shading of the late afternoon sun in summer, especially where sleeping areas might be seriously overheated.

Figure 5-4 shows the shading patterns for June 22 and December 21, when the sun has its highest and lowest positions in the sky. The information you need to determine the shading patterns in your area are the angles of the sun at these times of day and year for your latitude. Then, with a protractor, mark a ground line and draw in the angle the sun makes, say, at noon on December 21, as shown in Figure 5-4a. Now, move away from the house far enough that a full-grown evergreen tree (in the diagram, the height of the house, but a tree could be taller or shorter) will fit in between these lines. An evergreen tree *at least* this far from the house wouldn't shade the south windows at noon. However, if you want to take advantage of the solar heating options through windows on the south side of your home as explained in Chapter 16, then you'll want the sun to strike the south side of your home from 9 A.M. to 3 P.M. The angle of the sun at 40° north latitude on December 21 at either 9 A.M. or 3 P.M. is 14°; and, as shown in Figure 5-4a, an evergreen must be stepped back a distance equal to at least four times the height of the south wall to prevent shading. It should be mentioned that deciduous trees (which lose their leaves in winter) can be located closer to the house without serious detriment to the amount of solar energy collected than can an evergreen, as long as they are not densely spaced like a maple or poplar woods.

For the west side of your home, you'll want to have shade trees for late summer afternoons, as shown in Figure 5-4b. Using the sun's angle at 4:30 P.M. on June 22 for your latitude, construct two parallel lines intersecting the top and bottom of the west wall at this angle. You'll then want to choose shade trees so that the crown of the tree fills the space between these two lines when fully grown. Note that the farther the tree is from the west side of the house, the taller it must be. Good summer shading on the west side can also be obtained from a dense high hedge of

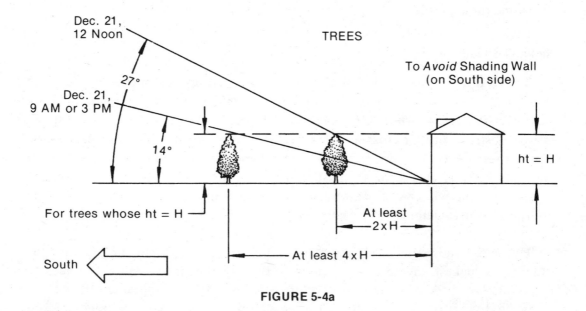

FIGURE 5-4a

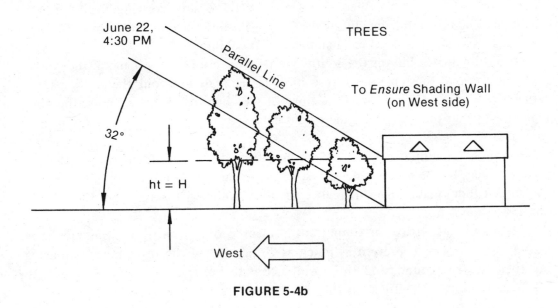

FIGURE 5-4b

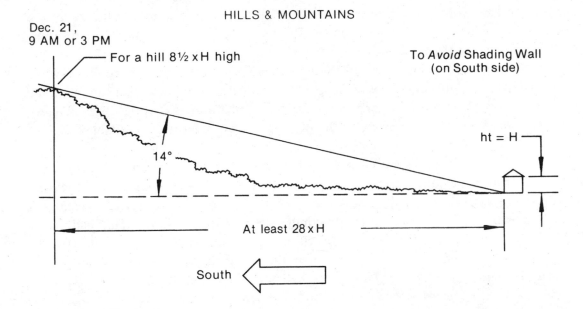

FIGURE 5-4c

evergreens, such as Canadian hemlock. However, this is slow growing. Faster growth can be obtained from tall deciduous shrubs like autumn olive or multiflora rose. The latter will also offer excellent food and cover for wildlife.

To ensure that a hill on the south side of your home doesn't block the winter sun, you'll want to step your house back far enough to avoid the hill's shadow between 9 A.M. and 3 P.M. on December 21 (see Figure 5-4c). Again, use the sun's angle at your latitude for this time of day and year to make sure that shading does *not* occur.

For those planning to build below the 40th latitude, *protection from* the sun (shading) may be the prime concern, especially in early or late summer. If you are at the 30° latitude, make up a sun angle chart for shading on June 22 at 4:30 P.M., for this gives good results.

Vegetation also plays an important role in cooling during the summer. The shading from trees can prevent as much as 70 percent of the sun's heat from being absorbed by the ground; and this, combined with the trees' transpiration process, will have a measurable effect on the air temperature around them. The air temperature will be from 10 to 15° cooler when the general temperature is 90°F. and 5 to 10° cooler when it is 70°F. Thus, it is important to be aware of the varying shade patterns offered by different trees, some examples of which are shown in Figure 5-5. Trees contribute several other "pluses" to our overall comfort. For example, their dense foliage will reduce air-borne sounds, and the viscous surfaces of the leaves will catch dust that might otherwise come into the house. Chapter 16 has more information on shading.

When you plot the winter and summer shading factors, you'll need to know the position of the sun for different times of day in different seasons. You can buy a sun angle chart that shows sun angles for each hour of the day from: Robert Bennett, 6 Snowden Road, Bala Cynwyd, PA. Be sure to specify latitude (available from 24° to 52° lat. in 2° intervals).

The sun's energy is stored outside your house as well as inside. The earth and rocks absorb the solar energy, and the more the slope is perpendicular to the sun's rays, the more heat energy is absorbed. Therefore, your house can be placed on a slope as in Figure 5-6 to help you maximize the heat storage in the fall. Shading and groundcover for the summer season will be important, though. Also note that the house and garden should be placed above the low point of the site. The low pocket will become a "cold sink." This is the area that is most affected by the early frosts and takes the longest to thaw out in spring. There is one lesson you can learn about solar energy on these outside slopes: the sun's energy is very diffuse and hard to collect. What follows naturally is the need to protect from the dissipating qualities of the wind what we can collect.

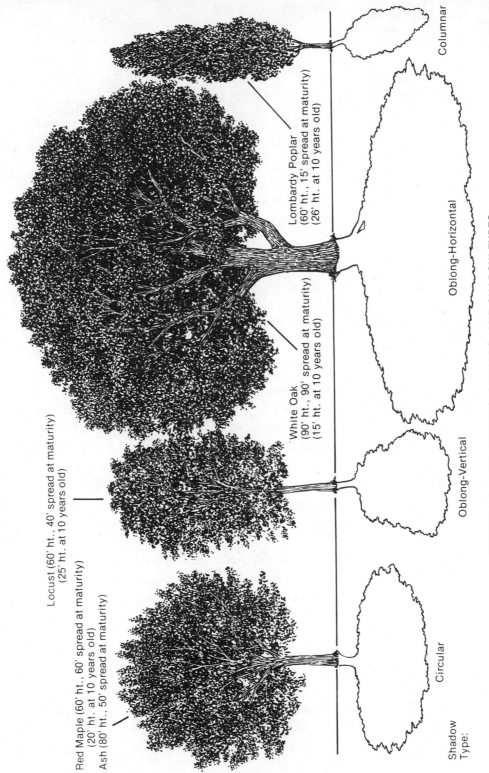

55

Columnar

Lombardy Poplar
(60' ht., 15' spread at maturity)
(26' ht. at 10 years old)

Oblong-Horizontal

White Oak
(90' ht., 90' spread at maturity)
(15' ht. at 10 years old)

Oblong-Vertical

Locust (60' ht., 40' spread at maturity)
(25' ht. at 10 years old)

Red Maple (60' ht., 60' spread at maturity)
(20' ht. at 10 years old)
Ash (80' ht., 50' spread at maturity)

Circular

Shadow
Type:

FIGURE 5-5: TREE HEIGHTS, SPREADS, AND SHADOW TYPES

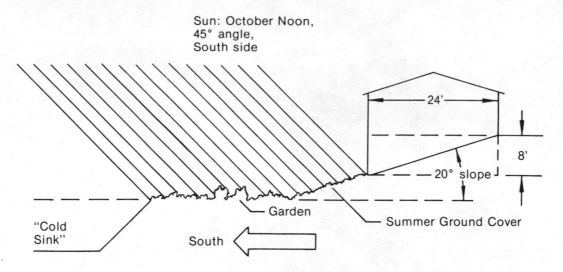

FIGURE 5-6: MAXIMUM FALL HEAT STORAGE

Wind

The wind is the burglar that comes from the northwest to break into our fragile structures. You can build features into your house that guard against the wind (described in Chapter 11) and can also create a more favorable situation for yourself with good site planning. By using natural rises in the land, or creating landscaped buffers between the house and the prevailing wind direction, you can slow up the wind sufficiently to minimize its effect. In areas above the 40° latitude, minimizing the effect of the wind during the heating season is a major concern. Figure 5-7 shows the protection effect of building a windbreak. More information will be found on this subject in Chapter 10.

If you are clearing a wooded area to build and plant a garden, or for other purposes, keep in mind that some trees may have adapted to being part of a wooded area and may not be able to stand alone. Therefore, stands or groves of trees should be considered rather than solitary ones; and these of course will also work well as windbreaks or shading.

Since objectionable winds occur in winter, the best type of tree to use as a buffer is an evergreen. If there are no natural stands, planting is required; for this, you have several alternatives. You could find suitable trees 6 to 8 feet high elsewhere on your site and transplant them as a hedge, or you could purchase young trees from a nursery stock. Small Canadian hemlock or Scotch pine will cost from $6 to $8 each. Older trees, 6 to 8 feet high, will cost from $24 to $32, or about four times

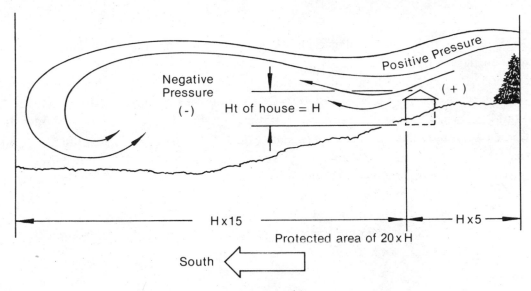

FIGURE 5-7: EVERGREENS USED AS A WINDBREAK

more than young ones. The young, small trees also must be protected more during the first few winters, since heavy snow and ice takes its toll on the weaker ones. Transplanted trees usually have to be cut back or pruned in order to assist the tree in recovering from the shock of uprooting, with the possible loss of nutrient-producing roots. With the right preparation and protection, however, you can expect 8 to 16 inches of growth per year, depending on the species.

Evergreens can be planted any time in the spring or fall, but spring is best since it gives the roots all summer and fall to grow and establish the plant. Dig the hole twice as wide and twice as deep as the root system (evergreens have long tap roots requiring deep holes). Prepare the base of the hole with a mixture of half peat moss and half coarse sandy soil. Spread the roots out carefully to allow room for normal growth. If the roots are in burlap, loosen the burlap. With the hole half full of dirt, fill it up with water and wait till the water drains out. Then fill the rest of the hole with more of the soil mixture, tapping as you fill so there will be no air pockets around the roots. Evergreens thrive in sandy soils. Mulch such as oak leaves, pine needles, or peat and evergreen food fertilizers should be used to supplement the soil, however. Shrubs and dense plantings should be kept a reasonable distance from the house so they will not inhibit good airflow. More open planting and grass near the house will provide a reservoir of cooler air to be brought into the house through openings designed to admit it. See Chapter 10 for more details.

The View

Keeping shrubbery and dense plantings away from the house not only provides for ventilation but ensures that a good view to the outside of the house is possible. Chapter 9 will explain how you can place windows to achieve the best view. However, if there is a private area just outside the house that you want to screen in, it would be best to bring the trees and shrubbery up close to that part of the house. In this case, you might want to put this area on the southwest side of your home so that the plantings will not only give privacy but, being close to the house, will also provide shade from the late afternoon summer sun. Then, with the plantings farther away from the southeast side of the house, you can use windows and vents in these areas for a view and ventilation purposes. See Chapter 16 for more details.

If we were to put together some of the ideas of this chapter on a hypothetical site, the result might look like Figure 5-8.

SITE UTILITIES

Utilities are a prime consideration in overall site evaluation. The need for water and the satisfactory handling of human, animal, and organic manure are ever present.

Water

The availability and quality of water can often be verified by consulting with helpful neighbors. Their experiences in finding water will usually be the ones you will have to contend with. If you cannot hook into a town's water supply, some water sources are springs, artesian wells, shallow and deep wells. Springs and artesian wells, though the least costly, are the least dependable. Springs are more apt to be contaminated than other sources, but if purity is maintained, the water is usually excellent to taste and soft (low mineral content).

Wells must be pumped mechanically or by electricity. A well is a dependable source of water that is normally hard (high in mineral content). Wells can be dug, bored, or drilled by the owner if the water table is reasonably shallow. A deep well, requiring professional drilling, can cost $7 per foot, and in some cases even more. Thus, a 300-foot-deep well could cost $2,100.

Waste

The problems of sanitation or waste handling are not simple. First of all, a good way to approach the subject is to avoid the word *waste* and become aware of the

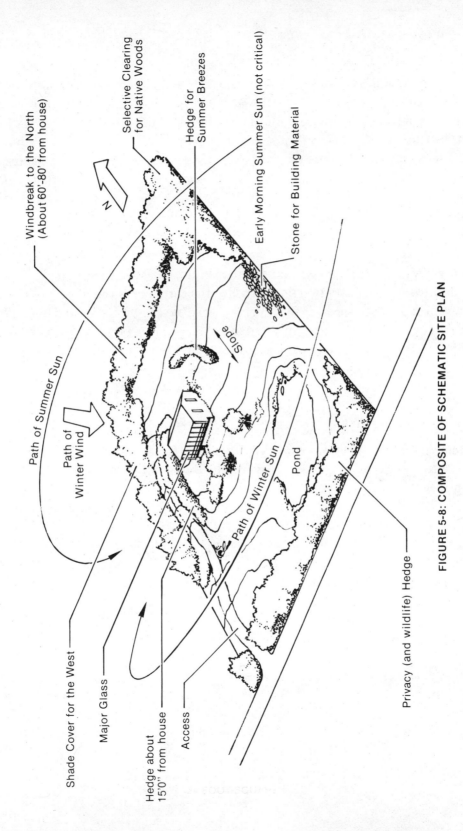

Selective Clearing for Native Woods

Hedge for Summer Breezes

Early Morning Summer Sun (not critical)

Stone for Building Material

Windbreak to the North (About 60'-80' from house)

Path of Summer Sun

Path of Winter Wind

Shade Cover for the West

Major Glass

Hedge about 15'0" from house

Access

Slope

Path of Winter Sun

Pond

N

Privacy (and wildlife) Hedge

FIGURE 5-8: COMPOSITE OF SCHEMATIC SITE PLAN

systems that recycle or utilize our sewage, kitchen scraps, and animal manure. One recycling approach is the method of methane digestion, which, as a bonus, produces a gas you can use for fuel, similar to natural gas (see the references for this chapter in the Appendix). Where family-sized amounts of human waste are available, you could also use the Clivus Multrum, or composting toilet, which forms a humus or garden fertilizer, in conjunction with your own composting of food scraps. More information on the Clivus and other water-saving toilet systems will be given in Chapter 15.

Most areas of the country have not adopted either the methane or the composting alternatives. Because of health laws that once made sense but now need reevaluation in the light of modern technology, often the septic tank/leach field system is the only "waste" disposal system permitted. If your land has poor percolation or none at all (rate of absorption of water into the earth) or is too close to "classified" streams or the watershed, then you may have a real problem, and seeking code approval for one of the modern alternate systems may be feasible. Also, read the plumbing section of this book carefully (see Chapter 15) and learn more about fixtures that conserve water and energy.

OTHER SITE FUNCTIONS

Remember the functions *other than shelter* that your site must perform when purchasing a piece of property, or laying out the location of your house. You need to consider three general areas:

1. Public use—This includes those features that set off the house, such as recreation areas, fences, etc.
2. Private use—Patio gardens, a swimming pool, etc., would come under this category.
3. Service use—This would include many things, from outdoor craft areas, storage areas, and animal sheds to considerations such as ease of getting packages of groceries into the house. The last consideration may make you think twice about locating your house on an off-the-road site or atop a steep hill with the carport below.

Most solutions, to satisfy all of these requirements, will be suggested by the site. If you find yourself saying, "If only the land were different," try to recall that the basic character or physiognomy of the land will speak very clearly, and it is up to you to listen. Once you've walked the land and studied its features, sit down with your family and draw up layouts for these public, private, and service uses until you come up with something that is pleasing and satisfies everyone's needs.

FREE GIFTS FROM THE LAND

One final area of site-house interaction involves the materials that may be found on the site for use in building. Putting these on-site construction materials to work can provide many fun-filled weekends for the whole family and is a wonderful way to get to know your property and its potential. I will describe some of these materials, their suitability for building, and a few of the techniques that you can use to integrate them into your construction. Further information can be found through the references for this chapter.

Earth

Earth has been used as a building material for as long as humanity has been making houses. Although earth is a material that is readily available and cheap, with good insulating qualities, there are limitations to its use because the process of making the building materials from earth is labor intensive. While it's possible, and many people do make their entire house from this material, I'd suggest a small project for a start, say, a retaining wall on some part of the property. The use of earth also has its problems due to its lack of strength, its vulnerability to moisture, and the erosive effects of external agents. The process of *stabilization* is a method of improving soil characteristics. For example, Portland cement can be added as a stabilizing agent. The stabilization process consists of taking soil from the earth, pulverizing it, adding to it a certain amount of cement, adding water until the optimum moisture content of the mixture is reached, and subjecting it to moderate pressure. This produces building material which, when set, possesses great strength. This material is able to bear a much higher work load than could be carried by soil without cement, and it is durable enough to withstand the continuous effects of atmospheric agents. The mixture is called *soil cement*. Three soil classifications are usually mentioned as suitable for soil cement construction:

Sandy Soil. This is soil that is stable when wet but unstable when dry. Sand has a high internal friction, but no cohesion and no plasticity. It does not contract when dry, and it is compressed almost instantaneously when pressure is applied.

Silt Soil. Silt is composed of small particles of soil. It has little cohesion or internal friction and is very difficult to compact.

Clay Soil. Clay has high plasticity and readily takes the desired shape. It is smooth to the touch and sticky when wet. Dry clay absorbs a considerable portion of water with a notable increase in volume. When it then dries, it returns to its original volume, but unlike sand, shrinkage produces cracking. Clay is virtually impervious and compresses very slowly when pressure is applied to its surface.

Before earth is used as a building material, a survey of soil samples from the site must be made, and physical as well as mechanical tests should be made to provide detailed knowledge of the soil's properties. These tests can be made by the local soil conservation service, university, or independent testing laboratory.

To prepare for a survey, cut a cross section about 6 feet deep below the topsoil, noting the types of earth encountered. A good practice is to contact the soil-testing agency first to determine what type of sample they will require in order to make their tests. You will want tests for the following characteristics: cohesion, compressibility, elasticity, and capillarity. From these, you can classify the soil as sand, silt, or clay.

Once the classification has been determined, the next step is to determine the cement and water proportions. The following are the normal percentages of cement to be added to each soil type:

Sand: 4.75% to 9.10%

Silt: 8.35% to 12.5%

Clay: 12.5% to 14.4%; 15% or more is not recommended.

For each soil type, several sample blocks using different percentages could be made and after a 15-day drying out period, the blocks should be tested for compressive strength.

The mixing process should be made as simple as possible by following these steps:

1. Prepare a level, hard, and nonabsorbent base—an area that can be kept consistently moist and clean.

2. Measure out the components in the specified proportions, using a receptacle of known volume and one which is easily handled. Deposit the soil first in layers no thicker than 3 inches.

3. The cement is added by uniformly sprinkling it over the soil.

4. Then, mix the components (by hand or with a mechanical mixer) until a uniform color is achieved.

5. When the dry soil and cement are thoroughly mixed, the mixture is spread out again and water is sprinkled on it from a watering can until the moisture is uniformly distributed throughout the mixture. The amount of water is determined ahead of time by the simple field tests given below:

> • The proper amount of water has been added if the mixture retains the shape of the hand without soiling it and if the mixture can be pulled apart without disintegration. An additional test is to drop the block from a height of 3 to 6 feet onto a hard surface. If it is properly prepared, it will disintegrate into a loose mixture similar to the original mixture.

● If there is too much water, the mixture will retain the shape of your hand but will stick to it and soil it. When the mixture is dropped, it will not disintegrate, but will flatten out.

● You have not added enough water if the mixture crumbles and does not retain the shape of the hand at all.

When the proper mixture is ready, it can be molded into the form desired for building. There are two basic ways in which soil cement can be used. One is "monolithic walling," where the material is placed in forms and compacted as the wall is erected. The second is to make "blocks," molded as individual units then joined together as with bricks, stones, or any other masonry units. Figures 5-9 and 5-10 illustrate both methods. A mold for making individual blocks is shown in Figure 5-11.

When the mixture is placed in a form or mold, the compacting process must be completed before the drying process is complete. This occurs within a 2-hour period. The drying and curing processes for walls and blocks are different. Monolithic walls must be covered with wet sacks for the first 3 days and sprinkled with a spray of water until the end of an 8-day period. Blocks, on the other hand, should be stored with maximum protection from sun and rain. They should be allowed to dry slowly and without drastic changes in temperature. Loss of moisture must be controlled for the first 24 hours by watering the blocks regularly. On the third day, they should be stacked or piled and the spraying continued until the end of 8 days. They can be used for construction 21 days after they are made.

Another way to make earth blocks is by using a machine press. This system offers the advantages of speed and uniformity of block size and strength. In addition, the equipment is relatively inexpensive, particularly if you share the cost with friends. Photos 5-1 and 5-2 show a block machine called the Cinva-Ram and a completed lower floor addition to a house constructed from these blocks. The materials cost for the 16 X 16-foot addition in Photo 5-2 was about $1.25 per square foot or about $300 total (in 1972). It was easily possible to make 40 blocks per hour with two people operating the machine.

More information on the Cinva-Ram machine and earth block construction techniques can be obtained from Bellows-Valvair International, 200 W. Exchange St., Akron, Ohio 44309. Their Cinva-Ram machine costs about $250 f.o.b. Akron.

Wood

A wooded site can offer a rich source of construction woods, depending on the type, age, and ability to mill the logs properly or use them in their natural state (to build a log cabin, for instance).

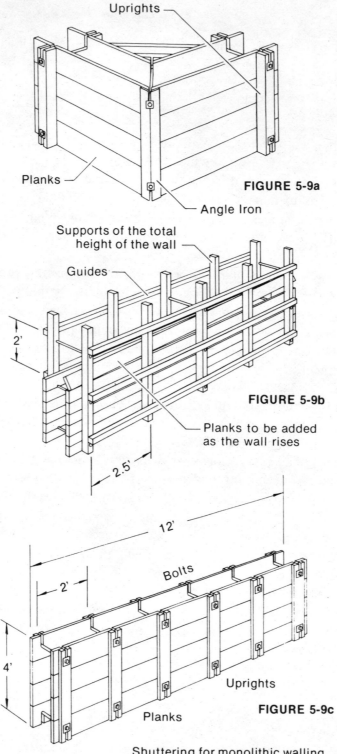

FIGURE 5-9a

Uprights

Planks

Angle Iron

Supports of the total
height of the wall

Guides

2'

FIGURE 5-9b

Planks to be added
as the wall rises

2.5'

12'

2'

Bolts

4'

Uprights

Planks

FIGURE 5-9c

Shuttering for monolithic walling
(Argentine Portland Cement Institute)

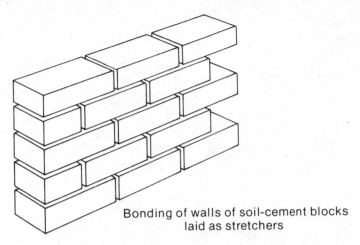

Bonding of walls of soil-cement blocks
laid as stretchers

FIGURE 5-10a

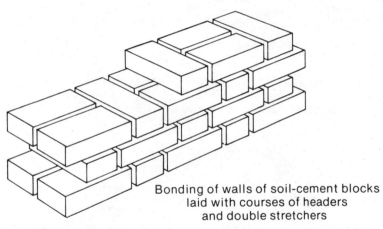

Bonding of walls of soil-cement blocks
laid with courses of headers
and double stretchers

FIGURE 5-10b

Types of Woods. There are very good field guides available to assist you in identifying the trees on your woodlot; we've listed some of these guides in the references for this chapter. At this point, however, let us point out that there are two basic classifications of woods: hardwoods and softwoods.

Hardwoods. These woods come from basically broadleaved trees. In fact, the term *hardwood* sometimes has no reference to the actual hardness of the wood, but refers only to the leaf shape. Some species in this category are ash, beech, red cedar,

FIGURE 5-11a

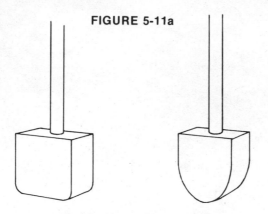

Rammers for compacting
monolithic walling

FIGURE 5-11b

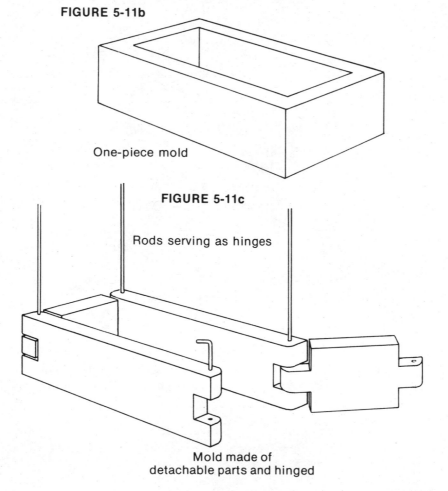

One-piece mold

FIGURE 5-11c

Rods serving as hinges

Mold made of
detachable parts and hinged

Photo 5-1: Earth blocks being made with a Cinva-Ram machine.

hickory, locust, and maple. Hardwoods are generally used for interior trim, flooring, furniture, and other purposes requiring durability and beauty.

Softwoods.   Woods known as softwoods come from the group of trees bearing needles or scalelike leaves; for the most part, they are evergreens. As with hardwoods, the term *softwood* has little actual reference to the softness of the wood.

Photo 5-2: A 16'x16' room addition to an old house where earth blocks were used as walls.

Species of softwood include Douglas fir, eastern hemlock, and Norway pine. They are used for structural framing and rough carpentry.

Selection of Lumber.   There are some facts you should be aware of as you survey your site for both clearing and usable wood sources. Lumber taken from trees killed by insects, blight, wind, or fire is undoubtedly as good for any structural purpose as that from live trees, provided additional injury from further insect attack, decay, or seasoning defects has not occurred. Tests on wood standing as long as 15 years after being killed by fire demonstrated that the wood was sound and as strong as wood from living trees. The important consideration is not if the tree is alive or dead, but whether the products are free of decay.

Decay.   Decay, even when sufficiently advanced to affect seriously the strength of a piece of wood, is often difficult to detect. Therefore, timber from dead trees requires more careful inspection. There are three places to look for decay. The first is in the sapwood, which is the light-colored wood beneath the bark; the second is in the heartwood, which is the darker colored wood from the sapwood to the center. These areas are usually affected by wood-destroying fungi. The early stages are often accompanied by discoloration of the wood, showing up as a somewhat water-soaked appearance. Later stages of decay can be more easily recognized because of definite changes in the color and substance of the wood. Here, the rotted wood may be white or brown. The brown, crumbly material is called "dry rot," though since no wood decays without moisture, the name is a misnomer. In living trees it is the heartwood that is more susceptible to decay. In logs and wood products the sapwood decays more readily.

The third area of decay to watch for is the surface. Here, fresh surface growths of decay fungi usually have a fluffy cottonlike appearance, or the appearance may be powdery in the case of molds.

The occurrence of decay-producing fungi may be greatly reduced by storing wood properly after it is cut. Logs or milled lumber should be kept off the ground and away from moist surfaces. The object is to reduce the moisture content in the wood, below 20 percent if possible. A common method for doing this is called air drying: lumber is stacked with spaces that allow complete air circulation, and the pile is located in a dry area and placed off the ground, crisscrossing the layers of wood with lathing strips 1 inch thick set between each layer. In addition, the top layers of the pile should not be allowed to be exposed to the sun and rain or warping and checking (splitting at the end) may occur.

These methods of keeping wood dry should be used whether you cut your own or buy materials from the lumberyard. If you are building a detached garage or other outbuilding, try to get this structure up first. Then it can serve as a storage place for sheathing, siding, studs, and joists.

Shrinkage. When you or your local mill cut up wood from your property, be careful about the shrinkage patterns. This shrinkage occurs as the wood dries and is most severe in a direction tangential to the annular rings; it is relatively unimportant in the radial direction. Figure 5-12 shows the effects of shrinkage on different shapes of wood cut from different areas of the tree's cross section.

Stone

Stone is available as a building material in many parts of the country. Rubble or common fieldstone from your property laid in walls whose thickness is 12 inches or more can be used as structural support walls. Stone walls have to be laid dry, and their strength is dependent on the tight or accurate fit of each stone.

If a stone wall is to carry a load, it is best to use stone in conjunction with mortar (one part cement, six parts sand, and one part hydrated lime). See Figure 5-13a. Figure 5-13b shows a result that could be obtained by installing the stone along with the concrete wall forms and pouring the concrete behind the stone. If the

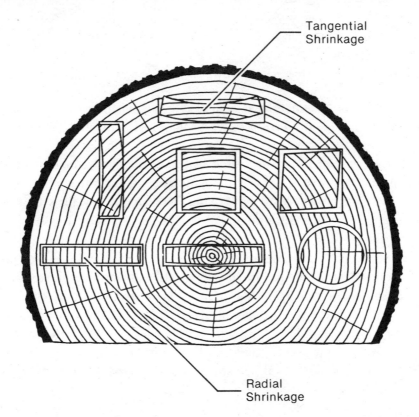

FIGURE 5-12: RECTANGULAR, CIRCULAR, AND SQUARE CUTS FROM DIFFERENT PARTS OF THE TREE SHRINK DIFFERENTLY

VIEWS FROM THE SIDE: STONE WORK

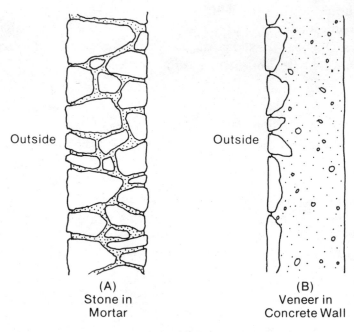

FIGURE 5-13

walls are laid up without concrete, it is desirable to "batter" the wall (e.g., make it wider at the base than at the top in order to help resist any horizontal pressure on the wall).

Be careful when you use stones such as limestone, shale, or sandstone, which have stratified seams. While it is easier for you to split and work with these materials, they are also more likely to split when laid up in a wall with their seams horizontal. For these materials, the seams should be placed vertically or the stone should not be used at all.

Although stone is a low-cost material, is fun to work with, and is pleasing in appearance, you will still have to insulate the inside walls of your home for comfort. The total resistance (R) to heat flow of a 12-inch stone wall is $R = 2$ (no more effective than a half inch of fiberglass insulation). This is about a tenth of what is really necessary (see Chapter 12). If you want to use stone on the inside of your home for the sake of appearance, build with a cavity construction, where a space is created between the outside and inside stone walls. The space can then be filled with insulation.

chapter 6

SMALLER HOUSES THAT UTILIZE SPACE AND MATERIALS MORE EFFICIENTLY

by ALEX WADE

PLANNING FOR THE MOST ECONOMICAL USE OF SPACE

Why does it cost so much today to build a house? Chiefly, houses have become increasingly wasteful of space and materials. Thus, at present, the only readily available economical choices for housing are trailers or apartments. Trailers, while frequently efficiently laid out, are not well insulated, often constructed of poor quality materials, and regarded by the lending industry as "temporary housing." As for apartments, these share many of the problems of trailers and, like them, are often too small for the average family. Some rational alternatives to these kinds of housing are clearly needed. This chapter will show you how to put economic and efficient house designs to work for you.

If you want to build a home today, what price would you expect to pay? Right now, the average cost of the typical detached ranch-style house in this country is above the $45,000 mark, well beyond the reach of the median income family and the rapidly dwindling supplies of mortgage money. Building costs vary widely, depending upon the area of the country and the capabilities of the contractor. For the most part, the cost of general living space ranges from $25 to $40 per square foot, and the kitchen and bath areas can run to double this amount. As can be seen from these figures, a reduction in the amount of space you build will result in considerable cost savings, particularly if these reductions are in the kitchen-bath areas.

Certain factors need to be studied if you want to bring down the costs of your new home or renovate a house efficiently. One major factor which is usually overlooked is the *shape* of the house. Geometrically, the circle encloses the most area with the least perimeter. Specifically, it requires 22 percent less wall area to enclose an efficient circular plan (the one we recommend in Chapter 19 was developed by the Forest Products Laboratory, of Madison, Wisconsin) than it would for a rectangular plan enclosing the same area. Though circular houses are difficult to design and have consequently never been popular, the Forest Products Laboratory house is very well planned and detailed. The U.S. government has built prototypes of this house and claims that it is possible to build it for *one-half* the cost of a conventional home.*

Since a square is also very efficient (12 to 15 percent more than the conventional rectangle) and much easier for ordinary contractors to build, we will confine most of our attention to square plans. (A couple of compact rectangular plans have been included for those who do not want a square house, though.) Figure 6-1 shows a design for a basic square house.

In order to obtain the required amount of space, most of our plans are organized *vertically,* with two or more levels. In addition to requiring less wall area to build, a two-story square house costs much less per square foot of floor area than a rectangular one of one story but equivalent area. The reason for this is that you are building only half as much roof and foundation per square foot of house. If you add attic lofts, you save even more. Therefore, we have given you these vertical designs to make the most efficient use of space; all of our designs have been carefully laid out to allow spaces to flow into one another so you can achieve elegance at low cost.

The best way to economize on house design is to take each space, study its actual requirements, and decide for yourself what's needed. If you want, you can save space in each area of your house. Or you can economize in one area, which will allow you to expand in another. Either way, you save. As we go along, take a piece of paper and work up floor layouts according to your own needs. Let's begin with the most expensive and potentially wasteful areas.

The Kitchen

The typical "modern" kitchen wastes quite a bit of space and hence many thousands of dollars. A very workable kitchen can be laid out in a linear, one-wall design 8 to 10 feet in length. Another space-saving design is for a compact U-shaped kitchen which can be installed in less than 50 square feet (see Figure 6-2). This sort

---

*Circular house plan in two sizes (804 and 1,134 ft^2) is available from the U.S. Government Printing Office, Washington, D.C. Order Plan No. FS-SE-5. Price is $2.00. See Fig. 19-22.

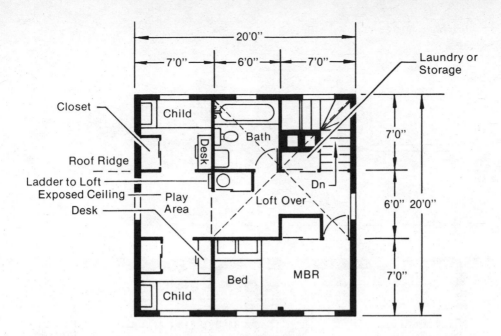

SECOND FLOOR PLAN

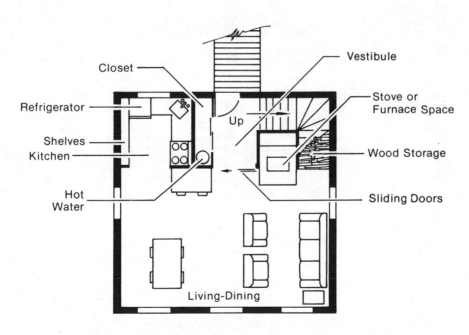

FIRST FLOOR PLAN

FIGURE 6-1

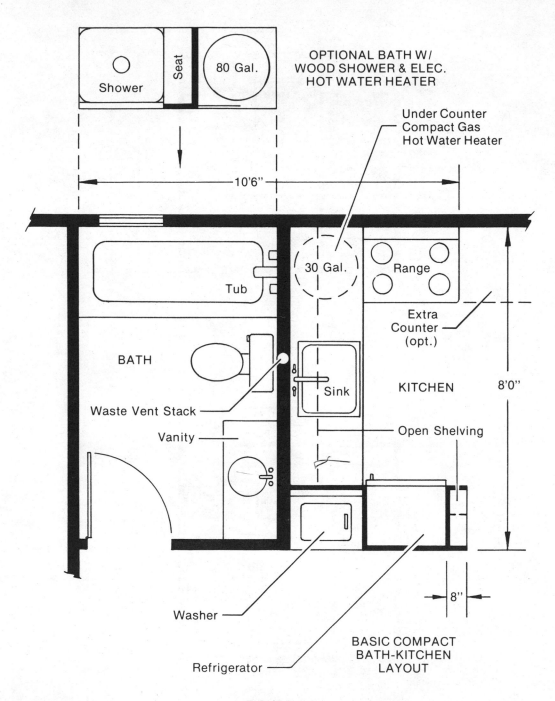

OPTIONAL BATH W/
WOOD SHOWER & ELEC.
HOT WATER HEATER

Under Counter
Compact Gas
Hot Water Heater

Shower

Seat

80 Gal.

10'6"

30 Gal.

Range

Extra
Counter
(opt.)

Tub

BATH

KITCHEN

8'0"

Waste Vent Stack

Sink

Vanity

Open Shelving

Washer

8"

Refrigerator

BASIC COMPACT
BATH-KITCHEN
LAYOUT

FIGURE 6-2

of kitchen saves a lot of unnecessary walking, too, yet still offers more than adequate work space. Space savings can be made by eliminating certain built-in appliances like garbage compactors, dishwashers, and the like. Also, by not building the range and refrigerator into the kitchen, it is possible to shop for used appliances, which are usually bargains. Older appliances, particularly gas ranges, were often of better quality than those produced today.

There are other ways to save money when you build your kitchen, too. Open shelves instead of expensive wall cabinets can be easily constructed by the owner for very little money. These are attractive, eliminate the danger of someone hitting his or her head on an open cabinet door, and provide ready access to the shelf contents. Next, the do-it-yourselfer can make handsome and inexpensive countertops by laminating scraps of wood such as hardwood flooring pieces. Plastic laminates should be avoided as they are expensive and amateurs can easily ruin large sheets of material.

The position of the kitchen within the house should be studied carefully with regard to the kitchen sink. For maximum economy of plumbing, the sink should be back to back with the bathroom plumbing or on the same vertical wall in a two-story scheme. The kitchen designs in Figures 6-1 and 6-2 are examples of this type of layout.

Bathrooms

The average tract house comes equipped with two and a half baths as standard equipment. In our basic layout, Figure 6-1, we propose a reduction to one bathroom with the fixtures located on the same wall above the kitchen. In the rectangular one-story version of the basic house, Figure 6-6b, we show two bathrooms. Note that this type of layout separates the bathtub and shower into two separate compartments that can be closed off, allowing more persons to use the facilities than in the standard all-in-one-room scheme. And it costs much less to split the fixtures into separate spaces than it would to provide two full baths.

Figure 6-2 shows yet another space-saving possibility. Using the half-bath concept, the hot water heater is integrated into the bathroom layout by substituting a shower for the tub. Showering can help you save on your water bills as well (see Chapter 15), and a really nice enclosure can be constructed of tongue and groove cedar boards. Sealant (silicone) should be applied between the joints of the boards as they are installed, and a coating of Woodlife or Cuprinol (clear) such as we will recommend for exterior waterproofing will help preserve the wood.

As plumbing is one of the most archaic and expensive of the building trades, this is an area in which you can make considerable savings. Keep in mind when laying out kitchen-bathroom areas that in addition to exorbitant costs for materials, there is a considerable loss of heat (and hence money) in transporting hot water over long distances. In remodeling work or other cases where long runs of piping are absolutely necessary, it will pay to thoroughly insulate the piping. Also, as both the main large discharge pipe and the main vent stack through the roof are connected to the toilet, it makes sense to place the toilet in the middle of the layout between the tub and the sink. The kitchen-bath design in Figure 6-2 is an example of this type of layout.

The Utility/Laundry Room

This space can easily be eliminated. These rooms usually house the furnace, washer, dryer, and hot water heater. In a really well laid-out scheme, all this equipment can be placed in space that is normally wasted. For instance, in a compact well-insulated house, you won't need an enormous furnace. A thermostatically controlled wood stove or a smaller under-the-floor furnace may be used; or a horizontally mounted forced-air furnace can readily be installed in a crawl space or under the upper roof (attic).

You can save on the space normally used for a washer and dryer also. A leading consumer testing organization has shown that it is cheaper to use commercial laundromat facilities than to own laundry machines. Thus it would be possible for some families to eliminate troublesome and noisy laundry machines altogether. Obviously, families with small children will desire machines on the premises. We have shown compact laundries in most of our schemes. These take advantage of washer and dryer units which stack vertically. Our laundry units have also been tucked into hall spaces so they can be adjacent to the plumbing.

Hot water heaters are another waster. A typical electric hot water heater uses twice as much money in heating a gallon of water as does a gas heater. Electric heaters are also much slower, which means that one must have a huge storage tank to compensate for their inefficiency. Obviously, we have to find some place to put this huge tank. For a comparison, a family of four can easily use a 30-gallon gas heater for their needs, while the same family would require an 80-gallon electric heater to provide similar amounts of hot water. Rudd Manufacturing Company and Rheem Company make compact 30-gallon gas hot water heaters which can fit under a standard kitchen counter. Your local plumbing supplier can order these units for you. Though these heaters must be vented and gas is in short supply, they should still be considered where possible. Our basic expanded kitchen-bath plan in Figure 6-2 shows one of these heaters fitted into an otherwise "dead" corner of the kitchen counter.

Because of fuel or code problems, a gas undercounter heater may not be practical in some cases. Local electric utility companies are now marketing a very high recovery electric undercounter model with the trade name "Day-Nite," which is set up for demand metering to make it more economical to operate. Finally, bear in mind that if you keep your plumbing system very compact, with all the pipes under the counter along with the water heater, everything but the bathroom will be protected from freezing; thus, if you leave the house for long periods over the winter, you can turn the heat way down. A small electric heater can prevent the freezing of pipes in the bathroom.

The Basement

If constructed, this space should be fully utilized. Basements tend to become storage areas for unused or unwanted items. In these times of drastic inflation, this is a minor luxury which you might want to consider doing without. As an alternate to the storage basement, the two-story basic house plan in Chapter 19 will show a basement that opens on grade level and provides extra bedroom space. This should be done, however, only on specially selected sites. A steep, south-facing lot with well-drained soil is ideal for a living basement. In this case, the exposed south wall admits light and air and the buried north wall protects against cold. (Also, Chapter 16 will explain how this basement construction can be used for solar heating.) However, this type of construction requires careful engineering to withstand earth pressures and moisture penetration.

Underground foundation walls should be constructed with a minimum of 12-inch concrete blocks, reinforced with horizontal block reinforcement every other course. Insert a #4 bar in the vertical cores every 4 feet and fill around the rod with mortar. Also, make sure that you buy three-core blocks so that the cores will line up. Then, waterproof the outside of the wall with membrane waterproofing, best applied by a roofing contractor, and install insulation as described in Chapter 12. Finally, install perforated drain tile and backfill the area next to the wall with porous gravel fill. This is an expensive solution, and those who construct in very dry areas with well-drained soil may get by with lighter construction. Remember, though, that walls are difficult and expensive to repair from the inside and rebuilding a cracked wall is *very* expensive. For the above design, we assume a residential foundation wall no more than 9 feet high and 24 feet long. For extra high or long walls or those supporting heavy loads of adjacent buildings or the like, we urge that you hire a professional engineer or architect to design the wall for you. If in doubt, *ask questions of a professional.*

The Attic

This space has been eliminated from most contemporary houses by the use of roof trusses. For a small house, though, trusses are completely unnecessary and wasteful of materials. We propose to reinstate the attic and make use of the space in a constructive way. The secret to gaining this space is post and beam construction. In addition to being quicker to build and more efficient of materials, the exposed beams make a beautiful ceiling. Our plans show ceilings open clear up to the roof in some areas to make small spaces visually larger. In addition, we show sleeping lofts under the eaves in several of the plans. If careful attention is given to insulation and ventilation, these lofts are warm in winter and cool in summer.

Later in this chapter and in Chapter 19, you'll find post and beam designs. Some make use of "shed-type" roofs which are very easy to construct. These roofs consist of a single sloping plane, generally with a slope of 4 inches vertically to each 12 inches of horizontal run. By sloping the roofs in this fashion, and by utilizing the post and beam system, you will be able to make good use of the spaces near the peak of the roof for sleeping and utilities.

Bedrooms

In addition to the sleeping lofts just mentioned, you can easily improve upon the space utilization of floor-level beds. In most tract houses, bedrooms are laid out in such a way that the bed has to be "floated" in the middle of the space, and as a result the area around the bed is largely wasted. We advocate building the bed into place with access from only one side and consolidating the walk-around space elsewhere. Moreover, this built-in bed can easily double as a storage chest. A hinged plywood lid can provide both a cover for the chest and a base for the mattress. (See Figure 6-3.) Build the bed a bit higher than normal (say, 28 inches) and it will be easier to "make" each morning. Also, a low, ranch-style chest neatly fits underneath.

Foam mattresses are best for this kind of bed. Two types are widely available. By far the cheapest and most common is plastic foam generally sold as "polyfoam" or "serofoam." These mattresses are usually very comfortable and hygienic; however, they may be hot in southern climates. In areas of extreme heat, ventilation holes should be drilled in the plywood and the mattress covered with a quilted cotton material. Much more expensive is the latex foam mattress. Latex foam is constructed with cores to permit the mattress to breathe. Either of these types is firm enough to provide comfortable support when used directly over a solid base. Leading hospitals, such as Massachusetts General, use latex mattresses because of their superior support of the back.

Children's beds can be similarly constructed. The beds can be placed in private

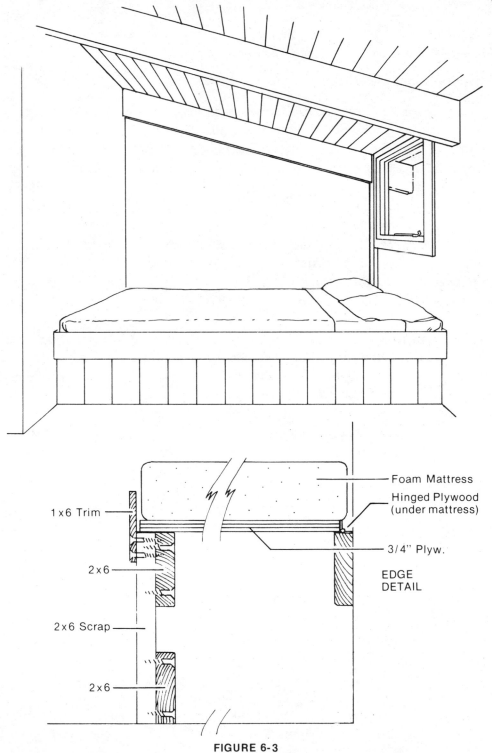

Foam Mattress

Hinged Plywood
(under mattress)

1 x 6 Trim

3/4" Plyw.

2 x 6

EDGE
DETAIL

2 x 6 Scrap

2 x 6

FIGURE 6-3

cubicles that open into a common play area. In the same way, part of the living room can be separated by sliding doors to provide guests with sleeping space.

The Dining Room

For best space utilization, this room should be combined with either the kitchen, the living area, or both. It should definitely not be a separate room in new construction as the very nature of a dining table and chairs is wasteful of space. If the table is placed in a separate room, enough space must be allowed to push the chairs away from the table. Also, enough extra "visual" space must be allowed to keep the room from appearing cramped. If the dining table is free standing within another living space, the area around the table becomes a visual and usable part of the other space. In this fashion, the dining table is likely to be used for many activities other than dining. Also, the table may be moved completely out of the way to enable expansion of the living area.

The Living Room

In order to reinforce our concept of attractive, open, flowing spaces, the living room should be combined with other rooms. The size of living room you can afford, if walled off by itself, might look crowded. You can easily avoid this by combining it with the dining room and/or kitchen. Combining these spaces into an open plan has other advantages also. Distribution of heat in winter and ventilation in summer are greatly facilitated. Another way to save money here is to build storage under the couches in the living room just as we did with the beds described earlier. By building this type of couch/storage space yourself, you can save a large amount of what is normally spent for furnishings.

The Garage

This is yet another dispensible luxury that has become a standard feature of American housing. At the most, consider a one-car carport which will help keep ice and snow off the vehicle in winter and can be used as an outdoor living space in mild weather.

To avoid the clutter of garden tools, bicycles, and other items that collect in garages and basements, we suggest that an outdoor storage unit about 3 feet deep by 8 or 10 feet long be constructed. This could be used as a windbreak for the carport/outdoor living area or as a privacy screen from neighbors or the street (see Figure 6-4).

Let's examine the results of our room-by-room economizing by comparing in a table the room sizes of our sample house with those of a smaller tract house (see

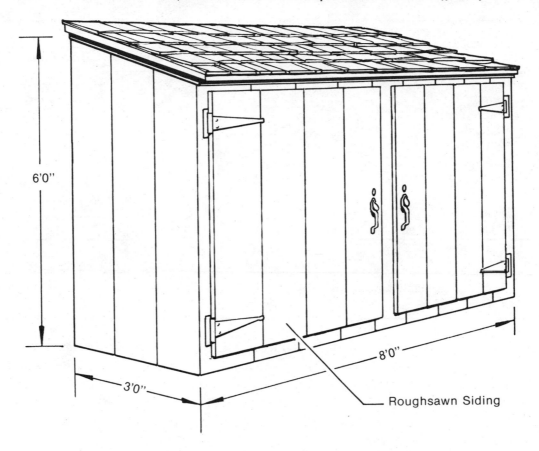

6'0"

3'0"

8'0"

Roughsawn Siding

OUTDOOR STORAGE/SCREEN

FIGURE 6-4

Table 6-1). We have also provided in the table an extra column so you can compare your own or any other design.

If you put all these ideas to work, it could save you about a third the cost of a new house.

It must be emphasized that building just "any old small house" can result in disaster. Typical houses constructed immediately after World War II had 900 to 1,000 square feet. Most of these houses were miserable because they were split up into numerous small-sized rooms. The result is most unpleasant. There is no reason, however, why a small house cannot feel spacious. We have concentrated in our designs on making the space required for sleeping more efficient. This is accom-

Table 6-1: *Comparison of Room Sizes*

Space	Tract House	Sample House	Your House
Kitchen	90	50	_____
Baths	100	40	_____
Laundry/utility	80	20	_____
Master bedroom	150	100*	_____
Bedroom #2	80	50*	_____
Bedroom #3	80	50*	_____
Family room	150	110	_____
Living room	200	160	_____
Dining room	120	100	_____
Circulation, storage, fireplace	150	120	_____
Total	1200	800*	_____

*This plan has an additional 96 square feet of loft space for sleeping, not included in these computations.

plished in the two ways already described: first, by building lofts in upper areas; second, by built-in beds in the bedrooms to make better use of space. Living-dining-kitchen areas are designed to flow together to make the spaces seem much larger. The result of all this is much more usable space than in the typical smaller house. We illustrate the principle with two 24-foot square houses. Figure 6-5b is conventional; Figure 6-5a follows the above principles. In addition, one of the houses shows how the use of cantilever construction can save a great deal of money. Both houses are pole-type. The conventional one requires nine poles, our alternate only four. The four-pole design is actually stronger and is much simpler and quicker to construct. With the loft space, this house is ample for a family of three children and may be preferred by some to the 800-square-foot design.

Our 800-square-foot design in Figure 6-1, however, should be ample for a family with two children, and adequate for one with three. If you have a larger family, you may wish to build one of the larger designs shown in Chapter 19, or design a house of your own. In Chapter 19, we also show a 20 X 24-foot "salt-box" variation on the 800-square-foot house, adding 160 square feet to the original plan. (The structural section shown in that chapter can be used for either house.) In addition, the two-story basic house plan in Chapter 19 shows an added basement for the 800-square-foot design, which brings the size up to 1,200 square feet, the same as a typical tract house. However, as the space is stacked into three levels, it is considerably cheaper to build.

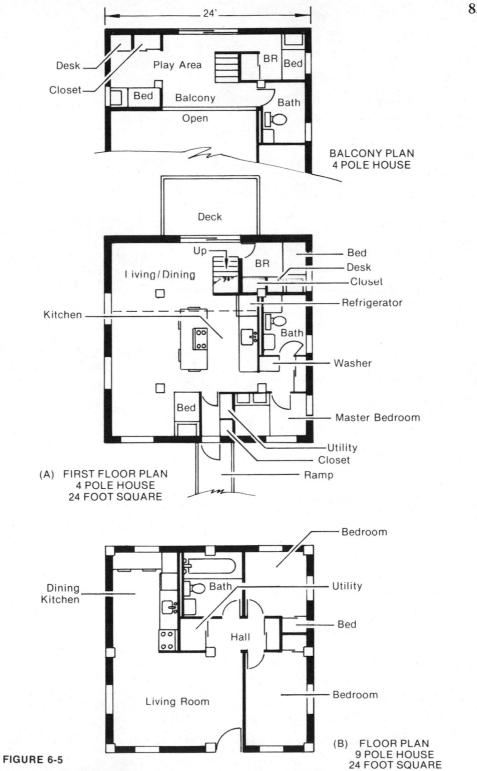

24'

Desk
Closet
Play Area
Bed
Balcony
Open
BR
Bed
Bath

BALCONY PLAN
4 POLE HOUSE

Deck

Up
Living/Dining
BR
Bed
Desk
Closet
Refrigerator

Kitchen
Bath
Washer

Bed
Master Bedroom
Utility
Closet
Ramp

(A) FIRST FLOOR PLAN
4 POLE HOUSE
24 FOOT SQUARE

Bedroom

Dining
Kitchen
Bath
Hall
Utility
Bed

Living Room
Bedroom

(B) FLOOR PLAN
9 POLE HOUSE
24 FOOT SQUARE

FIGURE 6-5

A final variation, which we detail under "Renovations" in Chapter 19, is a garage apartment that can be used as living quarters by older children or rented.

Another way to save money is to buy and renovate a smaller, older house by redesigning the interior with these plans in mind. Some ideas about how you can efficiently renovate a house will be found in the next chapter.

PLANNING FOR THE MOST ECONOMICAL USE OF MATERIALS

The American housing industry has been notoriously inefficient in comparison to the rest of the world in its utilization of labor and materials. Many excellent prefabrication and labor-saving devices and systems are available, but most of these are little used in this country. Information on these systems will be found in Chapter 2. We will concentrate in this chapter on ways of improving the efficiency of wood-frame houses primarily in on-site construction.

Most standard tract houses are designed entirely backwards from the standpoint of structural efficiency. Usually, the plan is selected with no thought to structure, and the structure is then forced to conform to this plan. Doing it the other way around is far more sensible. If a basic size and shape of house is first selected, a structural grid can be established. Graph paper with eight squares per inch is very handy for this purpose. A floor plan can then be worked into the grid, following the principles established in the first part of this chapter. In Figures 6-6a and 6-6b we give two examples of using the grid method.

The various structural systems described in this section can save you a lot of time and money in the construction of your house. Many of the post and beam designs presented here involve the installation of materials which double as structural elements and interior finish. This can save you work as well as give you a major psychological boost in seeing very rapid progress toward a completed house. These systems are stronger than conventional frame construction and will meet most building codes, particularly those written on a performance basis. Troubles might arise with local officials who are charged with interpretation of codes, or from very restrictive codes. (We tell how to deal with these problems in Chapter 4.) Now let's start our structural economizing from the ground up.

THE FOUNDATION, FLOOR, AND ROOF

Post and beam construction was selected for our examples as it can save considerably over conventional systems. Structural posts 4 or 6 inches square can be embedded directly into the ground to serve simultaneously as foundation and main

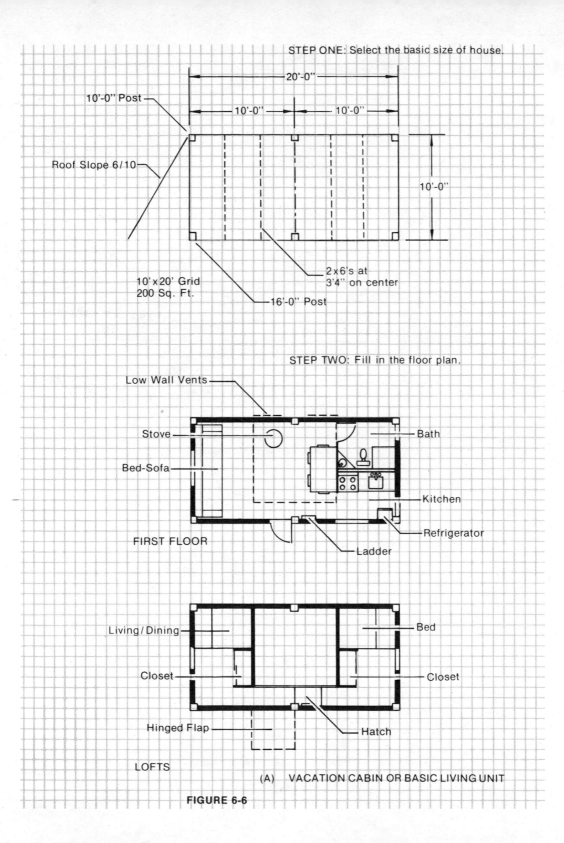

STEP ONE: Select the basic size of house.

20'-0"

10'-0" ← 10'-0"

10'-0" Post

Roof Slope 6/10

10'-0"

10' x 20' Grid
200 Sq. Ft.

2 x 6's at
3'4" on center

16'-0" Post

STEP TWO: Fill in the floor plan.

Low Wall Vents

Stove

Bath

Bed-Sofa

Kitchen

FIRST FLOOR

Refrigerator

Ladder

Living/Dining

Bed

Closet

Closet

Hinged Flap

Hatch

LOFTS

(A)   VACATION CABIN OR BASIC LIVING UNIT

FIGURE 6-6

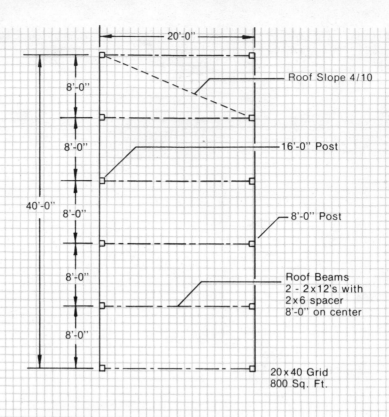

Roof Slope 4/10

20'-0"

8'-0"

8'-0"

16'-0" Post

40'-0"

8'-0"

8'-0" Post

8'-0"

Roof Beams
2 - 2x12's with
2x6 spacer
8'-0" on center

8'-0"

20x40 Grid
800 Sq. Ft.

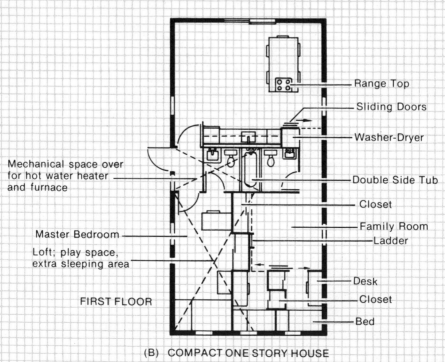

Range Top

Sliding Doors

Washer-Dryer

Mechanical space over
for hot water heater
and furnace

Double Side Tub

Closet

Family Room

Master Bedroom

Ladder

Loft; play space,
extra sleeping area

Desk

Closet

FIRST FLOOR

Bed

(B) COMPACT ONE STORY HOUSE

FIGURE 6-6

Photo 6-1: Here's a post and beam framed house being built. This house has a shed roof, though pitched roofs are equally well suited to this kind of construction.

vertical structural members for the house. See Photo 6-1. These posts must be of "pressure-treated" lumber, though. The treatment process imparts a light olive green color to the lumber and makes it suitable for constant contact with the ground. Life expectancy of these treated posts is from 60 to 100 years or more.*

By means of these posts, sloping ground that would either be passed over for construction or ruined by extreme grading can be economically used. Outside decks can readily be constructed by simply cantilevering the floor beams out over the slope. The suspended floor of the enclosed structure must be heavily insulated or the space beneath the floor enclosed to keep out the weather; but, even so, the savings in land cost can still be very substantial. See Figure 6-7a.

Post and beam construction, despite its speed and potential for cost savings, has been little used in this country in recent decades because of contractor resistance. The primary reasons usually given are that it's expensive and difficult to supply the heavy timbers for the main beams and that special attachments are necessary for connecting the beams to the vertical posts. In addition, electricians complain that

---

*Be certain pressure treatment is certified by the American Wood
Preserver's Institute. See reference on pp. 374.

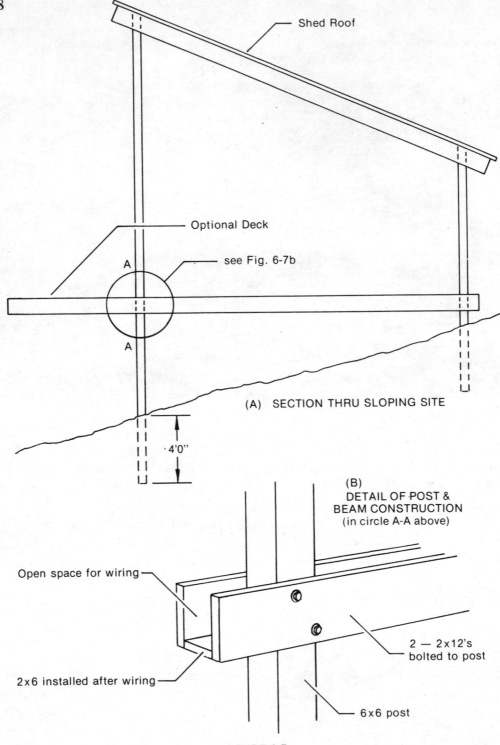

Shed Roof

Optional Deck

A

see Fig. 6-7b

A

(A) SECTION THRU SLOPING SITE

4'0"

(B)
DETAIL OF POST &
BEAM CONSTRUCTION
(in circle A-A above)

Open space for wiring

2 — 2x12's
bolted to post

2x6 installed after wiring

6x6 post

FIGURE 6-7

they have difficulty finding a place for wiring. Note, however, that by clever detailing we have avoided these drawbacks. The structural beams are made of standard dimensioned lumber (i.e., 2 X 8's, 2 X 10's, or 2 X 12's) and fastened to the sides of the posts with lag screws. The bottom is filled in later with a recessed 2 X 4 or 2 X 6, thus allowing the wiring to be placed. See Figure 6-7b.

These beams work equally well for floor or roof construction. Two-by-six tongue and groove, V-jointed fir or pine may be used directly over the beams to make a floor or roof deck. Moreover, few beams are needed. The maximum span for 2 X 6 decking used as a floor is about 6 feet, and the same span may be used for roofs in even very heavy snow areas. A greater span can sometimes be used for roofs, but must be designed for a specific loading. If the decking system is used for a roof, 2-inch foam insulation board can be used *on top of* the decking. This will enable you to keep the attractive wood beams of the ceiling uncovered.

USDA Handbook #364 illustrates this type of post and beam construction very thoroughly. Study the details and design tables carefully before selecting sizes. Some codes prohibit the use of used or roughsawn lumber (unless kiln-dried to a certified moisture content) for these main structural members. The structural members shown on our detailed plans in Chapter 19 are designed for *the specific loadings noted on the plans.* If you vary the sizes or the loadings, *be sure* to resize the beams accordingly. This will be particularly true if you plan to use slate or tile for your roof (heavy materials) or if you are building in an area of extremely heavy snowfall. Some areas will have unusual problems such as earthquakes, landslides, subsidence, and the like. In these locations, as well as in a few others with very strict codes, you will be required to submit plans with the formal seal of an architect or engineer before beginning construction. This may mean that you'll have to do some legwork to find a professional in your area who is knowledgeable on post and beam construction. Since the system is much stronger than conventional systems, however, you should have little trouble getting approval.

If you choose to use a more conventional roof system, we can still offer some improvements. Conventional tract houses normally use trusses or trussed rafters, which usually consist of a 2 X 6 top member with a 2 X 4 bottom chord, forming a ceiling joist. As well as limiting the use of the attic space, this system provides minimum depth for insulation. We propose to substitute 2 X 10's for the 2 X 6's and eliminate the bottom of the truss altogether, substituting lofts in some areas. See Chapter 19 for illustrations of our two-story house. The space between the joists (24 inches on center) should be filled with two layers of fiberglass insulation, one 4 inches and the other 6 inches thick—a total of 10 inches of insulation. Sheetrock can then be applied to the bottom surface of the joists.

Photo 6-2: An 800 ft.^{2} post and beam house professionally designed and built for $18,000. It sleeps as many as eight people at one time.

The post and beam system can also be used on flat land with a floor slab installed on grade. A very economical method of slab construction that eliminates conventional batch-mixed concrete and hence is well suited to remote sites can be constructed as follows: Pressure-treated sills are attached to the outside face of the structural posts and are partially embedded in the ground around the perimeter of the building. The earth floor of the building is leveled and tamped, and a 2-inch (minimum) layer of styrofoam or urethane insulation is applied directly over the ground and against the sills. Three inches of sand is then placed over the styrofoam. The sand should be moistened and leveled. (See Figure 6-8a.)

Brick is then tamped into place as a final finish. The joints between should be grouted *after* the brick is set in place. In this way, the spacing and pattern of the brick can be carefully controlled and the brick moved or rearranged as necessary. After the brick is set to the desired pattern, mortar is applied dry between the bricks, swept into the joints with a broom, and then thoroughly wet down with water. Repeat this if necessary to completely fill the joints. After the mortar starts to set, any excess should be washed off the brick surfaces. By the way, used brick is particularly suitable for this process. A wide variety of patterns such as herringbone or basketweave may be used, and the result is a handsome and permanent floor. (See Figure 6-8b.) If you desire to seal against moisture or brighten the brick color, a sealer may be applied. Two of my favorites are boiled linseed oil and polyurethane varnish.

Certain portions of the country, particularly in New England, are quite remote from the treatment plants that prepare pressure-treated lumber; hence, costs are high and supplies are sometimes limited. For these areas or for folks who would prefer a concrete slab, we offer a very economical method. Simply pour a grade level slab and turn down the edge. (See Figure 6-9.) This system is actually considerably stronger than conventional footings, because the load is distributed over the entire

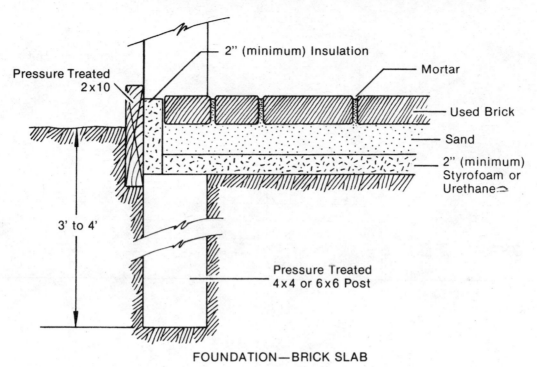

FOUNDATION—BRICK SLAB

FIGURE 6-8a

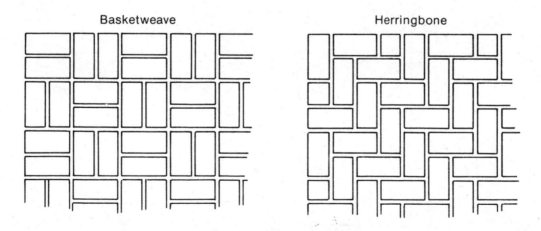

FLOOR PATTERNS

FIGURE 6-8b

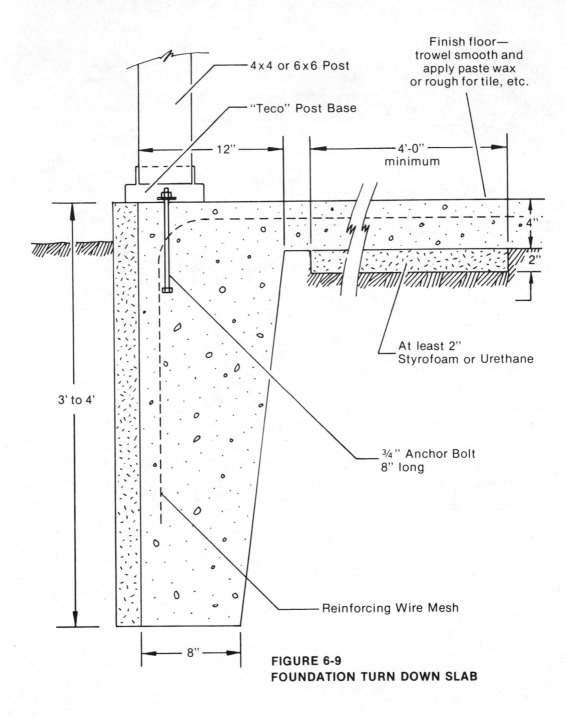

4 x 4 or 6 x 6 Post

"Teco" Post Base

Finish floor—
trowel smooth and
apply paste wax
or rough for tile, etc.

12"

4'-0"
minimum

4"

2"

At least 2"
Styrofoam or Urethane

3' to 4'

¾" Anchor Bolt
8" long

Reinforcing Wire Mesh

8"

FIGURE 6-9
FOUNDATION TURN DOWN SLAB

floor area. Moreover, if this slab is properly insulated, its mass will absorb heat and help maintain an even temperature in the house. To do this, it's very important to provide heavy insulation both on the outside face of the slab and under the slab. (See Chapter 12 for more detailed information.) For a very economical house, the concrete can be troweled smooth (preferably with a machine trowel) and simply waxed with old-fashioned paste floor wax. Floor tile or sheet vinyl can also be easily applied over the slab. Either post and beam or conventional construction can then be built atop this type of slab. However, the posts no longer need to be pressure treated.

WALLS AND PARTITIONS

For our wall construction, we reach back into early American history to retrieve and update a construction technique known as "Yankee Frame." This system was basically a nonstructural wall used with post and beam construction. Exterior walls consisted of 1-inch roughsawn planks applied vertically from a structural member at the floor to one at the roof. These planks were covered on the outside with horizontal siding or sometimes just vertical battens to cover the joints. The inside finish was usually lath and plaster. Primarily because of the poor insulating qualities of the system and the emergence of mass-produced lumber, it gave way to 2 X 4 stud construction which has changed little to this day. The development of efficient and reasonably priced rigid foam boards such as expanded polystyrene (styrofoam) make the Yankee Frame construction system practical once again. For an all-year home, you can put the insulation boards on the wall with mastic. Some of these boards come with an interior finish already attached, so the job is easy. However, styrofoam or urethane foam insulation *must* be covered with 5/8 inch of fire-rated sheetrock. Otherwise, forget using it, because of its dangerous flammability. For a summer cabin, roughsawn boards can be left exposed.

Roughsawn boards and beams can save a great deal of money. Native roughsawn lumber (both boards and beams) is available directly from local mills in many parts of the country. Few mills have facilities for kiln drying the lumber and, of course, this process uses fuel and increases the price. Some mills stockpile their lumber and allow it to air-dry, frequently charging a bit more for the product. Properly stacked roughsawn lumber has ½-inch spacers placed between each course of boards at about 3 feet on center. Check to see if your mill stores boards in this fashion. If you must buy green (freshly cut) boards, arrange to store them as described earlier for at least 3 and preferably 6 months. For other details, see Chapter 5. If the material must be installed green (not dried), make sure to use the board and batten pattern, as it was devised partly to hide the resulting shrinkage. Green lumber has poor nail-holding power, so use serrated (screw-type) nails or, in the case of main structural members, lag bolts. If desired, ¾-inch plywood with battens spaced at 16 inches on center can be used in lieu of the roughsawn to cut labor time considerably; however, the plywood will cost quite a bit more.

For those who would prefer not to use the Yankee Frame type of wall, here is an alternate system for building a wall while using post and beam construction. Use 6 X 6 posts set 6 to 8 feet on center as shown on the accompanying plans and those in Chapter 19. Then frame between the posts with 2 X 6's set *horizontally* at 2 feet on center. See Figure 6-10. This kind of framing is strong and can be well insulated. Since the roof and floor loads are being directly transferred to the structural posts, these walls carry no load other than their own weight and the wall's self-support is adequate. Moreover, the 6-inch thickness of the wall enables you to take advantage of the 6-inch-thick fiberglass insulation normally used for roofs.

You can improve on the usual ways of covering this framing too. Exterior sheathing should be insulating sheathing (Strongwall or Insulite) in 4 X 8 sheets set with the 8-foot dimension horizontal. Plywood should be used at the corners of the building for extra rigidity. For an exterior finish, I'd recommend roughsawn siding applied vertically with battans over the joints. For the inside finish, I'd suggest gypsum board (sheetrock) applied in large sheets of 4 X 12, 4 X 14, or even 4 X 16, again with the long joints running horizontally. This is the most efficient method of installing gypsum board. More information about interior and exterior sheathing will be given in Chapter 8.

You can save on the cost of constructing partition walls too. An inexpensive and attractive alternate to stud partitions faced with sheetrock was used at the same time as Yankee Frame construction. Again, roughsawn boards either 1 or 2 inches thick are set vertically. The edges are joined by shiplap or tongue and groove, and horizontal members at the top and bottom of both sides anchor the boards into place. If 1-inch boards are used, chair rails should be added for stiffness. Two-by-six tongue and groove decking may also be used for these partitions and any leftover scraps will work nicely on the countertop designs we mentioned earlier. Electrical wiring should be carefully planned to avoid these partitions, though, as running wiring in solid wood is difficult. Just use inexpensive baseboard raceways where you need to. See Figure 6-11.

MATERIALS

The golden principle of American business is to "charge what the market will bear"—not the value of the product plus a fair profit. This principle can be turned to

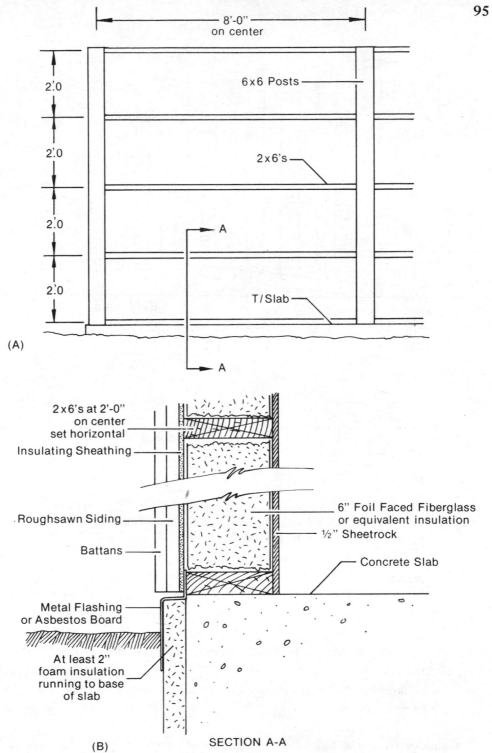

8'-0"
on center

6 x 6 Posts

2'.0

2'.0

2 x 6's

2'.0

A ➤ A

2'.0

T/Slab

(A)

2 x 6's at 2'-0"
on center
set horizontal

Insulating Sheathing

Roughsawn Siding

6" Foil Faced Fiberglass
or equivalent insulation

½" Sheetrock

Battans

Concrete Slab

Metal Flashing
or Asbestos Board

At least 2"
foam insulation
running to base
of slab

SECTION A-A

(B)

FIGURE 6-10

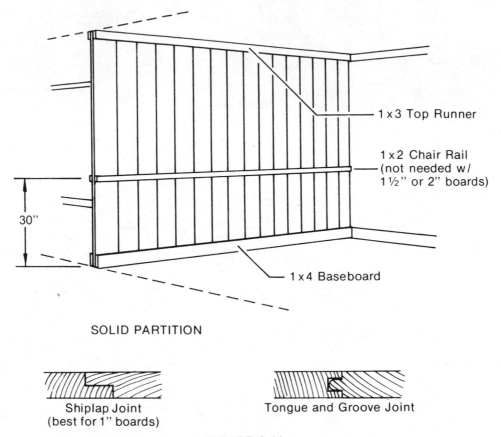

1 x 3 Top Runner

1 x 2 Chair Rail
(not needed w/
1½" or 2" boards)

30"

1 x 4 Baseboard

SOLID PARTITION

Shiplap Joint
(best for 1" boards)

Tongue and Groove Joint

FIGURE 6-11

your advantage when building a new house. Your local lumberyard probably sets its prices for the handyman who comes in from time to time to buy one board. If you make up a detailed materials list and approach several yards in your area for a package estimate, you may be amazed at the price differences. Be careful, though, to compare prices very carefully, as there may be a substantial difference in the quality of materials as well. In all your construction shopping, take a look at the supplier's clientele. If they are primarily construction workers and professional contractor types, you have probably found a good place to shop. If the place is overrun with families and laid out to look like a department store, run the other way. It wouldn't hurt to buy your whole package of materials from this sort of outfit, provided their prices were suitable, but don't go back for small purchases, as they count on making their profit from these. Plan *very* carefully to buy as many materials as possible *in bulk.*

Catalogs from the large mail-order houses can be a useful tool for comparing prices of some construction materials. Sears Roebuck is the largest merchandising firm in the world, and therefore tends to have the widest selection. In some rural areas, Sears actually does contracting work. They are also to be recommended because of their "return if dissatisfied" policy. You may have to be a bit firm to make it stick, however. The large mail-order houses maintain a wide variety of reasonably priced tools for construction, and those like Sears offer competent service departments for appliances and the like. In dealing with any of these mail-order houses, it will pay to remember that they deal in high volume, with low profit margins, and quality control is sometimes lacking. Remember also that repair parts have to be ordered by mail and can take a long time in arriving. If anything is wrong with what you have ordered, *return it immediately and ask for a replacement.* Don't count on getting replacement parts in a hurry.

Another very fine source of comparison prices and an excellent source of supplies is the Pease Lumber Company of Hamilton, Ohio. They put out a complete catalog of building items which can be most useful. They also market a wide variety of precut house packages, some of which are excellent and economical. These include precut domes; well-designed, small, efficient basic living units; and several models of more conventional houses. Prices start at well under $10,000 for a complete package home. Their central location makes them accessible to virtually the entire eastern half of the country.

What if you want to build just a small cabin, or what if your site is too remote for conventional lumberyards to deliver reasonably? Several companies make and market portable rigs that turn a chain saw into a small sawmill. In this way you can cut your own lumber for virtually nothing. The portable rigs run from $50 to $150. One manufacturer is Granberg Industries, 200 South Garrard Boulevard, Richmond, California 94894.

The use of the structural techniques we have just outlined can save as much as 40 percent over conventional construction methods. In addition to saving you money, these structures will be considerably stronger than conventionally framed buildings. Be careful about construction costs, though, as they vary widely between different areas of the country and even in the same town. For instance, an imaginative contractor who is looking for ways to build a better house more reasonably might build a post and beam structure for a reasonable price, while another contractor who will just figure his standard "price per square foot" and pocket any savings would bid twice as high for the job. It will pay to shop carefully and consider doing part or all of the job yourself.

chapter 7

RENOVATION AND RECYCLING OF OLDER BUILDINGS AND MATERIALS

by ALEX WADE

In this chapter we present other ways of obtaining decent housing at a reasonable cost. Our "throw-away" society is producing abandoned structures at an alarming rate. Frequently these buildings can be obtained for a fraction of their actual value and many of them could not be duplicated today for any price. It may even be possible to locate old structures which are about to be abandoned. If you can catch them before they are actually abandoned, many additional expenses may be avoided. Be very careful when you buy, though, as there are usually good reasons for the desertion of buildings. See the checklist (Table 7-1) at the end of this chapter for critical items to be verified before purchase. It may well be that the buildings were just too big and expensive for a single owner to maintain. If this proves to be the case, think about subdividing a larger building into two or more units, provided the local codes will permit multifamily dwellings. A cooperative venture of two or three families can help spread the financial and administrative burden of the renovation, or you can make the whole investment yourself and think about renting part of the house later. Also, it may be practical to start by renovating a small corner as an efficiency apartment to live in while the main portion of the house is being renovated. Then when you are all finished, you have the efficiency apartment to produce rental income.

Just how does one locate this ideal recyclable structure? It may not be easy to find just what you want, so be patient. A likely starting place is the delinquent tax listing in your city or county newspaper. This will give you an idea of who is in trouble and might be willing to sell. In a city, one can usually locate the property by the street address. In the country, it may be more difficult. The tax assessor and the post offices in the area may be of help. If you are looking in the country, by all means buy the United States Geological Survey (USGS) maps for the area you are interested in. They may be obtained from the Hammond Map Company in New York City or the geology department of most state universities. Get the large-scale maps. They show individual houses in the countryside and even indicate which ones are abandoned. You can go on many nice drives and hikes on back roads you would otherwise never have known existed. In this way, you can also tell whether the property has streams, lakes, forests, and other natural features. If you want a really good look and are in too much of a hurry to spend weeks tramping back roads, take your maps and rent a plane for an hour or so and fly over your area of interest. Take along binoculars and a camera. Be sure to mark your maps carefully so that you can find the property once back on the ground. Try to find several interesting pieces of property, as you may have to rule out some of them after reviewing the checklists.

RECYCLING AN OLDER BUILDING

Once you've found a likely candidate, then what? Three levels of recycling should be considered with an existing structure. First, let's assume that a structure is in an advanced state of decay. Such a building can be carefully demolished, saving all materials possible. Also, old foundations are often quite solid and can be either partially or totally reused. For example, old stone foundations can make a beautiful fireplace or stone wall, even if they can't be reused in place. A variation on demolition would be to tear down extraneous wings and additions, which are often of inferior construction, and use the salvaged materials for repair of the main structure. In your search for property, you might also find someone who owns an abandoned structure and will sell the building but not the land. Then you could either move the structure to your new property if feasible or tear it down for materials. A friend of mine took an old foundation for a chocolate factory with lovely stone arches, and added recycled materials from a couple of old houses. The result was a magnificent home for less than $5,000 (see Photo 7-1).

A second level of renovation would involve gutting or removing all interior finishes and nonstructural walls, and starting the interior from scratch. This procedure has become the norm for renovating city townhouses and can be applied as well to a country house with badly deteriorated finishes. Use *extreme* caution not to remove any supporting walls. Remove all the surface finishes first so that you can examine how the structure is put together. Be very careful of old multistory masonry structures as lateral interior partitions may be giving support. If you are in doubt, have an architect or engineer examine your structure before you complete demoli-

Photo 7-1: A $5,000 owner-built home constructed on the foundation of an old chocolate factory.

tion. By gutting the whole interior, you will be able to fully insulate your house and devise your own layout. An economical beginning for a project of this sort would be to construct a house within a house such as the "basic living unit" shown in Figure 6-6 of Chapter 6. Your house within a house can either be a freestanding unit within the shell of the old house or a small corner of the house left intact and outfitted as an efficiency unit. If you have the 11- or 12-foot ceilings common to old houses, this unit can be easily constructed. You can arrange your space carefully and end up with a finished retreat in which to live while renovating the entire building. This living area can be used as a rental unit after the main house is completed.

A third approach to renovation is necessary for the large older structure which is still in good condition. Many of these houses are very elegant, with high ceilings, ornate plaster, and paneling, which seem to preclude insulation and/or remodeling without destruction. Again, the "house-within-a-house" approach provides us with a solution. By designing a basic living unit within the larger structure for use during extreme winter months, the charm of the original house can be enjoyed in warmer weather without necessitating bankruptcy from the heating bill.

Try to locate your living unit on the south wall of the existing structure so that you can take advantage of solar heat gain. Also, be careful to use proper insulation. New installation techniques permit expanding foam to be blown into the walls. Do not use vermiculite or rock wool for this purpose, as was formerly done. These materials are permeable to vapor and there is no way to install a vapor barrier when insulation is blown into the walls. This vapor will rot the framing in short order, so stick to the foam. (See Chapter 12 for types and more specific methods of insulation.) Remember that you can also insulate interior partitions with this system. In this fashion we can provide an insulated corner in our big house for winter use. In Figure 7-1 we show one floor plan for a typical large house with two large rooms utilized for winter quarters. Obviously, tastes and individual old houses vary greatly. We hope that some of the suggestions on "thinking small" presented in Chapter 6 will enable the homeowner to construct his or her own living space within an existing shell.

What if the problem with your old house is that it is too small rather than too large? For these houses, it is frequently much less expensive to renovate attic space than to add to the original structure. An attic suite or loft beds in an attic space could be a good solution for children. A separate floor of their own, even if it is in the attic, seems to appeal strongly to them. Also, children enjoy ladders and the odd nooks and crannies which are frequently a part of attic living. A careful study of our plans, particularly of the built-in bed arrangements, may enable you to make better use of the space you already have.

You should be very careful to guard against making needless additions to an older house, though. If after you have drawn plans and studied carefully, you still feel that the house will be too small for you, we offer some additional suggestions. First, try adding a bay window. This window can be cantilevered out from the basic structure without adding to the foundation. Also, if the top of the bay is glazed, the window will admit lots of light into what may otherwise be a dark and seemingly cramped interior. The bay window shown in Photo 7-2 cost less than $300 to construct.

Another method for gaining space in an existing house is to build a deck or porch for use in warm weather. Decks tend to be better than porches as they are cheaper to construct and don't cut out light from the interior of the house. USDA manual #364, which we have recommended elsewhere, gives excellent details for

102

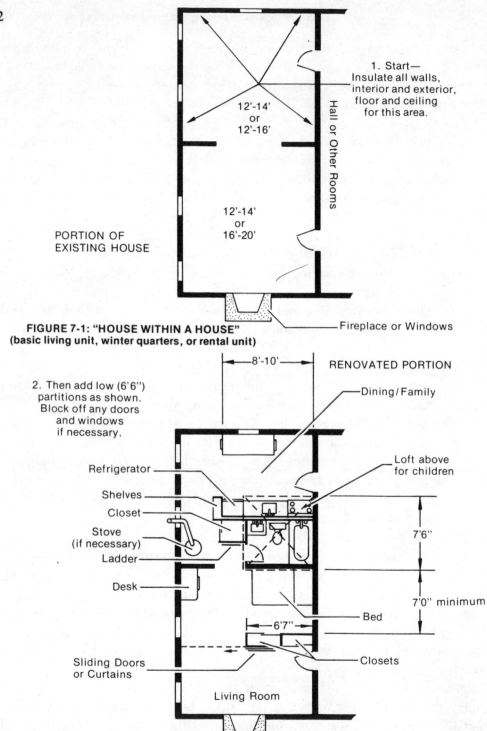

1. Start—
Insulate all walls,
interior and exterior,
floor and ceiling
for this area.

Hall or Other Rooms

12'-14'
or
12'-16'

12'-14'
or
16'-20'

PORTION OF
EXISTING HOUSE

Fireplace or Windows

FIGURE 7-1: "HOUSE WITHIN A HOUSE"
(basic living unit, winter quarters, or rental unit)

8'-10'   RENOVATED PORTION

2. Then add low (6'6")
partitions as shown.
Block off any doors
and windows
if necessary.

Dining/Family

Refrigerator

Loft above
for children

Shelves

Closet

Stove
(if necessary)

Ladder

7'6"

Desk

7'0" minimum

Bed

6'7"

Sliding Doors
or Curtains

Closets

Living Room

Note: If more than two people live in a renovated portion of a house as above, odor
build-up may be a problem *if* ventilation is lacking. Opening windows a bit, or using the
carbon filters mentioned in Chapter 10, will solve this problem.

Photo 7-2: The bay window added to the side of this house allows more natural light inside and provides extra room for house plants.

decks and porches. Also, see Chapter 17 for greenhouse construction. It may be that you will want to add one to your older house.

In working with any structure, one must plan carefully. Accurately measure the building and make an exact drawing showing the locations of all partitions, doors, windows, etc. Again, graph paper with eight squares per inch will be useful for trial layouts. First, lay out your existing floor plan on the paper; then sketch a variety of layouts with more compact sleeping areas. You will be surprised at the space savings. Use caution not to disturb walls that bear loads or contain plumbing (provided the existing plumbing is worth saving) as these changes can be quite expensive. "Before" and "after" plans for a second-floor renovation show two very different solutions (see Figure 7-2).

RECYCLING MATERIALS

Our throw-away society produces other benefits. Sound old buildings are constantly being torn down to make room for superhighways, tacky supermarkets, and the like. If you are clever, you can buy these materials for a fraction of their real value. If you catch the building being torn down, you can approach the contractor and offer to buy materials directly. Otherwise, he is probably selling them to a commercial salvage house. These salvage houses tend to stock everything imaginable. Best buys are windows, doors, trim, and plumbing fixtures. Large beams and staircases can be real money savers if they happen to fit your house design. For example, frequently, old houses were sided with cedar boards. These boards can be easily

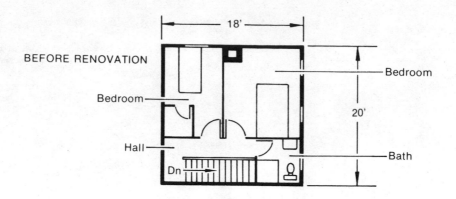

BEFORE RENOVATION

18'

20'

Bedroom

Bedroom

Bath

Hall

Dn

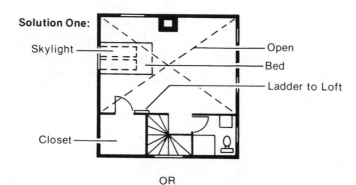

Solution One:

Skylight

Open

Bed

Ladder to Loft

Closet

OR

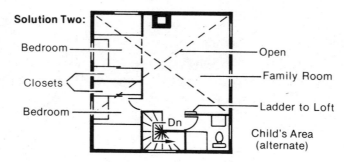

Solution Two:

Bedroom

Open

Closets

Family Room

Bedroom

Ladder to Loft

Dn

Child's Area
(alternate)

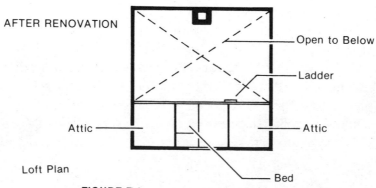

AFTER RENOVATION

Open to Below

Ladder

Attic

Attic

Loft Plan

Bed

FIGURE 7-2

reversed and reused for a natural finish, and time has already weathered them for you. A careful survey of the salvage materials available *before* you finish your plans will be very valuable.

Plumbing fixtures are one of your best buys at salvage yards. Old cast iron roll rim tubs can be very elegant if sunk into the floor. Make sure that the finish is still in reasonably good shape. Good commercial cleaners can be used to remove stubborn stains. Tub 'n Sink Jelly, manufactured by Woodhill Chemical Company, Cleveland, Ohio 44128, works very well. Plumbing fittings (faucets, traps, and the like) are good, too. Also, the older faucets were more attractive than many of the items marketed today. Don't be too concerned if the materials have been nickel plated and the plating is wearing off. You can either remove the remainder of the plating with abrasive paper for a nice brass finish or take the materials to an automotive plating shop for a refinish. I prefer the former and it's much cheaper.

Another area of savings at salvage houses is hardware of all sorts, particularly door and window hardware. Good modern hardware is extremely expensive. The plated steel hinges for sale at the corner hardware store will start to rust before you can get them on the door. Old hinges of cast iron or solid brass are best. Sometimes you will be lucky and the hardware will still be attached to your doors or windows. Be prepared to take your time and find materials that are harmonious with one another. You may also want to make some changes in your design to accommodate some of your finds.

POINTERS FOR BUYING PROPERTY

There are many important points to be considered when buying an old or abandoned house, particularly if it is way out in the countryside where it may appear to be a rare bargain. We will examine these step-by-step, starting with the most important: ownership.

Ownership

Can you legally assume title to the property? This will require the services of a lawyer to make sure that the party selling to you actually has a clear deed and the right to sell. Be prepared to wait a long time and put plenty of pressure on the lawyer, as they are notoriously slow at this type of thing (big money is elsewhere). You should also hire a surveyor to check the legal description of the property to make sure that it agrees with the apparent physical boundaries. Frequently it doesn't.

Access

Once you own the property, can you legally and physically get to it? In all weather? Many old pieces of country property are way off the main roads. For privacy and beauty this may make them seem very attractive. However, you may have to negotiate a right-of-way through neighboring property; and you may or may not be successful. Also, check to see that your access is of legal width. Utility companies usually want 20 feet and older properties sometimes had less. This problem can be gotten around by having the power company install a primary "trailer" service on a pole outside the right-of-way and then installing your own service line (see "Electricity and Telephone" below).

Road building can be very expensive. If the surface is reasonably hard, a modern four-wheel-drive vehicle can usually get in and out. However, the vehicle itself is expensive. If the road is reasonable, a good front-wheel-drive passenger car with mud-snow radials will probably make it with little trouble. Keep in mind that the road will have to be plowed or shoveled where winter snows are a problem. Also, fuel and supplies may have to be brought in during the winter. On the other hand, if you are looking for a summer place or a weekend retreat or are the hardy sort, a bargain piece of property way off the beaten path may be just the ticket.

Tenants

Legally, tenants can be a real hassle if they are already living in the house you plan to buy. Many cities and towns have laws protecting these tenants—laws that severely restrict your rights as the owner. In some cases it may be virtually impossible to remove the tenants. Usually, if the house is a single-family dwelling and you plan to live there yourself, you will have no trouble, except maybe with the actual eviction. Be extremely careful about purchasing a multifamily dwelling with tenants if you plan to remodel. Under these conditions, it would be very wise to consult a lawyer.

Water

Drilling a well is very expensive. Moreover, getting a drilling rig to a reasonable drilling site can require road building. Is there a good well on the property? Has the existing well become contaminated? There are possible ways around these problems. If there is a stream or large spring on the property, you may be able to install a water purification system for drinking water and use these supplies without drilling a well. Check to see that they are adequate in the dry season and that the local health officials will let you use them. This is particularly important in dry parts of the country where you might have to haul drinking water.

Waste Disposal

Is the soil suitable for a septic system? Heavy clays are a disaster. To determine the suitability of soil for sewage disposal, you should have a trained professional take a percolation test. However, you can get a rough idea by digging a hole about a foot deep. Pour it full of water, let set for 20 minutes, and then refill. If the water level can be visibly seen to decrease you are probably okay. If the water just sits there, you could be facing a $5,000 septic system. There are many proven, sensible alternates to a conventional septic system, including the Swedish Clivus Multrum and other less-expensive composting devices. See Chapter 15 for references on these. Unfortunately, most public health officials still frown upon them. In some areas you can use a well-constructed privy if you don't have cash for the septic system right away.

Electricity and Telephone

One reason for the abandonment of rural property is its remoteness from a power line. Power companies usually charge $300 or more per pole to run service onto your property. Poles are set about 300 feet apart, so it can add up to real money. As mentioned earlier, you can run your own secondary service for much less money. Be very careful in setting your own poles to get an OK from the telephone company, though, as they are much stricter than the power companies. Of course you may want to forget the phone. If you do, you can avoid the expense of running utility lines and install a small generating plant. This will be a more expensive source of power than that supplied by the commercial companies, and maintenance of the unit is a problem. However, putting a line all the way into an otherwise perfect piece of property may be a prohibitively expensive first cost, and this on-site generator can help spread the expenses out.

Structure

Now to the very important task of evaluating the structure to see just what should be done with it. If it's quite large and in very poor shape, partial or total demolition is the obvious answer. A small house in relatively good shape, on the other hand, should be kept as intact as possible. Since there are many good books on appraising houses for rot, termite damage, and the like, we will confine our remarks to major decisions concerning the structure. For additional reading, see the list of references for this chapter.

Major structural damage can cause a building to be almost worthless. If you are in doubt of your ability to decide yourself, by all means hire an architect or engineer to make a determination for you. Banks frequently have appraisers who are experi-

Table 7-1: *Checklist for Buying Property*

Item (property check)	OK	Needs Attention	Major Problem
1. Title to property			
2. Right of way			
3. Road			
4. Well			
5. Other water source			
6. Septic system			
7. Alternate method of sewage disposal			
8. Tenants			
9. Electricity			
10. Power plant			
11. Telephone			
Item (building check)	OK	Needs Attention	Major Problem
1. Roof—sag and condition			
2. Walls—bows and settlement			
3. Floors—settlement and condition			
4. Foundation—cracks and leaks			
5. Windows			
6. Termites			
7. Rot			
8. Wiring			
9. Plumbing			
10. Heating			
11. Chimneys			
12. Insulation			
13. Stairways			

enced in these matters. If you plan to get a loan, ask the bank's appraiser to look at your house. He may be able to save you the expense of hiring another professional. What follows are the major structural items you should look at:

A. Major structural damage is usually easy to detect. Look at the ridge line. Is it straight or is there a large dip in the center? If there is a dip, sight along the side walls: do they bow out at the top or in the center? If you have both of these problems, especially with a masonry structure, you are in serious trouble. A wood structure may be salvaged by pulling the sides back into place, but it's an expensive project that requires a professional. Unless it's really just what you want otherwise, you will be better off to tear it down or try for another piece of property. On the other hand, old structures frequently have sagging floors and differential settlement around chimneys, but this doesn't necessarily mean that there's anything wrong with the basic structure. If the sag is not too visually disturbing, it's best to leave it alone. If you are removing most of the interior partitions, it's usually a simple job to jack the floors back to level position.

B. On old masonry structures, which are usually good buys, look carefully for large cracks or signs of major settlement. Repointing brickwork is a lot of work, but it can always be done after you have completed the interior. Remember that a masonry structure is usually worth more money than a wood structure and also carries lower insurance rates in many areas.

C. Finally, unless old plaster is in excellent shape and the roof shows no signs of leakage, you will be best advised to remove all plaster and start over, particularly if you plan major partition changes and have to wire and plumb the house. Also, this will allow you to examine all structural members and replace any that may be necessary.

chapter 8
INTERIOR AND EXTERIOR SKINS
by ALEX WADE

Once you've chosen to build small and efficiently, or renovate an older structure, there are other savings you can make. The right choice of materials to build with can help you save money, and you'll be able to achieve an elegance and beauty for your home as well.

In this chapter we'll look at those materials which are the most economical, efficient, durable, and attractive for surface finishes on both the exterior and interior of your house.

ROOF COVERINGS

The roof is the most important single surface of your house—so don't skimp here. Standard asphalt shingles are a very sensible choice. Self-sealing shingles, 235 lbs., are generally stocked at most building supply houses. Do not use lighter shingles, and get heavier ones if you can afford them. Also, buy only major brands of shingles. Major roofing supply companies such as Bird, J-M, and GAF stand behind their product and will usually replace defective materials. Next, choose your color very carefully, remembering that the roof is a major visual highlight of your house. Solid black and solid white should be avoided because of poor durability. On the other hand, avoid shingles with wide ranges of color as these will look "jumpy" and cheapen the appearance of your home. Such shingles tend to be cheaper for the manufacturer to make because they don't require much quality control, but they are a poor buy.

Another excellent roofing material is called "roll roofing." As a roof surface it is cheaper and more easily applied than shingles and is known to the trade as "double coverage mineral surfaced asphalt" or sometimes SIS. It consists of a material similar to asphalt shingles, but manufactured instead in three-foot-wide rolls, half of which is surfaced with granules like standard shingling. See Figure 8-1. This material is applied in strips with each strip overlapping the last by one-half of its width; then the overlap is cemented with a cold applied cement. The finished product is more durable than a shingle roof, although maybe not quite as attractive. It is also much faster and easier for the beginner to apply. *Warning:* this material must be unrolled at least one day before application and allowed to straighten out, otherwise it will buckle.

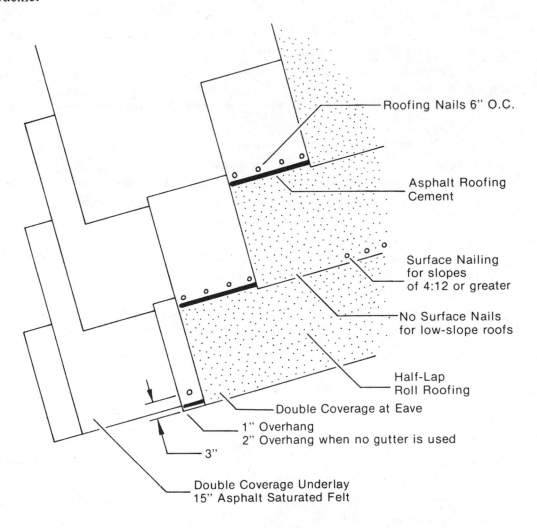

Roofing Nails 6" O.C.

Asphalt Roofing Cement

Surface Nailing for slopes of 4:12 or greater

No Surface Nails for low-slope roofs

Half-Lap Roll Roofing

Double Coverage at Eave

1" Overhang
2" Overhang when no gutter is used

3"

Double Coverage Underlay
15" Asphalt Saturated Felt

FIGURE 8-1

Another possibility is a Terne metal (lead-coated steel) roof. Terne roofing is usually used only for monumental public buildings because it requires a great deal of hand labor to apply. However, if you're willing to spend the time and effort, the materials cost is about the same as for good-quality asphalt shingles. The reward that you get for all your efforts is a very high quality and permanent roof which is good for the life of the structure (asphalt roofs have to be replaced every 15 to 20 years). Also, if you live in an area where water collection is important, these Terne roofs are great for collecting rainwater. They are known in the trade as "standing seam roofs" because they have a vertical ridge every 20 inches or so. Full information on standing seam roofs may be obtained from Follansbee Steel Company, Follansbee, West Virginia. As an historical note, the original standing seam roof on Thomas Jefferson's Monticello house is still in good shape!

Cedar shakes also make a very beautiful and durable roof. Unfortunately, they are a fire hazard and your insurance company will charge accordingly. However, cedar does make a handsome and durable siding material, inside or out. I'll comment more on cedar boards for siding later.

Now, for a tight seal, metal flashings are required at valleys, dormers, chimneys, skylights, and any other changes in direction of the roof. Flashings tend to be leak points, so make your roof as simple as possible. Although it is commonly used, aluminum is a very poor material for flashing, so use copper if possible. Terne metal is also an excellent flashing material, though it may be difficult to obtain unless you are near a large sheet-metal supply house.

In high-quality construction, flashings are also installed over window and door heads. The reason for these flashings is to catch any water which may have entered the wall and divert it back out again. If you use a good-quality, durable wood for your window frames and caulk very thoroughly, you can eliminate this expense. However, if you want to do a really fine job, use the flashings. Copper or Terne should be used if possible. Don't waste your time with aluminum or galvanized metal because they will deteriorate in no time and then cause lots of trouble by diverting the water into the wall that they are supposed to be protecting. Broken flashing is a frequent trouble spot in older houses, and you should examine carefully for this. See Figure 8-2.

While you are planning your roof, there are a few structural considerations to keep in mind as well. Keep your roof slope rather steep (4/12 minimum). This will cause the water to drain off more rapidly and make the shingles last longer. Also, the space you gain under the roof can be very valuable and costs virtually nothing. See Chapter 6 for uses of this space.

On the other hand, flat or very low-pitch roofs offer some opportunities and

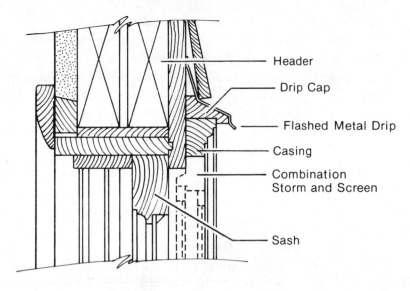

Header

Drip Cap

Flashed Metal Drip

Casing

Combination
Storm and Screen

Sash

FIGURE 8-2

they have their special problems. Porch roofs and low-slope roofs on old apartment buildings are the most typical examples. In certain hot, dry areas of the South and Southwest, dead flat roofs are also popular. This type of roof is much easier to construct and can be beneficial for both heating and cooling, as a flat roof will collect snow in winter, which acts as added insulation, and can be flooded in summer to provide evaporative cooling. Unfortunately, they are uncommon and thus rarely used. Another category of flat roof is the deck, which may be simply a section of flat roof covering a first-floor wing or bay window, or a section of roof that is cut out to form a flat area. All of these flat and low-pitch conditions require special attention. For very low pitches (which do have some slope, say 1/12 or 2/12), the roll roofing mentioned earlier is an easy solution for the homeowner. For a large, dead, flat roof, it is best to hire a professional roofer to install a "hot" roof. This consists of hot tar with five layers of building paper built into it. Make sure that the roofer uses coal tar rather than asphalt, particularly if you plan to flood the roof for summer cooling. Also, make sure that the flashings extend up the parapets far enough to allow you to flood the roof with 2 or 3 inches of water.

Now, if you have a small area of flat deck on an old house or you want to build a flat roof into your home (particularly one which can be walked upon), there is a very fine product for your use. It is a neoprene compound very similar to the material used to make automobile tires. This liquid neoprene is manufactured by DuPont under the trade name of "Hypalon" and has a consistency much like paint.

It can be applied by brush or roller in multiple coats until a film is built up. Moreover, it works well over plywood or concrete, and also adheres well to metal. Joints in plywood or cracks in concrete should be sealed with neoprene caulk and then covered with 2-inch-wide fiberglass tape (available from a marine supply house). Do not use the adhesive tape usually available with the Hypalon roofing system, as it makes a weak joint.

This system requires much hand labor. It is generally unknown except to very large, big-city roofing contractors who charge a small fortune to install the system. Properly installed, this system makes an exceptionally durable and permanent roof surface that can be used as a traffic deck. It is also an excellent system for resurfacing small decks on existing houses, waterproofing concrete foundation walls, or any other waterproofing operation. The major source of materials is Gates Engineering Company, Wilmington, Delaware. Much time, care, and patience is required to produce a watertight Hypalon roof. Many successive coats must be applied with great care, as the tiniest pinhole will leak profusely. If you use it on a relatively large roof, you should also make sure that you slope the surface to drain the water (minimum 1/8 inch per foot).

If you have time, the results of a Hypalon roof are well worth the effort. You will have a permanent roof surface for relatively low cost; and chimneys, exhaust fans, skylights, etc., can be flashed into the roof much more easily and securely than with conventional systems. Also, the final coat can be any color you desire. A word of caution: if you mention this system to a local roofer, he probably won't know what you're talking about. Roofing materials of a conventional nature haven't changed much for at least 50 years, and roofers get very set in their ways. So if you want to use this very fine system, you will probably be on your own.

As a final structural consideration on all roofing systems, think about where the rainwater will go after it leaves the roof. A roof overhang diverts water away from the house and eliminates the need for gutters except in a few special cases. Gutters, unless they are very expensive, are a major maintenance item. The only case where I would recommend them is where you have a basement with a major roof surface draining onto a ground surface that sloped *toward* the house. Try to eliminate such a condition by grading if possible. If you can't, there are two options for solving this problem. If you want to use gutters, the details in Chapter 19 show a simple built-in gutter which is economical to construct and durable. If you don't use gutters, foundation drainage is very important (see Chapters 5 and 6 for details) and it will also be necessary to install gravel at the foundation to prevent splashing from making a mess of the siding. Of course, you may want to collect this rainwater. In areas where air pollution is not a problem, it is exactly the soft water you'd like to use for washing clothes.

SIDING

Siding has the greatest visual impact of any of your surface materials, so choose carefully! It also represents the largest surface area to finish or maintain, so the type of finish is a most important consideration. We shall begin our examination of siding materials by telling you what materials *not* to use. Vinyl and other plastics are going to be very expensive because of the oil shortages. Moreover, their patterns tend to be poorly designed and plastics can shrink out of shape with age. Metal sidings, particularly aluminum, should also be avoided. Metals transmit heat very quickly; and because aluminum requires enormous quantities of energy to produce, it will increase in price, too. Further, the so-called "permanent finishes" on these metal products will eventually deteriorate.

All of these products were developed in an attempt to improve upon wood; thus far they have failed. Now that we have told you what to avoid, let's look at what should be used, in descending order of cost.

Cypress

The finest possible material for siding is cypress. It needs absolutely no finish whatsoever and will weather to a beautiful silver gray. Cypress is also naturally resistant to rot and termites and can be milled to any desired pattern. It can be applied readily in board and batten as described in Chapter 6. You can also have the doors and windows milled out of the same material and never have to paint or refinish anything—ever! Why, if this material is so good, isn't it widely used? It's relatively expensive and so scarce it must be gotten on special order (except in a few areas where it is native). In fact, it is wise to order well in advance of the time you want to use it. If you are comparing prices, don't forget to deduct the costs of stain, wood preservative, or paint.

Cedar and Redwood

Cedar and redwood have a resistance to rot and termites similar to that of cypress and they are easier to obtain. The major disadvantages are that they can weather black when exposed to water, and the appearance can be rather uneven because of different exposures. This can be corrected by applying a coat of clear Woodlife or Cuprinol at the time the siding is installed. This will cause the siding to turn a light buckskin tan, eventually fading to gray tan. The color is permanent and the preservative treatment need not be repeated. It may be easier to apply the preservative before installing the siding. Since all preservatives are intended to repel rot and termites, they are toxic. Be *very careful* not to contaminate the ground or water sources. In addition, there is a serious consideration to be made before using redwood: chiefly, this is a limited natural resource, a fact you should think about. The cheaper grades of redwood, by the way, also tend to be better looking than the expensive Clear Heart grade, but you may have to special order to get these.

Native Pine

Much cheaper than any of the preceding is roughsawn (unplaned) native pine. The vertical board and batten pattern described in Chapter 6 is ideal for this material, particularly if the pine is not well dried. (The batten hides any shrinkage that may take place.) A permanent finish that will never need attention can be achieved by dipping the boards in creosote before installation. This will produce the brown color frequently seen on utility poles. Unfortunately, brown is the only choice. However, clear Cuprinol, mentioned before, also comes with an added stain in a wide variety of colors. However, this material must be renewed every 4 or 5 years to maintain the weatherproof qualities of the pine wood. With any of the preservatives, give special attention to the end grain and make sure to touch up any cuts made after the wood has been treated.

Plywood

Another alternative is known as Texture 111 plywood. This is a plywood with grooves cut 4 inches on center to resemble planks. It comes in fir, redwood, and cedar. The fir is cheap but needs frequent coats of stain. The redwood is of good quality but depletes our limited supply of redwood trees. The cedar would be the best choice but probably will have to be special ordered in most places. All of these materials are normally 5/8 inch thick and come with a roughsawn finish that takes the various preservatives we discussed earlier. The use of these plywoods will enable you to eliminate sheathing completely, as these materials do a fine job of sheathing and siding all in one operation. Moreover, if you are hiring people to do the building work, it will pay to consider Texture 111 seriously, as the time it saves will considerably cut labor costs.

Hardboard

A final siding possibility is recycled wood chips or hardboard. These scrap wood chips are molded into various patterns of siding with a baked enamel finish. The best known of these are marketed under the Masonite trade name, and top-of-the-line products are guaranteed for 25 years. This material comes in a wide variety of patterns, including a very realistic-looking shingle design; and since these are rather thin, they make an excellent material for re-siding an old house because the existing trim can be left as is.

INTERIOR WALL SURFACING

In Chapter 6 we described various means of constructing partitions between rooms. Here, we will concentrate on the interior face of exterior walls. Gypsum board (sheetrock) is the most economical and logical material available for this. See Figure 8-3. Since the material is so inexpensive, you may want to do a really first class job and use two 3/8-inch layers. The first layer is securely nailed in place. The joints of the second layer are staggered and the material installed with panelboard adhesive and a few nails. Full details on the system can be obtained from a major sheetrock manufacturer such as Johns Manville or Gold Bond. For a single layer of sheetrock, 1/2 inch or 5/8 inch should be used. (Less careful builders use one layer of 3/8 inch.)

Some codes will require 5/8-inch fire-rated material in certain special locations, such as stairhalls and party walls. Check carefully. Waterproof sheetrock should be used in areas subject to heavy moisture, such as in bathrooms and behind kitchen sinks.

Spackling sheetrock joints is definitely *not* a job for an amateur. It is one of the very few areas of your house where you will be far better off to hire a professional. Make sure you hire those who do only this for a living—though sometimes they also paint. Do not trust a carpenter or a best friend who says, "Sure, I can spackle; don't waste your money."

Now that you have covered the walls, think carefully about how you will paint them. Cheap paint is a very poor bargain. Unfortunately, in our crazy marketplace cheap paint costs almost twice as much as good paint. Major paint companies such as PPG, Glidden, and Dutch Boy sell their paint commercially through stores that sell only paint. In five-gallon cans, good commercial grade interior latex paint runs around $5.00 per gallon. Your friendly neighborhood hardware store, on the other hand, will peddle "off-brand" or "house brand" paint of inferior or unknown quality for double the price or more, or stock name brands at premium prices. If you are using white walls, always buy pure white—it stays clean-looking far longer than all the near-whites. The recent proliferation of off-whites can make this a hard task, so check very carefully. (PPG #630 is genuine white.)

Now, using the sheetrock as the basic covering for the interior of outside walls, you can vary the effect by changing the pace a bit here and there. If you want a particularly rich effect that will help make your house unusually attractive, just bring some of the siding indoors! You can use up any leftover roughsawn siding on the interior walls. See Photo 8-1. The roughsawn also makes a nice refinish job for an older house. Don't overdo the wood, however, as it can make the interior dark and gloomy. I would recommend against using the numerous 1/4-inch prefinished

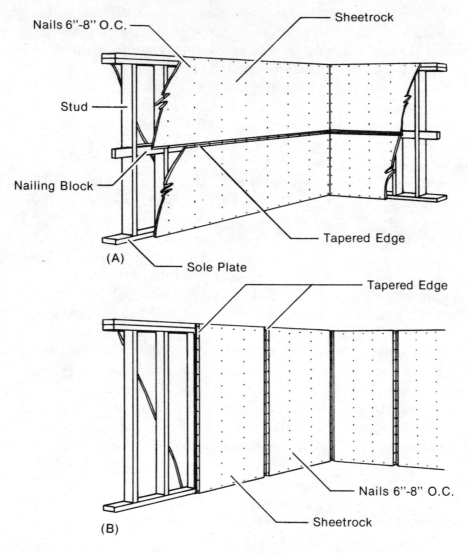

Nails 6"-8" O.C.

Sheetrock

Stud

Nailing Block

Tapered Edge

(A)

Sole Plate

Tapered Edge

Nails 6"-8" O.C.

Sheetrock

(B)

FIGURE 8-3

Installing sheetrock gypsum board on walls.
A, Horizontal application; B, vertical application.

Photo 8-1: Attractive use of roughsawn boards as interior sheathing in an owner-built home.

panelings. Only the very expensive ones are any good, and they tend to be more expensive than real paneling. Why buy a fake when you can have the real thing?

You can make use of other effects, too. If you have used our slab-on-grade method of construction described in Chapter 6, you may be able to add a very nice luxury touch, as you have already installed the necessary support. In many areas of the country native stone is to be had just by picking it up. You can easily use it to surface the inside of one of your walls, or even to build dividing partitions between rooms. The mass of the stone will absorb heat during the day and radiate it back to the space at night, particularly if the stone is located so that sunlight hits it during the day. For other ideas along this line, see Chapter 16. Even if you can't find native stone, by doing the installation yourself, the stone walls will add considerably to the appearance and value of your house.

You might also consider building a fireplace and chimney of the same material, or maybe a stone wall with a fireplace as a part of it. If you place a large mass of masonry such as a chimney in the middle of the slab, however, it will have to be thickened and reinforced to take the load.

What about those final touches? Modern interior trim pieces tend to be expensive. Leftover batten material from the exterior, on the other hand, can be stained and used for door and window trim, as well as baseboards, at considerable savings; and it can be just as attractive. Here's how to do it. Always stain or finish windows, doors, casings, and all trim *before* installation. Be sure also to paint your walls before installing the trim. Finally, whenever possible, finish both sides of a board. This will cause moisture to be more even and help prevent warping and splitting.

Two companies which make fine products for finishing are Minwax and Cabots. Both make a wide variety of oil stains. Minwax, in addition, makes a particularly fine satin finish polyurethane which is great for floors, countertops, furniture, and the like. They also make an extremely good finish called "Antique Oil," which is rubbed to produce a fantastic effect and can be used on almost anything indoors. Vermont Weatherboard, Inc., Wolcott, Vermont, makes a chemical that weathers wood to that elusive silver gray; it's very expensive, though.

INTERIOR CEILINGS

If you followed our recommendations in Chapter 6, most or all of your ceilings were installed with the basic structure exposed. All you need is a coat of sealer to finish the job. Otherwise, use sheetrock and paint. All my comments concerning sheetrock for walls apply equally to ceilings. If you need a fire rating for the ceiling to satisfy building codes, use 5/8-inch fire-rated sheetrock. Do not be tempted to use any of the various "acoustical" ceiling board and tiles on the market, as they tend to sag, shift, and get dirty.

If you select carefully, do all the materials buying yourself, and hire a contractor to do only the shell of your house, you can easily save half or more of the cost of a conventionally built home! If you do the mechanical work, you can save even more. Be careful in the selection of your building materials. You can go on saving in an attractive and elegant home as long as you live there.

Section III

LOWERING YOUR OPERATING EXPENSES

chapter 9

BETTER WINDOW PLANNING: BALANCING ENERGY AND ESTHETIC NEEDS

by WILLIAM K. LANGDON

We wish to maintain an environment in our homes which is nearly constant, even though the environment outside is continually changing. While a good shelter protects us from this outer environment, windows allow for a controlled exchange with the beneficial parts of the outside. Effective windows are really environmental valves, giving the light and view we desire but turning off the wind and rain. See Figure 9-1.

Someday, we may have large dome frames to contain a house and garden with covering skins which are responsive to the constant changes in the environment—skins made of a fabric that thickens at night to hold the day's heat within and then draws thin again on a winter's day to allow more heat to enter; a fabric that becomes partially opaque in the heat of the summer to shade the space from excessive heat, and with pores that open to allow free ventilation to cool the inside. Moisture levels might even be controlled by such a fabric.

Although there is no such fabric available today, this imaginary material would make an excellent window—such a fine window, in fact, that its placement or size would have few limitations, allowing a house and garden to be a single hemispherical window. The more we are able to equip the windows we use with such controls, the more freedom we will have in their use while still maintaining an energy-efficient home. Windows let light in, allow one to view out, and, with operating sashes, let air in and out. Curtains, shades, venetian blinds, and shutters all help to regulate the movements through this valve. Still, windows are often faulty as valves, for they let desired heat out or excessive heat in. We need a window that shades direct sunlight

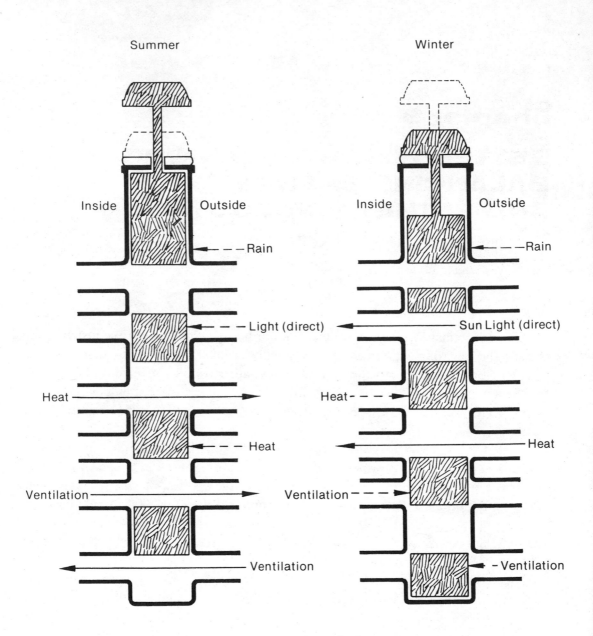

"ENVIRONMENTAL VALVE"

FIGURE 9-1

in the summer but allows the sun's penetration in the winter. We need a barrier to excessive heat losses on cold winter nights and heat gains on hot summer days.

Windows are by far the biggest heat guzzlers in most homes. On a per square foot basis, they generally lose five to 10 times more heat than do the walls or ceiling of a home. If you have a small, reasonably well-sealed home, adding storm windows or double panes to your windows will generally do as much good as adding 2 inches of insulation to your walls and ceiling. If you both add storm windows and insulate your home well (at least 3½ inches in walls and 6 inches in ceilings), you will still find the windows to be a major source of heat loss. Although heat losses through all surfaces decrease tremendously as you go from uninsulated, single-pane construction to well-insulated, double-pane construction, the percentage of your home's heat loss going through your windows rises from 25 to 35 percent. As you further insulate your walls and ceiling, again you lower the over-all heat losses, but the percentage of this remaining heat lost through the windows continues to rise.

When a ship is in good shape except for one bad leak that is pushing the bilge pump to its limit, it is a waste of time to be scrubbing the deck! Instead we need to fix the leak. And so it goes in a well-insulated home: double-paned glass just isn't leakproof enough. The truly energy-conscious homeowner will employ a movable insulating shield over his or her windows for additional thermal control. This shield can be a specially sewn curtain or hinged insulating shutters. We'll explain the details of this in Chapter 11; but for now, let's concern ourselves with *not building unnecessary windows that waste heat.*

PLANNING EFFICIENT WINDOWS

When planning a home's windows you *must* find a balance between too much and too little window area. Windows answer some basic human needs, and having too few of them will make a home dark and uncomfortable. Too many windows, on the other hand, not only will up the initial cost of a home, but will continue to wear your pocketbook thin with increased monthly heating bills. Generally, the amount of window area on the north wall should be kept to a minimum if you want to build an energy-efficient home, as you get no direct sunlight into your home from these windows. The east and west walls deserve caution. To avoid overheating problems in the summer you must be sure that whatever glass is placed on these walls can be adequately shaded, even with the sun low in the sky. The south wall should have the largest area of windows. Here you can capture the sun's heat in the winter but can easily shade windows on this wall with an overhang. To provide an *adequate but not excessive* window area you need to examine in detail your needs for both light and view. Some windows may be just for bringing in light, but before considering these you should begin by planning the windows that are for viewing. These "viewing" windows may bring in all the light that is needed.

THE NEED FOR VISION TO THE OUTSIDE

Having a view to the outside answers a whole variety of psychological needs. We don't like feeling claustrophobic or excessively "boxed in." This reaction often occurs when our senses are denied the rich interplay of light and shadows occurring outside. We miss these natural colors and hourly shifts in the position of the sun. Whether in city or country, much is happening just beyond that outside wall that is worth seeing. A well-planned home has windows carefully located so that one can capture desired external events; it has visual proximity between the kitchen and an outside play area so that one can watch children while preparing dinner; it allows the sun in the breakfast room to help warm a dreary morning; and it gives a glance of warning as neighbors approach, to allow for those last-minute preparations.

These are some of the kinds of communication from the outside that a window offers to the inside. But vision through a window is two ways, and those on the outside find a personality in a home just from how the windows are placed. In the evening, a home turns inside out to communicate with the world by the light it radiates. Some of this communication is unwanted, but visual screens are easily drawn over windows. A lighted living room sometimes will cue a passing friend to make an impromptu evening visit, while a darkened house indicates that it is an inopportune time for such a visit.

The placement of windows in any room should depend on both the nature of the site and the function of that particular room. In crowded urban sites you have more need for large glass openings to private gardens or courtyards than in rural sites where you are already immersed in nature. While a whole wall of the urban home may be a windowless shell to isolate the home and garden from the noise and intruders of a busy street, the country home usually has smaller windows to all directions of the surroundings.

No scheme for window placement or "fenestration" should be independent of the building site. To begin, you can consider the highlights of a particular site which you want to be able to see from your home's interior. In which directions do you feel crowded by streets, other homes, or an unpleasant view? Where do you want to create a private outdoor space, visually screened by shrubs and outdoor walls? These factors not only affect window locations but often dictate the whole arrangement of rooms in a home.

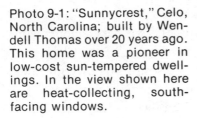

Photo 9-1: "Sunnycrest," Celo, North Carolina; built by Wendell Thomas over 20 years ago. This home was a pioneer in low-cost sun-tempered dwellings. In the view shown here are heat-collecting, south-facing windows.

Most homes existing today were designed primarily around the relationship between the site and the street, giving us a home which merely has a front and back. The front faces the street in a careful and formal way while the back opens to a more private area within a freer and more spontaneous format. This back portion is often screened by a garden, a small wooded lot, or, in densely populated areas, a fence. However, another consideration you should make about the building site is how the house will relate to the sun. To design an energy-conserving home, you need to balance the location of all the rooms so that they function in a unified way with energy considerations—so that the heat-collecting south window wall is in the living-dining area, if in that area you want a panoramic view. As difficult as it may be, we must get away from the notion that the home relates strictly to the street: we must bring back the importance of the sun's orientation.

Photo 9-2: "Sunnycrest" interior. These same south-facing windows provide ample natural lighting to the interior.

To fully utilize the sun in a home, we encounter a design approach which is more difficult than merely placing predesigned houses along a winding street. To find a satisfactory window relationship between a room layout and a site is a simple one-to-one relationship between two factors. However, if we bring the sun into this relationship, we now have three elements. On some sites these three elements work well together, but on others they aren't so compatible. See Figure 9-2. In Chapter 16 we'll look at how many windows you should have on the south side of your house and how to shade them with an overhang in the summer. For right now, see if you can come up with a scheme which places most of the windows on the south side of your home.

Lots with a private view to the south lend themselves well to a design with windows that provide both a solar heat gain and an open view to the outside. Building a sun-tempered home on such a lot should be quite straightforward. Let us examine what can be done on a more difficult site, a small city lot where a busy city

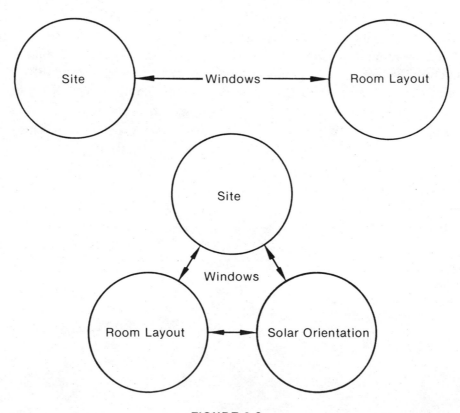

FIGURE 9-2

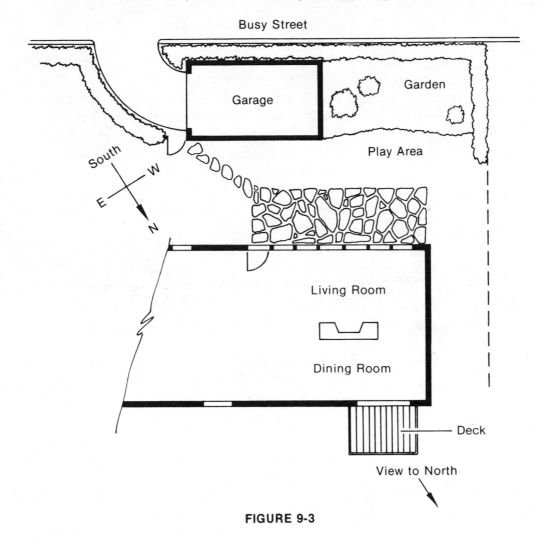

FIGURE 9-3

street is to your south and where a river valley on your north offers an exciting view. (See Figure 9-3.)

Does such a site mean that you deny yourself either the sun or the view? I don't think so, but it does mean that you are going to have to be a little more careful if you are going to get both, and there are limitations to what can be done on such a site (as you cannot change the orientation of the sun). Your major windows should still be on the south wall, but you will want to create a pleasant private space between your house and the road. At the street you can build a garage and wall to screen the public from view. If this wall is of stone, the heavy material will also screen the street's noise. Then by stepping the house back far enough so that the sun can get over the wall and into the south windows, and by planting and landscaping this private courtyard effectively, you can create from the barren features of this site an exciting outdoor space. It's a space to entertain friends within, where you can watch your children play safely away from the busy street, and a space through which you can get a solar heat gain. What about the view to the north? This view can be a distant surprise in contrast to the secure feeling from the courtyard that the south wall faces. You certainly don't want, nor do you need, a window wall on this north side; but a single floor-to-ceiling window or sliding glass door would be really exciting here, opening to a patio or deck where this full panorama could be gathered. However, a north wall window, even of this size, should be covered at night with shutters or a special curtain to keep heat in. (See Chapter 11.)

In any simple and economical building shape, not every space can have proximity to the best parts of a site. Usually the more public living, dining, and kitchen areas have larger windows and relate more strongly to the outside than the private bath and sleeping areas. In the bedroom, you may want a horizontal band of windows with a sill high enough so that you can dress behind it with no more than your head showing to one passing by outside. However, if the bedroom faces a private space, you may prefer a floor-to-ceiling window, and some bathrooms even open to enclosed patios where you can freely sunbathe.

At this point you should sit down with a pencil and paper and start to draw up designs for the windows in the house you want to build or renovate. Once you have decided where you want to put the windows for viewing the site and using the sun on the south side of the house, you can begin to fill in the details of these windows. Pick the top and bottom heights of your windows carefully. Windowsill heights vary considerably throughout a house, but to prevent cluttering the outside appearance of a home, you should usually avoid more than two different heights on any single wall. In Figure 9-4, some standards are given for windowsills and headers.

1. The standard header height or distance from the top of a window to the floor is 80 inches. It is desirable at times to raise this header even higher; but again, for appearance sake, you should try to keep this height fairly consistent on any given wall. More on esthetics later.
2. A clerestory window for light should begin above eye level (never put a sill right at eye level). Often such a window begins just above the door and window header height (80 inches) and runs up to the ceiling.
3. For privacy when dressing, 40 inches is a useful standard, but the effectiveness of this height varies with the relative height of the public area you are screening the dressing area from. Forty inches will also be above the height of most

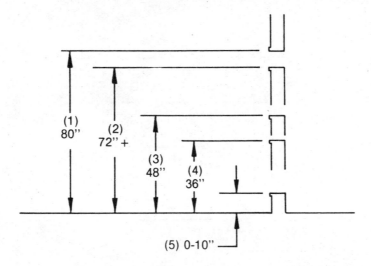

FIGURE 9-4: HEADER-SILL STANDARDS

chests of drawers. Furniture heights should be considered when determining sill heights, not only in the bedroom but in all parts of the home.

4. Federal law requires all glass less than 36 inches from the floor to be tempered or otherwise "unbreakable," i.e., plexiglass, wire glass, etc. These types of glass cost over twice as much as ordinary double-strength glass, so running a sill no lower than 36 inches is a good pocketbook consideration. A variation of this at the cost of using tempered glass is to have a sill at 30 inches in dining areas, which allows a fuller view but stops chairs from slamming into the glass. If windows extend below 30 inches, a chair rail should be installed at the 30-inch height anywhere the glass is endangered by sliding chairs.

5. A full-length window can extend all the way to the floor. A variation of this is to run a sill at 10 inches, which allows a baseboard heater to go beneath it. Whether fixed or sliding, a full-length window at one or two carefully chosen locations brings a lot more excitement into your living area than the typical "picture window," which often has little purpose other than as a sales gimmick.

For reasons of economy and visual security, it is often desirable to break up large windows with horizontal dividers. This allows a fixed glass above with an operating vent sash below or an ordinary glass pane above with a tempered pane below. It also provides a chair rail where needed. Be careful when planning these dividers, though, so that they don't block a horizontal view when someone is either standing or sitting (Figure 9-5). One such division can be placed between 53 and 57 inches off the floor, which is between your vision when standing and sitting. Anoth-

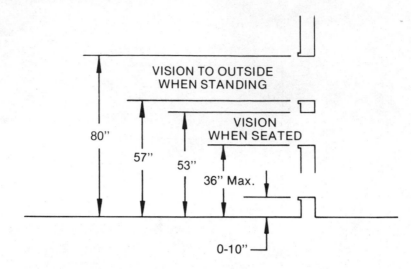

FIGURE 9-5: HORIZONTAL DIVIDERS

er divider no higher than 36 inches is below your vision when seated. This divider can double as a chair rail.

Planning the layout of windows should accompany the planning of a home from the start. Still, it is never too late to better the design of an existing home. An old home may have windows placed awkwardly or may not be receptive to recent changes in the site. With the same consideration to site, interior activity, and solar orientation you can add, replace, move, or remove existing windows to make your home more effective. The information in Chapter 11 will explain how this can be done, but first let's look at how to use windows for natural lighting.

WINDOWS TO PROVIDE NATURAL LIGHTING

Equally as important as the ability to see outside is the natural lighting which windows allow to enter a space. Because of wide fluctuations in outside illumination or light levels, interior illumination cannot be planned for with any large degree of certainty. Still, certain window shapes and locations give certain characteristic results. Examine the design for windows that you have put together so far. Try to maintain its best features while taking the following factors about natural lighting into account:

Nearly all the useful light in a room comes from above. The light coming from low windows has to first be reflected off the floor, and then once again off the ceiling, to be much good for anything other than showing one where to step. For this reason, the reflectance levels of walls and ceilings are very important. A room

Photo 9-3: Sandy Mush Primary School, Leicester, North Carolina. This small school is sun-tempered, utilizing recycled window panes and a dramatic clerestory. It was built entirely from volunteer help on a budget of $5,000, which included the cost of two bathrooms, septic tank, and electrical work.

maintains higher illumination levels when the walls are a light color and the ceiling is white. Openings for light placed well above the eye level bring us this daylighting directly and allow this light to penetrate to the very back of the room. Light entering a room near the floor, on the other hand, illuminates only that portion of the room adjacent to the window and leaves the back part of the room quite dim. Running windows up past the standard 80-inch header height helps by allowing the penetration of light from above. Another aid to natural lighting in any room is venetian blinds. They reflect the incoming daylight up to the ceiling and shower the whole room with light; so does a long horizontal window placed high on a wall. It provides more light than the same window turned vertically, although the vertical window does offer a better view.

Both skylights and clerestories are windows designed exclusively for natural lighting from above. A *clerestory* is a horizontal, clear glass panel placed on a wall well above our line of vision to allow full daylighting of a space. A band of glass just a foot wide lets a great deal of light enter a room. The *skylight* is a window mounted into the roof of a building for the same purpose. Skylights are usually made of frosted plastic that diffuses the incoming light, although some are still of clear unbreakable glass (wired or tempered). Skylights can be used where a roof doubles as a ceiling, while clerestories are dependent on a supportive architectural form. Thus, skylights are more adaptable than are clerestories. However, this flexibility is

Photo 9-4: This home-office was built from the remnants of an old log cabin. The clerestory shown provides good lighting to the interior.

achieved only by skylights shedding water, and there are many leaky skylights on the market today. Check to make sure that the manufacturer of a skylight is reputable and that their product has proved durable. The right kind of skylight *can* be a good way of getting light from above. Also, with both skylights and clerestories, since they are placed in a warm area near the ceiling and also in an area hard to reach in terms of a curtain to retain heat on winter nights, it's imperative that you install equipment that will lose as little heat as possible. The only way to do this is to have the glass fixed in place and use *three* panes. The use of triple glazing will ensure that the heat lost at night is balanced by the light and heat gained during the day.

You should strive for a uniform illumination level from daylighting in any room, as large contrasts in light levels cause eye discomfort. Your eyes cannot adjust simultaneously to two contrasting light levels, and as a result they can feel strained. This "direct" type of glare is quite common where a traditional series of separate small "punched out" window openings creates intense bright areas on the otherwise

Photo 9-5: This residence in Asheville, North Carolina, has two clear glass skylights on the roof. Skylights such as these can be made of tempered or wire glass and add much excitement to the interior.

FIGURE 9-6: GROUPING WINDOWS REDUCES DIRECT GLARE

dark wall they are placed in. The discomfort caused by this can be relieved by grouping these windows together and thereby eliminating the dark areas between them. See Figure 9-6. For the same reasons, corner windows or conventional bay windows do a poor job of increasing daylighting uniformly. However, by placing windows along two walls, you can help to even out the light in a room. If you are planning to have large south-facing windows in any room that is deeper than it is wide, you should plan on some additional lighting from the back, from above, or from the side to balance out the lighting in this room.

Glare is something to be avoided wherever possible. Direct glare can be avoided by a uniform lighting level. Reflected glare is caused by light mirroring off a work surface at the same angle it enters. See Figure 9-7. Fortunately, the work surfaces of

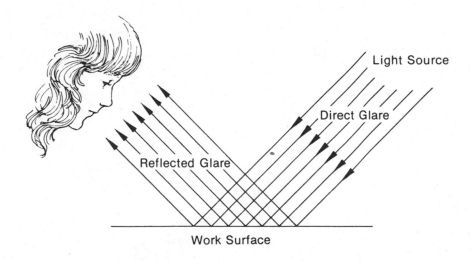

DIRECT & REFLECTED GLARE

FIGURE 9-7

a home are more flexible than a school chalkboard where reflected glare is a common problem. If you're annoyed by reflected glare when working at the kitchen table, let's say, all you need to do is to shift your position to relieve the glare.

Many building codes require the glass areas in the walls of any room to be at least 10 percent of the floor area. Use caution when you apply any such rule of thumb to something as variable as natural lighting. This 10 percent is probably an adequate window area where most of the glass is at or above eye level. If your glass areas are near the floor, however, you should consider having a window area equal to 20 percent or more of the floor area. Since winter heat loss and summer heat gain are big problems with windows, try to keep your glass areas away from the floor. This way you can reduce fuel bills because of smaller windows and still have the light and view you need. An area of glass at eye level equal to 15 percent of the area of the floor is a good balanced design figure to work with. Then, a greater area can be used for south windows (see Chapter 16) and less area in north-facing windows.

To design your home to provide adequate daylighting will not just spare you the waste and expense of electricity. In the comfort of the natural colors which daylight provides, you will find less heat per unit of light to overheat your home in the summer than is produced by either fluorescent or incandescent lighting systems.

OPERATING AND FIXED SASHES

So far we have ignored a very common function of windows: their ability to open for ventilation and sometimes for an emergency exit. Nothing in the nature of a glass panel is inherently related to providing ventilation. Operating sashes do have a few advantages, but they are the common means of residential ventilation largely due to tradition. In certain places you may want to specifically design your windows to allow the enjoyable experience of "opening a window" and peering through to the outside with a fresh breeze blowing across your face. However, while light best enters from windows that are above eye level, vents usually are most effective when placed below this level, near the floor. The next chapter will deal with this subject of ventilation in detail, but for now let's make sure that our viewing and natural lighting designs don't interfere with ventilation.

Having planned your window areas independent of ventilation considerations should not present problems, as you can still provide adequate vent openings. In Chapter 11, Figure 11-7 shows details for a vent flap made by sandwiching styrofoam between two layers of plywood and hinging this panel out from the top. By supplying your fresh air from such vents near the floor, you are free to glaze the window areas with fixed glass. Another similar scheme which is quite common today is the use of fixed glass at eye level with an operating sash directly below, usually one of the awning type. You can cut costs considerably by using fixed glass for eye-level viewing, which can often be obtained second hand, and constructing separate vents below.

The function of operating sashes as fire exits is not something to be ignored, however. Most building codes have a section entitled "means of egress" or "exit requirements" which spells out the number of exits required for each type of building and the allowable "dead end" distances from each exit. With single-family residences, these requirements are usually quite lenient, and by using a little common sense you should more than comply. The second floor of any home should have at least one exit directly to the outside. A small deck or platform by this exit will make it more effective and a second exit at the opposite end of this floor is often a good idea.

In the home you are planning, avoid dead ends more than 30 feet away from an exit on any occupied floor. Exits can be operating sashes of adequate size for the size and agility of the occupants. In other cases, they can be small, well-sealed doors or partial doors constructed to match the interior and exterior surfaces. With such a door you need not interrupt an economical window scheme utilizing fixed sashes.

Window sashes are available in wood, aluminum, and steel. Wood has several advantages but, to resist decay, should receive a good coat of wood sealer or preservative. Metal sashes lose considerably more heat than do wood sashes and are cold to touch in the winter, often with moisture condensing on them. Aluminum is corrosion resistant and makes a good storm sash where it is exposed to the weather but insulated by an inner wood sash. Some new, but expensive, sliding glass doors have a breach, or a plastic-filled space between an inner and outer frame and thereby insulate better. Steel frames are strong, but they corrode easily and insulate poorly.

In the list that follows, six different types of operating sashes are described. See also Figures 9-8a to 8f. The first two types slide a glass panel in front of another one and subsequently allow only 50 percent of their areas to ventilate. In the last four, glass panes are hinged either in or out and thereby open the whole window area. However, they may project in or out in such a manner as to be an obstruction.

1. *Double hung*—This is the most common window in homes today, although one finds it questionable why this is so. The dividing bar between panes is often at eye level and an obstruction to the view. It's not a particularly good view window or vent but makes a good "punched out" opening where such is needed.

2. *Horizontal sliding*—This is another common window today. When purchasing, check the tracks to make sure the window slides well and check to be sure it can be adequately weatherstripped. Some horizontal sliding windows are hard to seal.

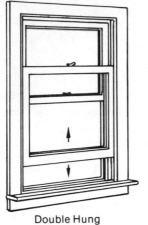

Double Hung
FIGURE 9-8a

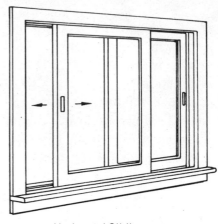

Horizontal Sliding
FIGURE 9-8b

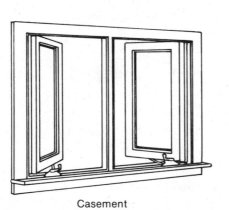

Casement
FIGURE 9-8c

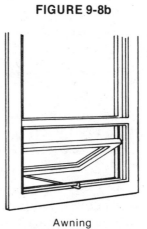

Awning
FIGURE 9-8d

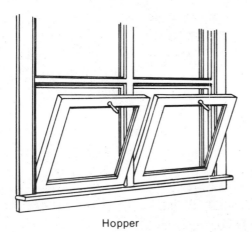

Hopper
FIGURE 9-8e

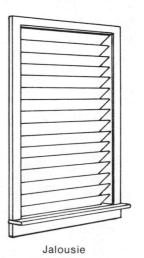

Jalousie
FIGURE 9-8f

3. *Casement*—This is the oldest of the operating sashes and one of the easiest to get a good seal with. Beware of the hazard to those walking by such a window when it's projecting to the outside.
4. *Awning*—This is a good window to use below fixed glass. It will shed rain when open but also easily collects dirt.
5. *Hopper*—The hopper is similar to the awning but it hinges from the bottom inward. It may obstruct furniture in a room.
6. *Jalousie*—In southern climates this is a good window because of its ability to direct air flow. For northern climates, check if it will seal adequately and whether it will accept storm sashes.

SUMMARY OF PRINCIPLES

In this chapter, so far, we have seen a method described that determines where windows should be most useful in a home you are planning. Before we become immersed in how to bring these windows together to create a unified, pleasing appearance for your home, let's review this window planning method:

1. Windows cannot be placed indiscriminately, or house heat and comfort will be sacrificed.
2. When balancing your window needs, try to place major window areas on the south wall and keep window areas on the north wall to a minimum.
3. Begin by planning window areas which will allow vision to the outside. Plan your house so the south windows which offer a good view are in rooms that will use this view.
4. When assessing the placement of window areas, think in terms of standard, repetitive window sizes and uniform sill and header heights.
5. Consider how well your need for natural lighting will be met by the windows you've already planned for viewing. Where needed, plan clerestories and sky-lights to bring in extra light from above. Also, reconsider your "view" windows to see if your needs in some rooms can be met by high, light-giving windows without these "view" windows.
6. Determine each room's need for ventilation (with help from the next chapters). Study the various types of operating sashes available and apply these to those window openings where your needs justify the expense. In other places use fixed sashes (which are cheaper) with a system of separate vent openings.

ARCHITECTURAL QUALITY

Our discussion so far has centered on a window's utility, its logical and pragmatic considerations. Does this mean that the builder should feel strait-jacketed by these concerns, that in response to these new energy standards, our homes are to become dull and institutionally identical? Not at all! Our homes now have an opportunity to respond to their environment as never before. Just as each site, each human need, and each climate is different, every home can be unique. But let this uniqueness come from within. So much energy is wasted by people trying to build homes that *look* entirely different from any other, and the end product is usually inferior to a home which simply performs its function well. Examine the homes in your area. Try to find ones that make a good use of the site, climate, and local materials. Find a way of building which is natural to your region and use it unashamedly. Keep the appearance of your home simple and you can avoid the costs of all those frills that

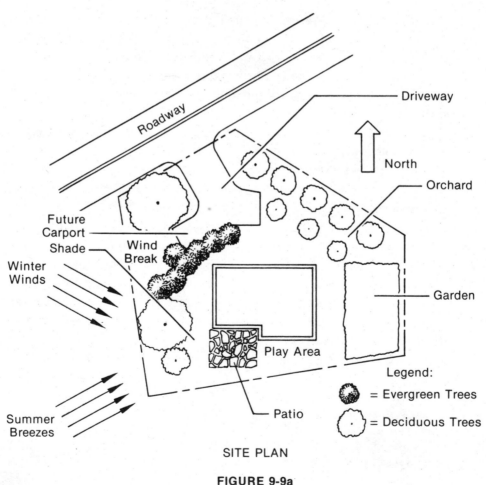

SITE PLAN

FIGURE 9-9a

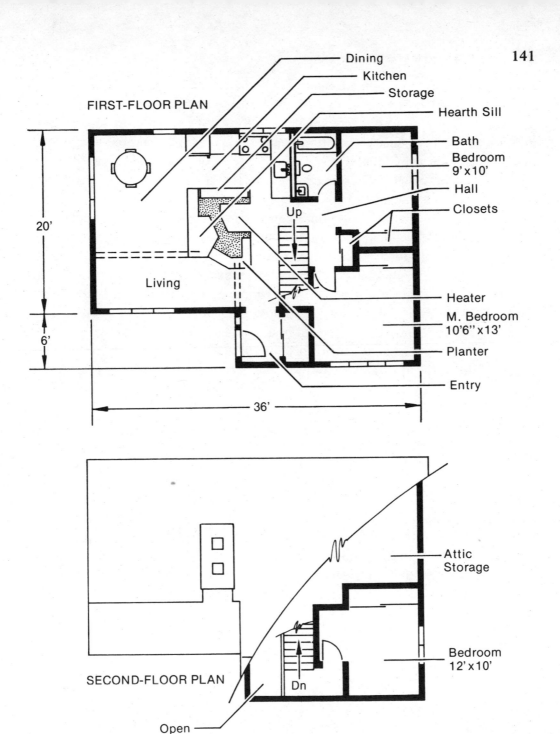

FIRST-FLOOR PLAN

Dining
Kitchen
Storage
Hearth Sill
Bath
Bedroom
9' x 10'
Hall
Closets
Heater
M. Bedroom
10'6" x 13'
Planter
Entry

Up

Living

20'

6'

36'

SECOND-FLOOR PLAN

Attic
Storage

Bedroom
12' x 10'

Dn

Open

FIGURE SHOWN WITHOUT
ANY VENT OPENINGS

FIGURE 9-9b

do nothing. Why use "colonial" shutters stamped out of aluminum to remind you of a period in history when aluminum didn't even exist? Instead, why not try to devise a system of insulating shutters which actually open and close to the weather and times of day? To design with natural materials in a simple and honest way is a good start toward a healthy-looking house exterior, but you also need to carefully balance the arrangement of window areas to other exterior surfaces.

Windows are the single most expressive element of a home to the outside. To build an esthetically pleasing home you need to position your windows so they relate as a unified composition to the outside, even though their individual placement in any single room is individually planned. If you have a floor plan sketch of where you want windows and an approximate idea of their sizes and heights, now is a good time to assign trial sizes to these windows. Standardize all near-similar windows—for the fewer the number of window sizes, the simpler your work will be all the way around. If using the post and beam system from Chapter 6 with posts 6 feet on center, pick a standard width (including frame) that divides this evenly (2 or 3 feet). If the bathroom can take only an odd-size window, consider using a skylight instead. This will make it easier to find a good elevation for this wall and most small bathroom windows are too high and too small to be good for anything but bringing light in anyway.

Figures 9-9a and 9-9b show the site plan and floor plan for a small home that has been carefully positioned on the site. The major window areas face the south and the west. These are well shaded. The main entry is located out of the winter winds, and the kitchen entrance will be closed off during the cold months. The only windows on the north side are a few in the kitchen. The bathroom is illuminated by a skylight. This is a compact and efficient house combining many of the principles we have outlined so far in this book. Although not shown, a basement can be easily added to the plan and the upstairs can be expanded by adding to the roof to make a sleeping area that extends all the way to the back (north side) of the house. One interesting thing about this house is that it faces away from the road, and what would be called the "back" of this house faces toward the front of the lot. This is brought to your attention because if you orient your home to the sun, *south* will be the way it will face and that side will be the most exciting. If the street is on the north and you are energy sensitive, you will build a tight house with a nearly windowless wall on this north side. It shouldn't be a problem to have to go into your private backyard to enjoy the good looks of your home.

You can begin drawing elevations to your home with a ground line, the roof lines, eavelines, and end walls or corners. Then lightly pencil in floor lines, windowsill lines, and header lines. Add to these the doors, porches, posts, chimneys, and other projections from the basic house form. See Figure 9-10. Lightly sketch vertical

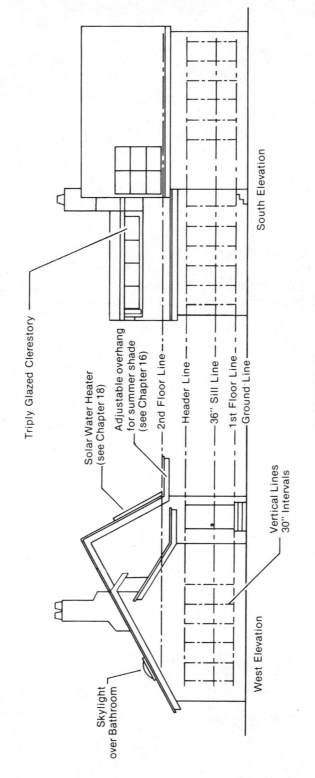

Triply Glazed Clerestory

Solar Water Heater
(see Chapter 18)

Adjustable overhang
for summer shade
(see Chapter 16)

2nd Floor Line

Header Line

36" Sill Line

1st Floor Line

Ground Line

South Elevation

Vertical Lines
30" Intervals

West Elevation

Skylight
over Bathroom

FIGURE 9-10

lines on the elevation at the regular intervals of this standard window width and see if you can get the whole elevation to work with this grid. Don't expect to find the best elevation on your first sketch. Draw several with different spacings of the vertical lines and different window patterns. Group windows where possible into continuous bands. These look better and give more even light on the inside. Also, try to balance these glass areas with the other surfaces on the elevation. If second-story or clerestory windows are in the plan, place them so they harmonize with the ones below. A clerestory usually appears as a strong horizontal band with mullions at rhythmic intervals showing structure.

Second-story windows are usually fewer in number than those on a first floor. Each should be placed directly above one of the first-floor windows so that the window's edges or centers are in vertical alignment. In a small two-story house, balance will be the main problem in your elevations; and sticking to symmetry will make things easier. The single-story house will be more a problem of finding a rhythm and horizontal continuity. Here you may want to vary the exterior surface in between or above the windows for a certain effect, or you can run a continuous sill across the wall between several windows to accent the horizontal.

Figure 9-11 and Figure 9-12 each show two sample elevations of a home. In Figure 9-11 you find a repetitive distance between windows which are grouped together into single units for each room. The standardization of window sizes creates a pleasant rhythm and a feeling of horizontal continuity between windowsills. In the elevation in Figure 9-12 you find a lack of balance, too many sill heights from too many window sizes, and erratic spacing between windows. The awkwardness of so many punched openings here could be eliminated by clustering these windows into a few groups.

Photo 9-6: This is a condominium project in Asheville, North Carolina. It is a good example of a balanced application of simple, repetitive units.

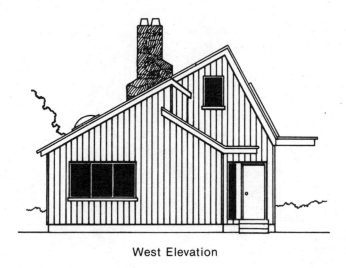

West Elevation

Window Mullion

Clerestory

South Elevation

FIGURE 9-11

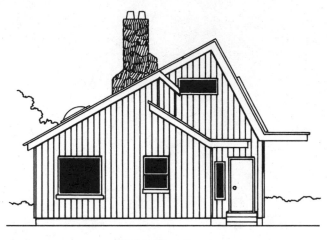

West Elevation

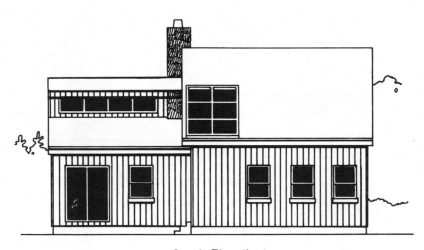

South Elevation

FIGURE 9-12

Photo 9-7: This solar house model was designed and built by Bill Langdon. The model was made from 1/16" chipboard with a ruler and utility knife. Scale: ¼" = 1'0". Shown is a self elevation with a large collector area and sunken greenhouses below. Note repetitive intervals of windows. (To make "windows" in a model, simply darken the window areas with black pen, or — as in this model — cut them out.)

As your elevations begin to work well by themselves, try putting them together. As you look at a home, you usually can see two elevations at once; so, as you near completion of your house plan, I suggest that you make a small cardboard scale model of your home at either ¼ or ½ inch to the foot. Paste or draw onto the model the basic sill and mullion lines of your home, lightly draw in siding materials (board and batten or whatever you plan to use), and then darken to almost black the glass areas. (Windows appear darker than exterior siding in daylight.) Now you can study the appearance of your home in three dimensions!

As you work through the preceding method, it needn't limit your design. Just as any musical performance is based on a score, let this method be a score for your home's exterior performance. Then at some special location you can break stride from this pattern and do something surprising and dramatic. For instance, you can use a standard header height but allow an entrance roof and window to project up out of this horizontal line, or in another location allow a certain group of windows to run with their own special theme. Be your own judge as to what is complementary and what you like.

While it's true that large windows are expensive initially and generally increase heating bills, carefully choosing your windows doesn't have to mean that you miss the excitement of the out-of-doors. A home is a shelter from the outside, a place of security and introspection. You don't need large glass window areas to allow exciting things to happen within your home, nor do these glass walls really "bring the outside in." Why not bring plants into your home along with native crafts and driftwood and bits of nature? By building a fireplace and planters into an interior wall, you can create a whole microcosm of nature to share within your dwelling. A well-designed home will at times beckon us to turn inward from the world outside, to reexamine ourselves, to gather with our kinfolk, and to center our lives within this shelter. Windows, properly placed, will emphasize this inner realm with light and color. A clerestory placed to shed light on a brick or stone hearth will not only dance light over the hearth's texture in the daytime, but will fill its mass with heat to reradiate from the core of this space during the night.

chapter 10

AIR MOTION AND VENTILATION

by CHRIS LOGAN

This chapter deals with the movement of air around and through your home. In any house design, trade-offs will be encountered, and this also applies to the subject of air flow. However, if you keep in mind the air flow patterns when deciding where windows, vents, doors, and interior walls are to be placed, your house will be cooler in summer and warmer in winter. This need for outside air varies throughout the year and depends on many factors, which include seasonal changes, local climate, the number of people living in the house (including smokers vs. nonsmokers), and the type of heating system. Generally, you want to *maximize summer air intake and minimize winter air intake.*

Before we start to quantify the required amounts of ventilation, let's see how air is introduced into a typical house. Outside air enters the structure as either ventilated or infiltrated air. *Ventilated air* is introduced by a mechanical means (i.e., a fan) at a controlled rate to accomplish a specific task, or by merely opening a window to air out the house. Ventilation occurs primarily in the summer as replacement air when hot air is removed, and to a lesser extent in winter during odor removal from kitchens and bathrooms. *Infiltrated air,* on the other hand, is more difficult to deal with since it enters the house at an uncontrolled rate through cracks, around windows and doors, through walls and the foundation, etc. In winter, infiltrated air is a major problem because it causes cold drafts and adds significantly to the heating costs of your house. A study in St. Paul, Minnesota,* showed that for a typical one-story insulated house 15 percent of the heat was lost through the walls, 13 percent through the roof, 5 percent through the floor, 27 percent through

---

*The Invisible Family Test House, St. Paul, Minn., Wood Conversion
Co., 1961

Photo 10-1: The wind pattern over a barrier is demonstrated here by blown smoke.

windows and doors, and fully *40 percent* through air exchanges based on one air change per hour for the house. This means that for every dollar spent to heat the house, 40¢ was spent heating outside air that managed to get in! This "one air change per hour" was based on the average insulated house. However, in an older home, air changes can be as high as two to three per hour. As you can see, then, two-thirds of the heat loss in a house is associated with doors, windows, and air infiltration. Much of this heat loss can be eliminated using easy-to-apply techniques and low-cost materials. In this chapter we'll concentrate on the principles that must be used to reduce air infiltration with cracks and through walls while ensuring that summer ventilation needs can be met. The next chapter will discuss the specifics for dealing with windows, doors, and vent openings for optimum comfort in both winter and summer.

HOW AIR GETS INTO YOUR HOME

Let's look at the various ways that air infiltrates a house. Since the wind plays a major role in air infiltration, it is important to understand the dynamics of outside air movement and its effect on a structure. On the windward side, a positive pressure is built up, since the air is compressed; while on the leeward side, a negative pressure is created due to the suction of air away from that side of the house. See Figure 10-1. This pressure difference is one of the driving forces that create air infiltration.

To analyze these conditions, we can consider air as similar to water flowing in a pipe. If the pipe is capped at one end, no water will "flow" regardless of the pressure applied. This same condition exists for air in a house; that is, there must be an inlet and an outlet for the air to travel through. Infiltrated air enters through cracks, walls, and spaces on the windward or positive pressure side and exits on the negative pressure side. The rate of air infiltration increases with increasing wind velocity; so with careful siting of wind barriers, you can slow down the winter wind reaching the house or speed up a summer breeze to increase comfort. Knowledge of the seasonal prevailing winds and winter storm patterns will indicate where you should plant vegetation or construct winter windbreaks and where to place your large and small windows.

Air infiltration also occurs if there exists a negative pressure within the structure. This condition is created by exhausting air for ventilation (i.e., the kitchen exhaust fan) or by venting a combustion unit to the outside. The oil burner or wood stove behaves in this manner; it takes air from an enclosed space, heats it in the combustion process, and vents it up the chimney. This results in a slightly negative pressure,

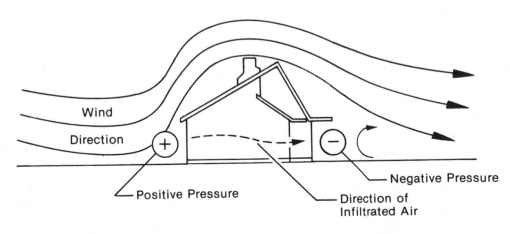

FIGURE 10-1: WIND FLOW PATTERNS OVER HOUSE

which causes the outside air to infiltrate the house. Probably the worst offender in this regard is the charming, old-fashioned fireplace. I am sure that anyone who has ever sat in front of a fireplace can vouch for the fact that a steady stream of air from the room goes up the chimney. In fact, an average-sized fireplace will draw about 3000 cubic feet of air per hour—about a one-third air change per hour through the house. This is *much* more than will be drawn by an oil burner or tight wood stove.

The third mechanism of air infiltration is due to the difference in temperature (or density) between the outside and inside air and the vertical height separating the inlet and outlet openings. An exaggerated example of this type of flow is the operation of a chimney. Air flows from the bottom to the top because of the large temperature difference and the height of the chimney. Even with a small temperature difference over a short vertical height, however, a significant amount of air infiltration can occur. Figure 10-2 illustrates the chimney effect and depicts how a house creates a similar air flow situation.

As we have seen, the three main causes of air infiltration into a structure are (1) wind movement which creates positive and negative pressure areas outside the house, (2) negative pressure within the house caused by various exhaust mechanisms, and (3) temperature differences between inside and outside air. Of the three, the first (air pressure variations outside the structure) and the third (temperature differences inside vs. outside) are the most significant.

You can put these infiltration forces to work for you. In most cases it's possible to minimize both your winter heating costs and your summer cooling costs as well. Your local climate conditions will be the governing factor. Let's first look at how to reduce the heating bill.

WINTER INFILTRATION

As mentioned earlier, heating costs can be substantially reduced by minimizing the infiltration of cold winter air into your home. To reduce infiltration to a minimum requires several things. First, you'll want to reduce the speed of the wind outside your house with a windbreak, if possible. While this is important, a full program of "tightening up" your house will have the greatest effect in reducing air infiltration and hence your heating bills. This involves the use of a vapor barrier to prevent air from coming through walls and the application of caulking and weatherstripping around cracks, the foundation, floor, ceiling, windows, and doors. The relative importance of any of these areas will depend on the type of construction you have. In this chapter we'll explore the general principles involving air infiltration and look at those areas which are often overlooked, like wind barriers and vapor barriers; in the next chapter, the use of caulking and weatherstripping materials to

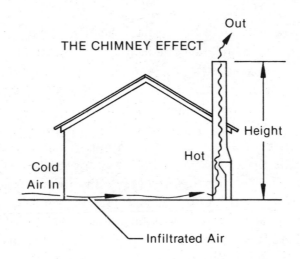

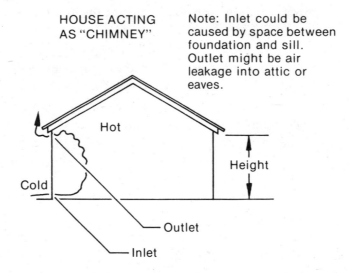

FIGURE 10-2: INFILTRATION CAUSED BY THE "CHIMNEY EFFECT"

solve the most common problems around doors and windows will be explained. Now, let's look at how to make use of effective windbreaks.

Reducing Winter Winds

To reduce the effects of winter winds, the first place to start is outside the house. Find out the speed and direction of the winds near your home. To determine the prevailing winds at your site, the best source of data will be from a weather-conscious neighbor or a local farmer. If you cannot find someone to provide this information, call the local radio station's weather department and ask for the direction and speeds of the winter and summer (seasonal) prevailing winds and the winter storm winds. Another source of data is the National Climatic Center, NOAA, Federal Building, Asheville, North Carolina 28801. Ask for the *Local Climatological Data, Annual Summary with Comparative Data* (15¢). Since wind speeds and direction are site specific, you will have to look over your area for wind barriers such as distant mountains, buildings, or trees that would alter the wind's direction. The data from the National Climatic Center have generally been collected at stations situated in major cities or at airports and should therefore be used *only* as a rough guide for a particular site. However, it is worth sending for, as it also contains lots of information that is quite useful in house planning, such as the amount of solar radiation per month in the area, cloudiness factors, temperature extremes, and relative humidity for each month.

Now, you want to reduce the speed of the wind reaching your house. By lowering the wind's velocity, you'll cut down on infiltration and hence lower your heating bill. This can be done by placing wind barriers (trees, garage, barn, hedges, or fencing) between the house and the prevailing winter winds. (If your winter and summer winds prevail from the *same* direction, read the section of this chapter regarding summer cooling to determine the correct method of dealing with this situation.)

If you plan to use trees or shrubs for your wind barrier, it is important to plant vegetation that will give you a dense growth and that will grow to a height equal to or greater than your house. The maximum distance from your house to the windbreak should not be more than five times the building height, measured from the leeward wall. See Figure 10-3.

The type of tree or hedge you plant will depend on the height you need and the local climate. A good source for advice regarding the type of tree to plant is the local Cooperative Extension Service of your state's land-grant university, or the Department of Agriculture's county agricultural agent, both of which you should be able to find in the phone book. Tell them the size of the area to be protected and the desired height of the trees. Also, if you plant close to the house, ask about root

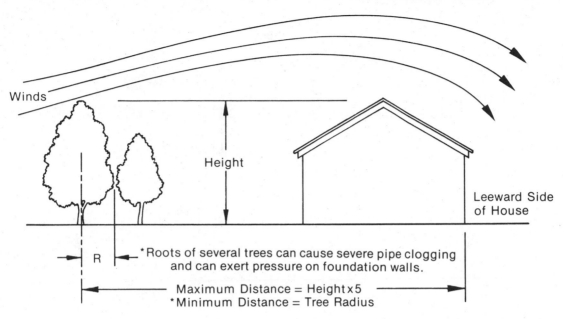

Winds

Height

Leeward Side
of House

R

*Roots of several trees can cause severe pipe clogging
and can exert pressure on foundation walls.

Maximum Distance = Height x 5
*Minimum Distance = Tree Radius

FIGURE 10-3: WIND BARRIERS

damage to sewage pipes and foundation walls. Basically, the problem is that tree roots (especially willows) will seek water and will find their way into a leaking pipe. As tree roots generally grow to the drip line of the outermost branches, make sure to give them enough room.

Reducing Air Infiltration

Next, let's go inside. The best defense against air infiltration is to construct a tight house. You may be surprised to find out that air infiltration occurs not only through cracks but also through the walls themselves! This is because the pressure on the windward side forces air through the tiny holes in wood siding and even masonry walls. The solution for reducing air infiltration in this case is an effective vapor barrier. While designed primarily to keep water vapor within the living space, the vapor barrier is also an extremely effective barrier against air infiltration. To ensure the application of a tight vapor barrier when building a new home, stress its importance with your builder, or do it yourself. Aluminum-faced insulation is a very good vapor barrier, but you must make sure that there are no raised areas or "fish-mouths" when the insulation is stapled to the studs. See Figure 10-4. You can also tape over the joint for a good seal. If polyethylene is used, make certain enough excess material exists at the ceiling and wall joints to prevent the polyethylene from being stretched or stressed when the interior covering is nailed in place. Watch out

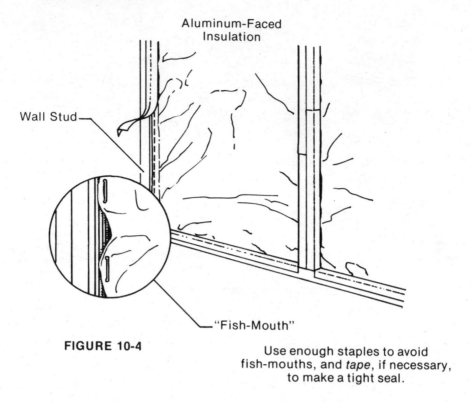

Aluminum-Faced Insulation

Wall Stud

"Fish-Mouth"

FIGURE 10-4

Use enough staples to avoid
fish-mouths, and *tape*, if necessary,
to make a tight seal.

for rips or splits in the polyethylene (use 4 mil minimum thickness) and replace or repair these with tape.

For rehabilitation or renovation work where you are considering adding new interior surfaces, you can apply a polyethylene film first, and then put on your new wall. Plaster walls are effective vapor barriers, so you might want to repair cracks in existing walls. If your plans call for sprucing up the interior with paint, certain good quality paints serve well as vapor barriers. Ask your local paint supplier for a paint with a low permeability value or low air transmission properties. Another problem in older homes is that vapor barriers often are not tight near ceiling light fixtures, around cellar doors, attic access doors, etc. This permits large amounts of air to infiltrate into the house through the basement or first floor, while warm air exits through the attic vents. Make sure to seal up cracks like these with caulking or weatherstripping. (See the next chapter for specifics on these materials.)

A good vapor barrier not only prevents air infiltration but also retains moisture in the house. Some people don't like the vapor barrier idea, though, since they feel that a house must "breathe." Actually, most houses "breathe" too much and are drafty. Additionally, any problems with moisture in the walls can be overcome by proper installation of the vapor barrier. The barrier should always be placed next to the *interior* wall (on the warm side of the insulation material), as this arrangement allows air to circulate in the outer wall, thus reducing moisture problems, and will still be effective in reducing infiltration of air into the house. Watch out, though, as some builders will try to put the vapor barrier between the sheathing and the stud, and this will lead to a condensation problem *in* the wall and in your insulation. Remember, you want the moisture in the living space, not condensing in your insulation!

Thus, the windbreak, the vapor barrier, and the weatherstripping and caulking explained in the next chapter help you to save money in several ways. They enable you to set your thermostat lower because the humidity in the room will be higher. This is because the comfort level in your house depends on the relative humidity as well as the temperature. The higher the humidity content in a room, the less moisture is given off by your body, and you feel warmer. So the first reason the vapor barrier helps is that by keeping house moisture inside, you can keep the thermostat down and less heat will be lost from the building. The second reason is that the vapor barrier reduces the amount of air infiltration into a house. Hence, less cold air must be heated. Yet a third reason is that the cold, dry outside air is excluded so it doesn't have the effect of lowering the humidity inside. Cold air has a much lower humidity content than warm air. If the infiltrated air isn't humidified, the indoor relative humidity falls below an accepted minimum level of 20 percent. At this humidity, a much higher temperature is required for thermal comfort. Taken together, the savings resulting from a full program of winterizing can be very substantial, some 30 to 50 percent of your heating bill!

The degree of tightness or air permeability has not yet been established by the building codes, and at present rests solely with the builder. Many people ask, "What if I make my house too tight? Will I be able to breathe?" The answer to this is that using conventional building materials and techniques you *can't* build a house so tight that lack of oxygen will be a problem! In fact, the amount of carbon dioxide is the controlling factor in the amount of air required for ventilation, and the tightest houses still let in more than adequate amounts of air. However, there have been houses built so tightly that *moisture* builds up inside the house and becomes a problem. This situation can occur in houses constructed to use electric, solar, or heat-pump heating systems; or in houses heated by a combustion process (coal, oil, wood, or gas), where the codes require direct furnace room ventilation via high and low wall vents so that sufficient amounts of air will not be drawn in and circulate throughout the house to avoid moisture buildup. So, if you are worrying about the possibility of making your house too tight, stop worrying! If moisture builds up, you can open a few windows a little, to let the cold, dry air lower the humidity for those rare times when this is a problem; or you might want to install a fireplace, as shown in Chapter 13. Then, if during the winter the house gets too humid, gather some firewood, light it, sit back, and enjoy it. Since your house is tight, you may want to open a few windows a bit to increase the fireplace draft. For example, if moisture in the kitchen is a problem, you can open a window, which will cause air to flow in through the kitchen and up the chimney. Working with the fireplace is more pleasurable than emptying pans from the dehumidifier and has other charms as well.

Another question that people ask is about airing their house out. Occasionally, houses will need an "airing out" for removal of kitchen, bathroom, or other odors that build up in a tight structure. Rather than introducing outside air for odor control, you can use activated charcoal filters that can be installed in the air returns of forced hot air systems, over the kitchen range, or in conjunction with bathroom fans. For other types of heating systems, smaller units are available for individual room use. These filters work because of the large surface area in the charcoal. A 1-inch-thick by 1-foot-square filter piece, for example, has an effective surface area of over 1 million square feet! It's no wonder they are so effective in removing virtually all household odors. The filter units are inexpensive, last about 6 months to a year (depending on use), and are easily replaced. They can be used to remove just the worst kitchen, bathroom, and tobacco odors or can be used to completely purify the air in the house at all times. In the latter case, they also remove dust when used in conjunction with furnace filters, so household dusting can be all but eliminated as well. Thus, these filters make a tight house clean, pleasant, and economical to operate. For more information on hot air furnace and individual room filters, contact Barneby-Cheney, Inc., Columbus, Ohio 43216, or your local heating and ventilating dealer. For more information on kitchen range hood and bathroom units check with companies like Sears Roebuck that sell household (kitchen and bathroom) appliances.

SUMMER COOLING AND VENTILATION

Summer cooling of your home by natural ventilation should be a prime consideration during your site planning and preliminary house design stages. Unfortunately, people's energies are usually focused on staying warm in the winter, and the opportunities for summer cooling are often overlooked. Recognition of the prevailing summer breezes and internal air flow patterns provides the information you will need to determine your summer cooling options. There *are* alternatives to expensive electric air conditioning which are quiet, ecologically sound, and easy to install and operate. Since summer heat loads vary throughout the country, we will concern ourselves here with the principles of natural ventilation which can be applied generally to various sites. If your climate is such that natural ventilation will not maintain the comfort level (such as in areas with almost *no* summer breezes), the principles discussed will still apply so that your operational costs for a supplemental summer cooling system will be reduced. In these cases, you can open the windows and vents we'll describe later to make the best use of the cooling breezes when they exist, and then use a fan or close the house off and use a room air conditioner on those days when the air is still outside.

Basically, the human body maintains a constant temperature (98.6°F) by giving up heat to its surroundings. This is necessary, since the body "burns" food for energy and must discard its excess heat. There are three ways in which heat is given off: radiant heat, convection to moving air, and evaporation of moisture from the skin. When the surrounding environment is too hot, too moist, or lacks adequate air movement, our body's heat-loss capability is reduced and we feel uncomfortably warm. Thus, a ventilating system should provide dry moving air at a cool temperature. However, we cannot always achieve these ideal conditions because of varying climatic conditions. If, for example, you live in an area where the outside air is hot and humid, the only cooling effect you can hope to obtain besides expensive air conditioning is by convection, which aids evaporation of moisture from the skin with high-speed air movement over the body. If in another situation the outside air is hot but with a low humidity, the body will lose heat through moisture evaporation from the skin at low air velocities. These climate conditions can vary considerably, so let's determine what we can do to obtain maximum cooling effects.

Shading Your House

Starting on the outside of the house, we first need shade trees to keep out the sun. Trees are the best natural cooling systems available, as they keep out the sun's heat rays, and the evaporated moisture from their leaves creates cool areas under their branches. Shade trees should be situated to protect the east and west sides of your house in the summer, since these sides are exposed to much of the sun's heat and are difficult to shade because of the changing location of the sun. Make sure that your contractor and bulldozer operator are aware of the trees that you want to retain. In monetary terms, the inconvenience and additional time spent in building by working around a good-sized shade tree will repay you in one or two summers. If you are planning the rehabilitation of an existing house, consider planting shade trees as a long-range solution for summer cooling. There are several fast-growing trees you can purchase, and the same Cooperative Extension Service at your local land-grant university or the county agricultural agent mentioned earlier will have free booklets and good advice on this subject. Give them a call.

Insulation and Ventilation

Once the house is shaded, the next major requirement is insulation. Yes, insulation also works for you in the summertime! What happens is that insulation increases the house's resistance to outside heat, so it takes longer for heat to get in. By the time heat does reach your living area, it is later in the day and the temperature outside is beginning to drop. Then cool evening air can be brought in to effectively remove heat from inside the dwelling. Put the insulating techniques described in Chapter 12 to work for you. They will make summer comfort that much easier to obtain.

The next step is to get these cool breezes into and through the house. Let's look at how to maximize the cooling effect with ventilation and where we want the air to flow. The factors involved with natural ventilation cooling are as follows:

1. the summer prevailing wind velocity and direction,
2. the location from which we take air into the house,
3. the location from which air exits,
4. the time of day, and
5. the outside air temperature and humidity.

The installation of high and low air vents is one effective method of achieving summer ventilation. The hot air in your house will rise and exit from the high vent, while cool air will be drawn in through the low vent. The location of vents is important. You should try to have your *intake vent on the windward side* of the house and your *outlet vent on the leeward side.* In many instances the direction of the prevailing winds will vary during the day and night. If this applies to your site, plan to locate most of your inlet vents on the *evening* side to achieve maximum evening air flow, as the major cooling of your house will occur at this time. These vents must also be operable so they can be sealed during the winter months and on hot sunny days. Details for such vents will be given in the next chapter.

Another consideration when placing vents is to minimize any impediment to the flow of air to the vents, such as shade trees or windbreaks. In the case of shade trees, the air will move freely to the vent as long as the crown of the tree is not in direct line with it. Make sure that the crowns of your shade trees are above these vents or else install the vents on either side of the crowns. If there are conflicts with windbreaks because winter winds and summer breezes come from the same direction, you will have to balance these effects. Generally, the summer ventilation needs will be most important, as you can eliminate the windbreak and still minimize winter air infiltration by using the other techniques in this and the next chapter. If a private outdoor patio area is desired and the windbreak then serves to enclose this area, use vents along the wall area outside the protected patio area for ventilation. Chapters 5 and 16 contain information on balancing your needs for shading, ventilation, and having a protected area.

Other important factors involved in achieving efficient air flow through a house are, first, the location of interior walls and, second, the size of operable windows and vents. This is because the speed of the air flow depends on the size and position of the inlet and exhaust openings (doors, windows, vents, etc.) and the speed of external cooling breezes. Natural cooling will occur when cool air is allowed to enter a house and replace warm air. Hence, by increasing the velocity of air passing through a structure, you increase the cooling effect.

Figure 10-5 shows the air flow patterns through a house and the effects of where walls and partitions are placed. As you can see in Figure 10-5a, the only region not getting some air movement is I. With Figure 10-5b, regions I, II, and III feel little air movement. Be careful where you place your interior walls. Avoid placing partitions and openings in such a way that the wind gets trapped in dead ends. If the layout in your house is not ideal for this cross-ventilation, add vents to move air into the "still-air spaces" of your house. Figure 10-5c shows vents added to the layout of Figure 10-5b, substantially increasing summer comfort.

Since air velocity through the house is needed for cooling, Figure 10-6 shows the effect of relative inlet and outlet vent sizes. Normally, maximum flow is desired, and

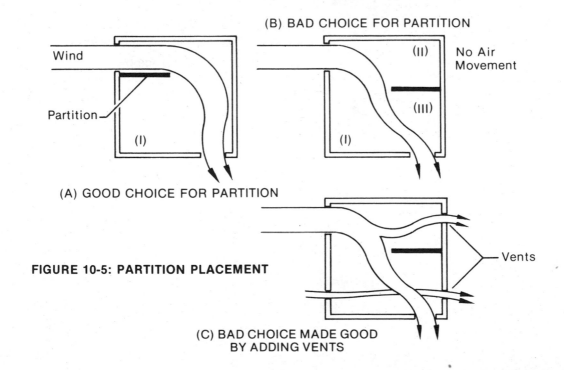

FIGURE 10-5: PARTITION PLACEMENT

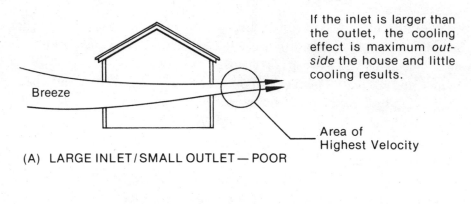

If the inlet is larger than the outlet, the cooling effect is maximum *outside* the house and little cooling results.

Area of
Highest Velocity

Breeze

(A) LARGE INLET / SMALL OUTLET — POOR

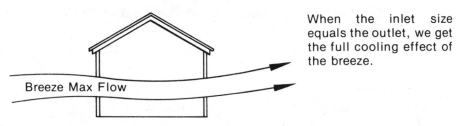

When the inlet size equals the outlet, we get the full cooling effect of the breeze.

Breeze Max Flow

(B) INLET = OUTLET AREA — FAIR

If outside humidity is *high*, the most cooling will take place by increased internal air flow velocity. This is done by decreasing the inlet opening while leaving a large exit opening.

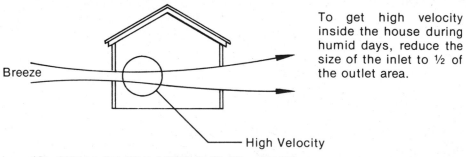

To get high velocity inside the house during humid days, reduce the size of the inlet to ½ of the outlet area.

Breeze

High Velocity

(C) SMALL INLET / LARGE OUTLET — GOOD

FIGURE 10-6

therefore the area of the exit opening should be *at least* as large as the inlet opening. The plan at the end of this chapter gives an example of how you can estimate the number of vents you will need.

The final consideration is the height of vents or operational windows on the wall. Figure 10-7 shows the effect of various venting configurations. Note that while the size of the outlet is important, its position on the wall is somewhat arbitrary because you can vent any part of the room by adjusting only the position of the inlet vents. To decide which arrangement is best for you, first decide what area you want to vent. For example, if you want to vent the ceiling area, use the vents shown in Figure 10-7a; if you want to vent the floor area, use the arrangement shown in Figure 10-7c. As you can see, there are several choices. Generally, Figure 10-7c works best since you benefit from the temperature difference (e.g., chimney effect) as well as from the wind movement, and the chimney effect can be increased even more by using high wall outlet vents and ceiling and attic vents.

The decision to use a fan will be determined by your local climate conditions. Generally, humid southern regions of the country require a high volume of air movement through the house since the major cooling effect is due to air velocity passing over the body. In regions with low humidity, on the other hand, cooling will take place by evaporation of moisture from the skin, so a lower air volume will produce a comfortable cooling effect.

One area you'll want to make sure to ventilate is the attic; in winter for condensation control, and in summer for reduction of day time temperatures. A trap door can be used to allow hot air in the living area to vent through the attic. The best ways to vent the attic is by using a continuous vent along the ridge and eaves or by installing turbine ventilators in your roof. You can get ridge vents from local building suppliers and turbine ventilators from:

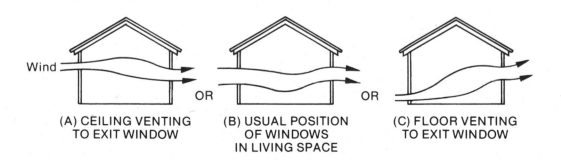

FIGURE 10-7: ANY PART OF THE ROOM CAN BE VENTED BY ADJUSTING THE POSITION OF THE INLET.

Turbine

Roof

FIGURE 10-8

The Sears, Roebuck & Co., mail-order catalog.

Louver Manufacturing Co.		Wind-Wonder, Inc.
P.O. Box 519		P.O. Box 36462
Jacksonville,	or	Houston,
Arkansas 72076		Texas 77036

An attic fan is also a good investment since it will reduce attic temperatures during those windless summer days when the sun is shining, and in the evening it can be used to bring in cool night air. See Figure 10-9.

The size of the fan will vary with the site, but for a rule of thumb, southern regions can design for up to one air change in two minutes; in northern climates, one air change per three minutes should be adequate. Thus, for a house in a southern climate with a volume of 10,000 cubic feet, the fan size can be found by dividing 10,000 cubic feet by 2 minutes, or 5,000 cubic feet per minute. This is the "ventilation rating" you want, *not* the free air capacity. The latter describes the fan's maximum output, that is, without any load. As you start to load the fan with resistance (i.e., ducts, louvers, screens), the actual air movement will be greatly reduced. A well-built fan will give you years of service with little maintenance, so look for quality when you buy. Look at the name plate or product literature, as you'll want a fan that meets what are called the ASHRAE or NEMA test standards. I suggest you get a two-speed fan, with the top rating equal to your calculated air flow. The lower speed can then be used during the cooler days and evening times. This will provide you with a thermally comfortable home, will save you money on electricity, and will be quieter in the evening because of the reduced speed of operation. As with all rotating electrical equipment, safety precautions are a primary concern. Check your local building codes for the electrical and fire regulations

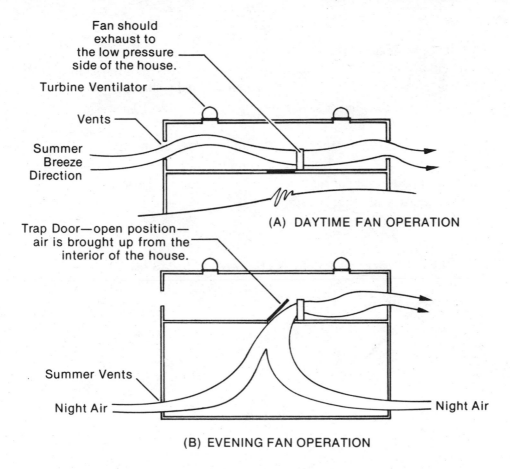

Fan should exhaust to the low pressure side of the house.

Turbine Ventilator

Vents

Summer Breeze Direction

Trap Door—open position— air is brought up from the interior of the house.

(A) DAYTIME FAN OPERATION

Summer Vents

Night Air

Night Air

(B) EVENING FAN OPERATION

FIGURE 10-9

regarding attic fans, and use safety guards when persons, especially children, might have access to the fan.

What do you do in situations where there are few summer breezes? In this case, it will pay to take as much advantage as possible of the chimney effect (the fact that heat rises) by using the floor-to-floor vents described in Chapter 13, together with low inlet vents and high outlet vents along walls, in the top floor ceiling, and in the attic. This way, hot air will rise up through the house, bringing replacement air into the house behind it. This "breeze" effect will give some comfort in these situations and can be improved upon by placing shade trees, a pond, or groundcover on the inlet side so that air is cooled by the evaporation of water from these areas before it is drawn into the house. On hot, still days when this method isn't enough to satisfy your needs, the attic fan mentioned earlier will bring the air speed in the house up to the comfort level.

PLANNING, WITH THE NEW ENGLAND AREA AS AN EXAMPLE

Since in New England the major concern is the long winter season, you'll want to be particularly careful to minimize winter infiltration (i.e., by planting evergreen trees on the northeast and northwest so that the cold prevailing winter and storm winds don't reach the house). See Figures 10-10 and 10-11. The double door entrance is on the east, away from both the positive and negative pressure sides of the house in winter. The southeast, south, and southwest are open to permit the sun to heat the house in winter. For natural summer cooling, a breezeway is preserved by this planting arrangement and the driveway location. Note that if the driveway were on the southwest, the "cooling" summer breeze would be heated *before* it reached the house.

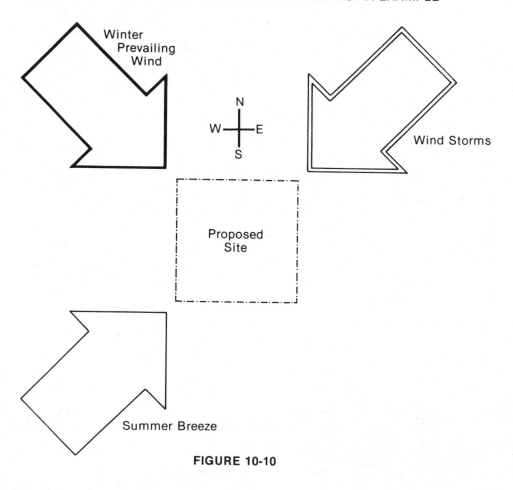

PLANNING, WITH THE NEW ENGLAND AREA AS AN EXAMPLE

FIGURE 10-10

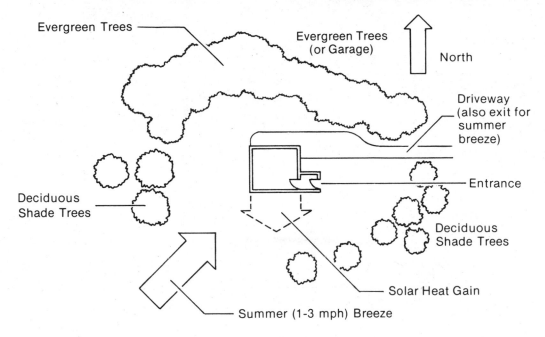

FIGURE 10-11

Now, let's look at the size and placement of vents for summer cooling. For a 1,000-square-foot house, the total volume is $1,000 \times 8 = 8,000$ cubic feet. Thus, for the New England area, a three-minute air change requires movement of

$$\frac{8000 \text{ ft}^3}{3 \text{ minutes}} = 2,700 \text{ c.f.m. (cubic feet per minute) through the house.}$$

Since the summer prevailing breeze is low, the winter wind barrier, acting as a funnel, will help to increase the air velocity through the house. To rely entirely on the summer breeze to move 2,700 c.f.m. through the house, we would need openings with a total area of 50 square feet, which is determined as follows:

$$A = \frac{Q}{EV} = \frac{2,700 \text{ c.f.m.}}{(.3)(2)(88)} = 50 \text{ ft}^2 \text{ of openings,}$$

where A is the required area opening in square feet, Q is the required air flow in c.f.m., E is the factor for the direction of the wind (.5 for perpendicular winds, .3 for diagonal winds), and V is the average wind velocity in feet per minute (ft./min. = m.p.h. $\times$ 88).

To rely on the temperature difference between the ceiling and the floor, we would need an area of 35 square feet:

$$A = \frac{Q}{9.4\sqrt{H\Delta T}} = \frac{2,700}{9.4\sqrt{(7)(10)}} = 35 \text{ ft}^2 \text{ of openings,}$$

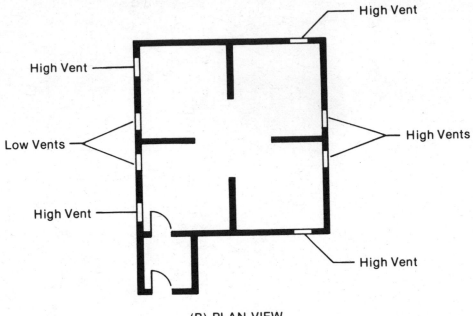

(B) PLAN VIEW

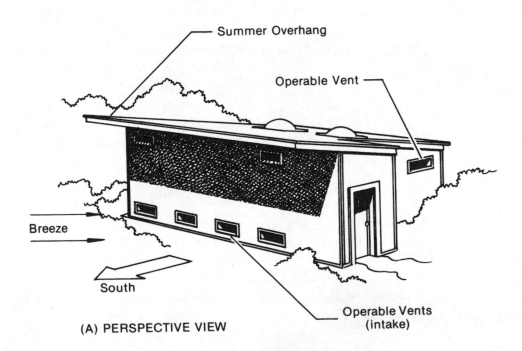

(A) PERSPECTIVE VIEW

FIGURE 10-12

**(WINDOWS OMITTED TO SHOW VENT LOCATIONS —
OPERABLE WINDOWS MAY BE USED AS VENTS.)**

where A is the area in square feet, H is the vertical distance in feet between vents, Q is the required air flow in c.f.m., and ΔT is the temperature difference between the inside ceiling and the outside shaded area ($T_{\text{inside ceiling}} - T_{\text{outside shade}}$).

For this particular site, I would rely on *both* the flow due to the temperature difference and that due to the summer breeze. See Figure 10-12. I would plan my outlets (exhaust vents) to be located on the summer evening's leeward side of the house.

With both the temperature difference and the summer breeze acting together, about 25 to 30 square feet of inlets should provide adequate ventilation. Most of this can come from windows, as outlined in Figure 10-7, but four low inlet vents of 2 square feet each and four high outlet vents of 3 square feet each will help tremendously. This way, most of the windows can be fixed so infiltration will be low in the winter, and yet the few operable windows and the vents can still provide adequate summer cooling. Two small high vents have been added on the breeze side to prevent heat build-up along the high part of the ceiling area, because of its slope.

This situation wouldn't change all that much as you go farther south. There, with the required ventilation rate higher, more windows and vents would be necessary. It would still be possible to get adequate cooling from natural ventilation, provided you take the effort come winter to do a good job weatherstripping the windows and vents.

Let's summarize some of the important aspects of this chapter:

For winter operation, *Minimize air infiltration* by:

1. Plant or construct wind barriers to protect your house from winter storms and prevailing winds.
2. Install a good vapor barrier which will keep out the cold air and keep in the humidity, thus ensuring a cozy house all winter.
3. Check around windows, inside cellar doors, and around ceiling light fixtures for cracks that will permit air to escape or enter. Then follow the instructions in the next chapter for caulking and weatherstripping.

For summer operation, *Maximize cool air intake* by:

1. Plant shade trees on the east and west sides of the house. If you are building a new house, plan your site with existing trees in mind.
2. Plan to utilize summer breezes (i.e., don't build a garage that will prevent the breeze from reaching the house). Also, the driveway collects a lot of heat, so avoid placing a hot-top driveway between your house and the summer breeze. However, a pond or shaded area upwind from your home will cool the air before it enters the house.

3. Make maximum use of high and low vents for the removal of hot air and the intake of cool air.

4. Keep heat out during the day and introduce lots of cool air at night. For maximum air flow through the house, the outlet vent area should be *at least* as large as the inlet vent area.

5. In addition to the above, put turbine vents in the roof; and, if you need to, install a two-speed attic fan.

chapter 11

THE SPECIFICS OF CONSTRUCTING BETTER WINDOWS AND VENTS

by EUGENE ECCLI
WILLIAM K. LANGDON
ALEX WADE

As we saw in the last chapter, 27 percent of the heat in the Minnesota test house was lost via conduction through doors and windows. In addition, of the 40 percent of the heat lost by one air change per hour in this test house, at least 20 percent was associated with infiltration around doors and windows. Thus, doors and windows in your home can account for some 50 percent of the heat loss. The infiltration problem occurs because of cracks where the door and wall meet the door frame and where the sash and wall meet the window frame. The conduction problem occurs because glass and most doors are not good heat insulators; hence, warmth goes through them very easily. See Figure 11-1. Consequently, you can see that careful selection of windows and doors and very thorough weatherstripping and sealing are of vital importance in lowering your fuel bills in winter and achieving summer comfort.

For the windows, several considerations are required. Here, the air infiltration around the window and the conduction of heat through the glass are even more important than with doors. Moreover, as pointed out in the last two chapters, the solution of these problems *must* be accomplished in a way that still allows us to provide the light, view, and summer ventilation we need. This chapter will explain the specifics for dealing with these problems, but first let's look at some of the important considerations for both new and older houses.

DOOR

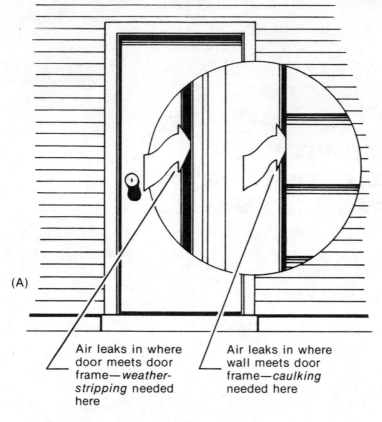

(A)

Air leaks in where door meets door frame—*weatherstripping* needed here

Air leaks in where wall meets door frame—*caulking* needed here

WINDOW

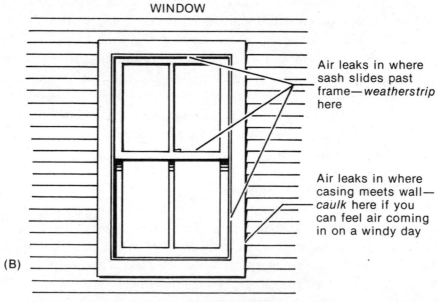

Air leaks in where sash slides past frame—*weatherstrip* here

Air leaks in where casing meets wall—*caulk* here if you can feel air coming in on a windy day

(B)

FIGURE 11-1

The two big problems of infiltration around and conduction through doors and windows require that you be aware of several options before you buy new equipment or do repair work on an older home. Basically, the options to look for can be broken down as follows:

Windows

In a new home, make sure that you minimize the window area, leaving just enough for light, a view, and ventilation, as outlined in the last two chapters. When you buy new equipment, make sure that windows are weatherstripped and that they are double glazed as well.

On older homes, make sure to weatherstrip and caulk very thoroughly around all the windows. Also, be sure to add storm windows. If the placement of windows is incorrect for the view, natural light, or ventilation needs that you have, close off some windows and install new ones to achieve the desired effect.

In both new and old homes, consider the use of fixed windows for light and viewing, and use operable vents for ventilation purposes. The fixed glass areas are less expensive than sliding windows. They also eliminate the infiltration of air around the window while still allowing adequate ventilation from the vents.

Finally, be sure to add insulating curtains or panels to cover the windows at night, as this is one of the largest sources of heat loss from the home during winter.

Doors

For new doors, make sure that they make a tight seal when closed and are thick enough to slow down the conduction of heat.

For older doors, make sure to weatherstrip and caulk to get a tight seal.

In both new and old homes, use only those doors which are necessary in the winter and seal off the others. Try to choose the door least affected by the winter winds and add an entrance foyer if possible. The entrance foyer will cut down on the infiltration around the door.

Why not consider all the options just outlined? Most of these projects are low in cost and easy for the homeowner to work on (or to check that your contractor does a good job on), and the savings are really substantial. If your heating and cooling bill is $500 each year, a few days' work on sealing up the house and perhaps $50 spent on materials can save you $100 to $150 each year thereafter. The work necessary and the cost of materials to change the position of windows is greater; but the view, the light, and the natural ventilation achieved will bring comfort and savings for the whole life of your house. Moreover, the cost-benefits and the added comfort that will result from putting the ideas of this and the last two chapters to work are just as important as insulation! Now, let's look at how we can put the problem-solving opportunities mentioned so far to work for us.

NEW WINDOWS AND DOORS

With doors, infiltration problems arise because air leaks to the inside of the house where the door and wall meet the door frame. Another problem is that heat conducts right through the door itself. If you buy new doors for your home, manufacturers like the Morgan Door Company make prehung, fully weatherstripped wood doors that will eliminate these problems. Generally, choose a wood rather than a metal door, as these conduct less heat to the outside; and in the wood line, choose a thick door for the best insulation. However, some aluminum doors will work well if filled with a urethane insulation.

Windows have air leakage problems where the wall and sash meet the window frame. If you buy commercial windows for your home, try to spend the extra money and buy windows of a quality equal to those sold by companies like Pella or Andersen. These companies make excellent wood sash and frame windows that are factory glazed with insulating glass and fully weatherstripped. Look for "maximum infiltration" standards which are stamped on new doors and windows. You'll want "AAMA Standard 302.8" or "ANSI Standard A134.1" with infiltration rates of no more than ½ c.f.m. at a 15 m.p.h. wind. This will ensure that the new windows and doors seal tightly and that you get the quality you pay for.

APPLYING WEATHERSTRIPPING

What can you do about air infiltration if you don't have the money for expensive, prefitted, factory weatherstripped windows and doors? You might decide to build your own, but then how do you weatherstrip them? Also, what can be done with the windows in an existing house? In both cases, the answer can be found at the corner hardware store: it is a gasket compression-type weatherstripping somewhat similar to those used on automobile doors. It comes with a peel-off back and an adhesive coating. Be careful with the coating, though, as the adhesive is usually of very inferior quality and the stuff falls off. This can easily be remedied by applying a coat of contact cement to both the weatherstripping and the surface to which it is to be applied. Make sure to buy the black weatherstripping with a fabric face, not the white or grey plastic foam variety, as the foam will damage rather easily. If you are buying for the whole house, it will pay to buy from a real hardware store that sells only hardware. These stores are usually found in large and medium-sized cities under the yellow pages listing "Architectural Hardware."

Compression-type weatherstripping, properly applied, makes a completely weathertight seal. It is much easier to apply than the traditional spring-bronze types, which are very difficult to seal at corners. This compression weatherstripping should be used on hinged doors, as well as on casement and awning-type windows. Casement windows work just like a hinged door and the weatherstripping is applied accordingly. Awning windows were described in Chapter 9 and, like casements, can be easily job constructed and weatherstripped. On sliding windows the spring bronze or felt-hair weatherstripping will work best. Also, make sure to install good latches on your windows. The latches you select should have a cam action that tightly forces the window sash against the weatherstripping. Simple, inexpensive latches of this nature are available at any hardware store.

It is also possible to make the windows for your home and then weatherstrip them. One money-saving route you should consider is to construct your own operating or fixed sashes from glass and hardware which are separately purchased. Most large window manufacturers assemble their products from the parts that smaller independent glass and window hardware manufacturers supply. The Blaine Window Hardware Co., Hagerstown, Maryland, prides itself on supplying all window and sliding-door hardware, weatherstripping, and replacement parts for any window made in this country. Their catalogue is worth ordering. Glass, particularly tempered insulating glass, if ordered in standard sliding glass door sizes (28 X 76, 34 X 76, 34 X 92, 46 X 76, and 46 X 92), costs less than half what the major companies charge for it. Try to find out from your local building supply outlet who makes glass for your part of the country. Two such companies are the Economy Glass Corp., 315 Columbus Avenue, Boston, Massachusetts 02116, and the Shaw Glass Company, Inc., 200 Turnpike Street, Stoughton, Massachusetts 02072. They sell in small quantities and ship orders to the New England area and the eastern United States, respectively. Photo 11-1 is an example of a homemade awning window made with one of the hinges from the Blaine Window Hardware Co. mentioned earlier.

Your problems with an old house with double-hung windows may be more complicated. If the windows are in good condition, a combination of storm windows and interior insulated shutters as described later in this chapter is the best answer. At the other extreme, if the windows and frames are deteriorated beyond repair, rip them out and replace them with new windows, either factory weatherstripped or job-built as just described. In the most typical instance, however, just the sash and maybe the windowsill need replacement. A large lumberyard with a millwork department can either order or make a replacement sash for you. You can usually get them custom-made to any size. The replacement sash can even be made in pane sizes to resemble the original design of the window. The new sash can then be installed in the existing frame with hardware from a company like the one mentioned before,

Photo 11-1: One of the simplest kinds of windows to construct and weatherstrip is an awning window, like the one shown here.

and the compression gaskets thus produce a fully weathertight window. If the windows are of a particularly nice design or if you want to restore them exactly, a large lumber company can supply exact replacements. Your best bet here is to have them grooved on the sides and install new spring-bronze type weatherstripping. This is an expensive solution, but well worth it if the house has otherwise fine detailing.

Many old houses had leaded or stained glass windows of particularly beautiful designs. By this time many of the wood sashes are flimsy and the lead is brittle and maybe a piece or two of glass is missing. Still, it is very handsome, but you can't afford to have it rebuilt. What to do? Acrylic plastic sheets commonly sold under the trade names of Plexiglas or Acrylite can be used to salvage these old sashes. Cut a piece of the acrylic just small enough to clear the window stops. Drill holes about 4 inches apart and screw the sheet to the outside face of the window sash. In this fashion, you achieve the extra insulating value of double glazing, strengthen the sash, and protect the leaded glass from the elements—all in one operation.

CAULKING AND SEALANTS

All of your careful work and expenditures on doors and windows will do little good if the wind simply leaks in around the outer frames of the openings. Caulking of openings and cracks in both old and new construction is probably the most important step after weatherstripping to assure yourself of a comfortable home. Caulking is also important to prevent penetration of water into the house during driving rainstorms. As we have seen earlier, houses develop negative pressure inside, and the tiniest pinhole will admit water.

The newest caulking compounds have vastly better adhesive qualities than the older, standard caulking compounds. These high-quality caulks are usually referred to as "sealants." Unfortunately, good quality sealants are not readily available to the homeowner, or to the typical small contractor for that matter, as the two-component catalytic sealants require special mixing equipment and are not suitable for small jobs. To make matters worse, the caulking compounds found in a bewildering array at your hardware store or lumberyard are usually worthless, or nearly so. Remember, it is a great deal of work to apply caulking, and the inferior stuff found locally will have to be replaced within a few years. Fortunately, there is a very fine sealant which *can* be readily obtained with a little hunting. It is called MONO and is manufactured by The Tremco Company, New York, New York. It has terrific adhesive qualities and will withstand a great deal of movement in the substrate, making it ideal for sealing cracks in old construction or between dissimilar materials. For a house of reasonable size, old or new, a whole case should be purchased. A case (24 tubes) costs about $50 to $60. The best source of supply is a large glass-storefront installer such as those who have franchises with glass companies like Pittsburgh Plate Glass. Or try the yellow pages under "Caulking" in the nearest large city. *Accept no substitutes*—there are none!

Some special techniques have to be used in the application of MONO. First, unless the outside temperature is 85°F or so, the material should be heated prior to application. This makes it much easier to apply and greatly increases its adhesive qualities. It absolutely *has* to be heated in cold weather to be applied at all. If an oven is available, this is the easiest method. If not, heat the tubes with a light bulb as directed on the carton. Be very careful in application because the superior adhesive qualities make the stuff extremely difficult to remove. The solvent for the material is methyl ethyl ketone, commonly available as a plastic pipe cleaner or from your drugstore. MONO can be used to attach glass or acrylic plastic directly to wood supports if quick, easy glazing is required such as with a greenhouse. It is also very useful for sealing troublesome skylights and chimney flashings, old or new. On an old house, it can be used to make temporary repairs to rusted-out flashings. Simply cut a patch and stick it on with the MONO. This may enable you to put off more expensive repairs for a year or two.

A second sealant which you may wish to consider for some jobs is silicone, which *is* generally available from local supply sources. Unfortunately, it is very expensive and cannot be painted. It is manufactured by General Electric and Dow Chemical in a variety of colors, including white and clear. The latter may come in handy for repairing leaded glass and stained glass and for sealing butt joints in glass and plastic. Your best bet with the silicone is to buy it from a glass supplier for about 35 percent off what you pay at the corner hardware store. In an emergency on small jobs, try to pick a caulk which says it contains either butyl or neoprene.

All types of caulking and sealants should be applied over clean, solid, dry materials. Large cracks (over 3/16 inch) should be filled to within 1/4 inch of the surface with a backup material such as oakum (available from plumbing supply houses) or polystyrene rope (available from glass supply outlets). MONO is much less critical about the size of a crack and about cleanliness than are other materials. However, the surface must be *completely dry* for any of them to adhere properly. See Figure 11-2.

Now, commercial contractors rarely do a thorough job of caulking, and many operations are done best by applying the sealant to the materials *before* they are assembled. Therefore, check carefully to see that the contractor you choose does a good caulking job. When the job is done, check for leaks with a garden hose, and recaulk if need be. It may save trouble later.

ADDING AND CLOSING OFF WINDOWS

If you want to use the ideas in Chapter 9 and find that your present home needs some changes in window locations and sizes, think about closing off some of the unnecessary windows and adding others where appropriate. This should not be too difficult if the exterior of the house is wood siding. If the exterior of your home is of brick or masonry, you will need to call in a skilled mason for help; this can be expensive. On any wall you are considering opening for a window, first check from the basement for any plumbing, electrical, or heating lines which might have to be rerouted. This may change your mind about where you locate the window you are thinking of adding.

Once you decide where a new window will go, the first concern is for safety on the job. If your opening will be much over 3 feet, you should pin up any overhead ceiling joists while you tear out the wall opening until you can get a header in place. To pin up these joists (see Figure 11-3), cut a 4 X 4 to a length slightly longer than

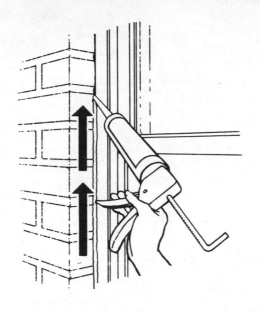

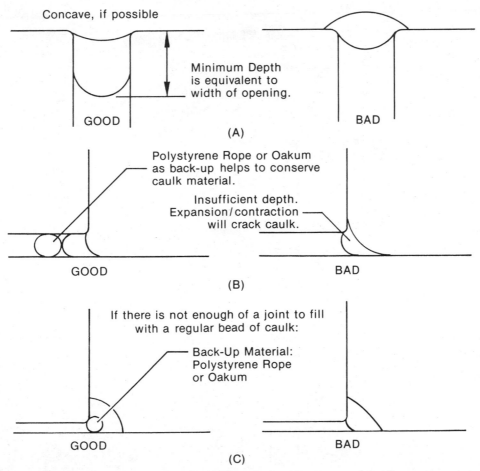

Concave, if possible

Minimum Depth
is equivalent to
width of opening.

GOOD

BAD

(A)

Polystyrene Rope or Oakum
as back-up helps to conserve
caulk material.

Insufficient depth.
Expansion/contraction
will crack caulk.

GOOD

BAD

(B)

If there is not enough of a joint to fill
with a regular bead of caulk:

Back-Up Material:
Polystyrene Rope
or Oakum

GOOD

BAD

(C)

FIGURE 11-2

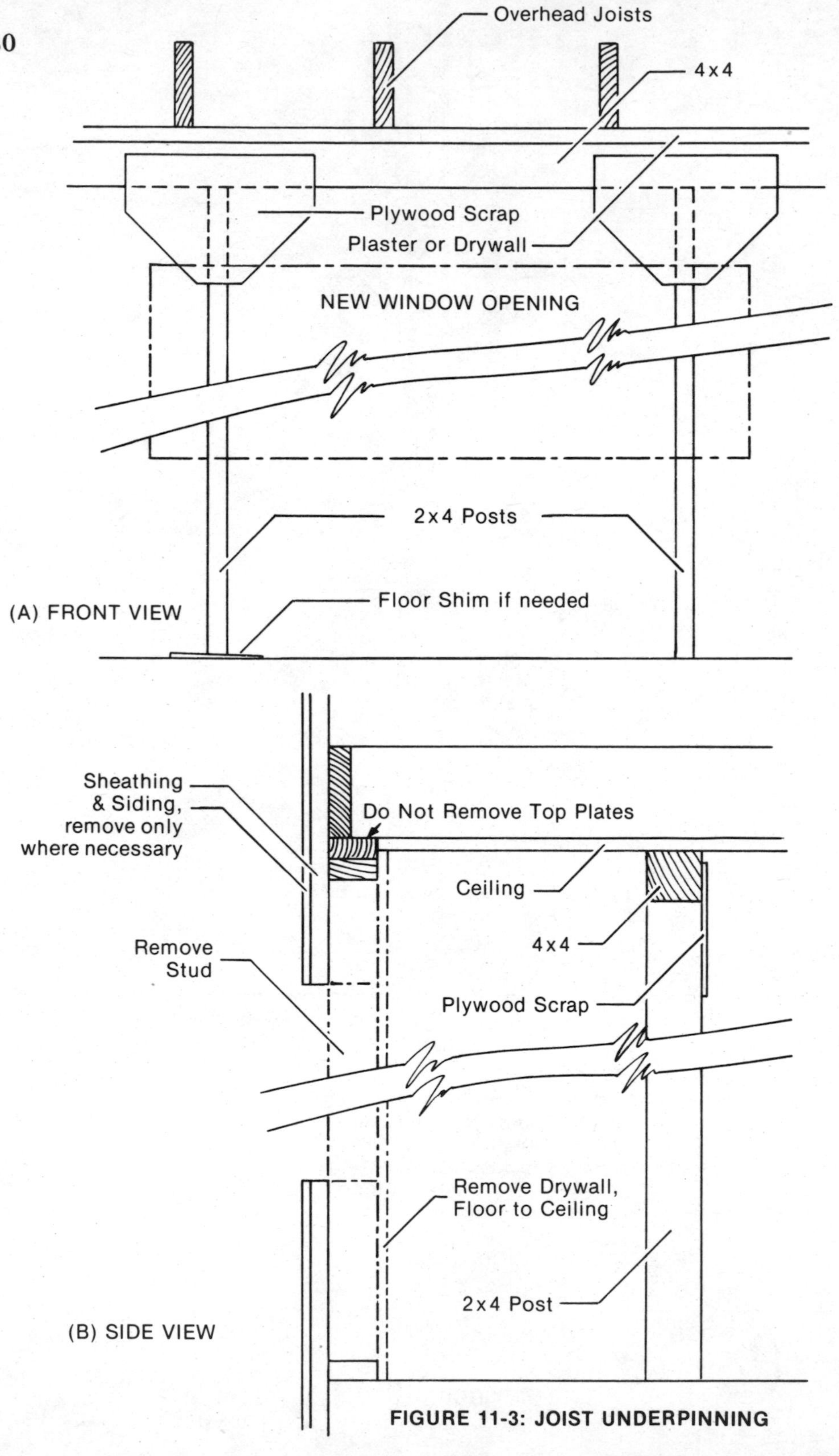

Overhead Joists

4 x 4

Plywood Scrap

Plaster or Drywall

NEW WINDOW OPENING

2 x 4 Posts

Floor Shim if needed

(A) FRONT VIEW

Sheathing
& Siding,
remove only
where necessary

Do Not Remove Top Plates

Ceiling

4 x 4

Plywood Scrap

Remove
Stud

Remove Drywall,
Floor to Ceiling

2 x 4 Post

(B) SIDE VIEW

FIGURE 11-3: JOIST UNDERPINNING

the size of the planned window opening. Measure the floor-to-ceiling height in front of your new window and cut 2 X 4 posts to precisely 3½ inches less than this height. Then, with a few scraps of plywood, attach the posts to the 4 X 4 plate and mount the 4 X 4 against the ceiling in front of the opening. The 2 X 4's should fit tightly against the floor but shouldn't be a forced fit.

Now you can tear out the existing studs in your wall and put in a header. The header should be made of double 2 X 6's for an opening up to 5½ feet, double 2 X 8's for up to 7 feet, and double 2 X 10's for anything over 7 feet. You should support the header with double 2 X 4's on the sides. See Figure 11-4. With the header in place, cut short blocks to fill in as spacers between the top of the header and the bottom of the existing top plate. Now you can remove the temporary support you made from 2 X 4's and 4 X 4's. Then you can go ahead and finish the rough framing as shown in Figure 11-4: mount your window casing in place, shim until level, and nail it to the frame from the outside. For fixed glass windows (Figure 11-5), you can make your own frame out of clear 1 X 6's which are shimmed and leveled to make an opening ¼ inch bigger than your glass size. The glass is fixed in place with ½ X ½-inch wood strips and silicone caulk is applied to the outside.

If you are adding windows to your home, you may find that some of the older ones are really unnecessary. As you renovate the old windows in your home, some may be in poor condition; and if these are on a north wall, you may want to remove a few of them. Removing an old window will be easier than adding a new one elsewhere. To remove an old window sash, first gently remove all the trim on both the inside and outside. In newer windows, the casing and sash are usually a single unit and should come out as one piece. In some situations this window unit will be set in under the exterior siding, and removing some of this siding will be necessary. Once the window is out, you have three options for its future. If the sash is in reasonably good operating shape, you could reuse it in your home or recycle it to a friend. If the window is just slightly deteriorated, you can mount it into a fixed sash opening. The third option, of course, is to discard an obvious piece of junk.

Now, with your window out, you have a gaping hole in your wall with rough framing around it. Structurally, this opening is secure, but you will need to frame in studs at 16 inches on center to attached interior and exterior wall coverings. For the exterior, you may be able to reuse siding cut out of a window opening that you are adding. Otherwise, shop ahead of time at lumberyards and used material yards to find something to match your existing siding. If your exterior is of brick or stone, it will be far simpler if you still use a wood panel here, matching exterior woodwork. Calling in a mason will be expensive, and the additional brick most likely won't quite match the existing masonry anyway. Wood infill panels will do, and you can paint or decorate them to complement the existing exterior of your home.

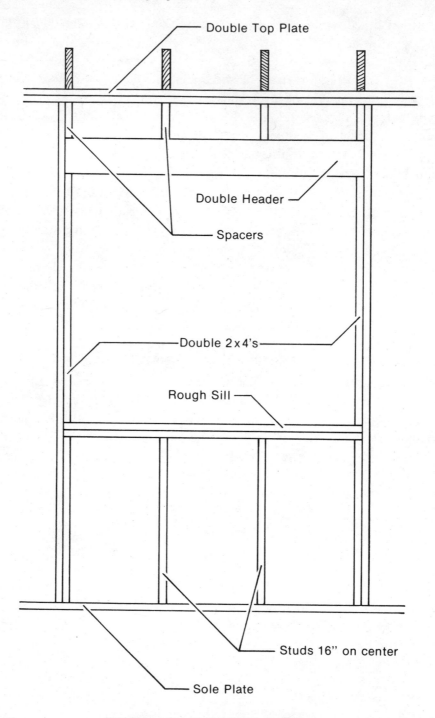

FIGURE 11-4: WINDOW FRAMING

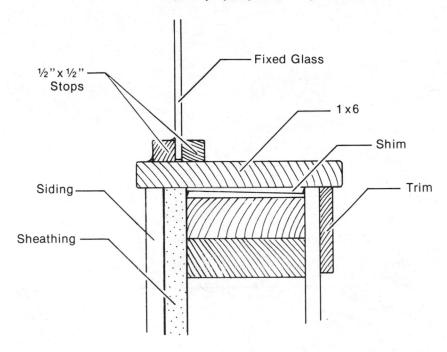

FIGURE 11-5: FIXED GLASS DETAIL

Now that the exterior sheathing and siding has been placed over this opening, insulate well between the studs (as much as possible) and then cover with dry wall. Spackle or panel the opening to match the existing wall coverings, and you should have a warm wall where there once was an old, leaky window.

Some north-facing windows in your home may be a liability in the winter but an asset in the summer. North-facing windows give little daylight in the winter to offset their drain on your heating bills. However, it is not uncommon for such windows to be quite valuable in the summertime—offering additional light and, more importantly, cross-ventilation. If such is the case, you can construct a set of removable panels which seal and insulate these windows during cold months.

These panels are made out of 3/8-inch interior plywood with styrofoam glued to their backs. This styrofoam should be available from your lumberyard in ¾, 1, and 1½-inch-thick sheets. The plywood is cut to a size just smaller than your window trim. After styrofoam is attached to it with an adhesive, the panel is screwed to the window frame at its edges. See Figure 11-6. Most window sashes are set back far enough to allow for 1 inch of styrofoam on this panel. Measure your window to see how thick this foam can be. If you have enough clearance to use 1½ inches of styrofoam, go ahead! It will take at least 1½ inches of styrofoam with two panes of glass to be as effective as an insulated wall.

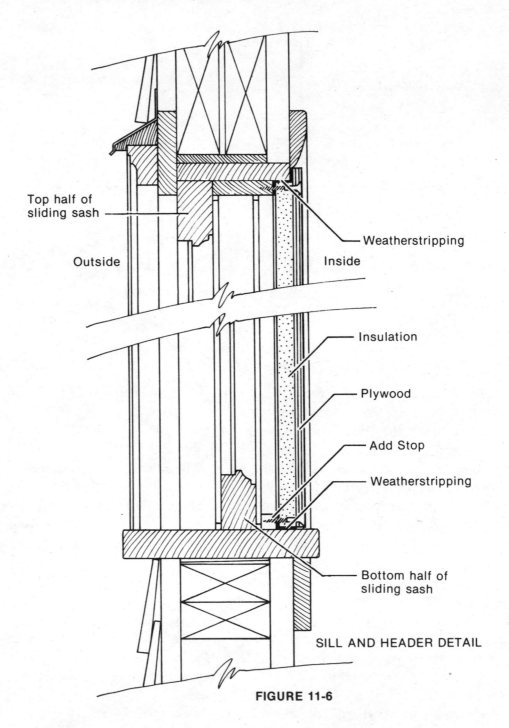

Top half of
sliding sash

Weatherstripping

Outside

Inside

Insulation

Plywood

Add Stop

Weatherstripping

Bottom half of
sliding sash

SILL AND HEADER DETAIL

FIGURE 11-6

When installing these panels, you should line the edges with felt hair or the black weatherstripping mentioned earlier to ensure a good seal. If this window has a storm sash, it will help to stuff the space between sashes with loose fiberfill. (See more on fiberfill later in this chapter.) The twice-a-year task of mounting these panels into the window frame during the fall and taking them down in the spring should be no harder than putting up and taking down storm windows.

ADDING VENTS

Ventilation in older homes invariably comes from operating window sashes (usually double hung) which were planned with no regard to prevailing breezes, or any of the other considerations outlined in Chapter 10. If your home actually does vent well for summer cooling, the forces of chance have served you well! If not, you will probably want to add some vents at strategic locations.

Figures 11-7a and 11-7b show a simple vent which you can make at home. With a styrofoam backing, it insulates well and, with weatherstripping, seals well, too. An insect screen can be mounted in front of it. This vent works best with an area between 1 and 3 square feet, as outlined in the last chapter. As shown in Figure 11-7a, a vent can be placed in either a 2 X 4 vertical stud wall or a 2 X 6 horizontal stud wall.

If you are removing a window that definitely aids in summer ventilation, this is a logical place to add one of these vents. If you are removing a window on the northeast side where this is also the leeward side to summer winds, the window should be replaced by a *high* vent, allowing the hot air to exit. A really effective location for adding a small, near-the-floor vent is on the windward side of a room which already has large openings on the leeward side.

As you study your home and local climate to decide where the best places are to add vents, also consider the type of wall you are going to have to punch through. While wood siding will be easier to work with than masonry, these openings are relatively small and venting even through a brick exterior should not be too difficult. However, you should avoid having to relocate heating ducts and plumbing lines.

With wood siding, you can work between the wall studs and not have to frame in a header as you do with larger window openings. With a brick veneer exterior, you will have to remove a few brick and add a lintel cut from 3½ X 3½ X 5/16-inch angle iron. Complete instructions on how to do small masonry jobs can be found in the *Reader's Digest Complete Do-It-Yourself Manual,* pages 446-460. With help from this guide and a little practice rearranging brick around these small vent openings, you may learn quickly and want to go on to add larger window openings in a

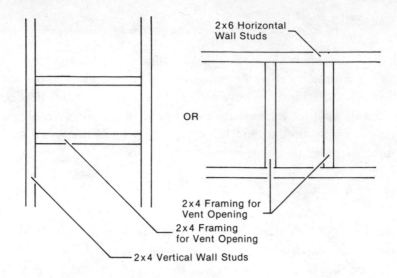

2×6 Horizontal
Wall Studs

OR

2×4 Framing for
Vent Opening

2×4 Framing
for Vent Opening

2×4 Vertical Wall Studs

FIGURE 11-7a

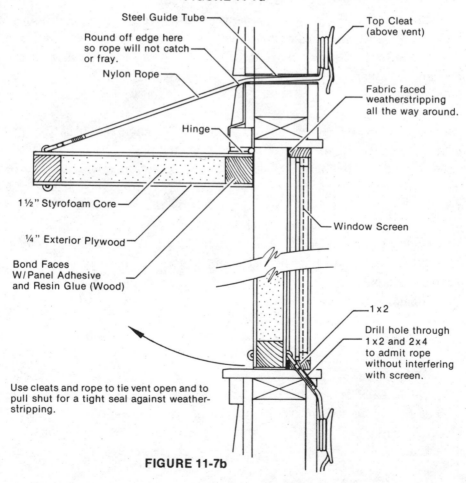

Steel Guide Tube

Round off edge here
so rope will not catch
or fray.

Nylon Rope

Hinge

Top Cleat
(above vent)

Fabric faced
weatherstripping
all the way around.

1½" Styrofoam Core

¼" Exterior Plywood

Bond Faces
W/Panel Adhesive
and Resin Glue (Wood)

Window Screen

1 × 2

Drill hole through
1 x 2 and 2 x 4
to admit rope
without interfering
with screen.

Use cleats and rope to tie vent open and to
pull shut for a tight seal against weatherstripping.

FIGURE 11-7b

masonry wall. This can be done, but seek the advice of a skilled mason and check your local codes. Undermining the support of a masonry wall with large window openings can be dangerous.

Figure 11-8 shows you which bricks have to be removed to make a small 10 2/3 X 16-inch vent opening (4 X 2 2/3 X 8-inch nominal bricks). To remove the first brick, carefully chisel out the mortar all the way around the brick with a hammer no heavier than 2 pounds. Avoid massive blows to the brick or you will crack the whole wall. Once the first brick is out, then carefully remove the other bricks, one by one, and set in the lintel with half bricks at the edges.

If your wall is of concrete block, you can remove one for an 8 X 16-inch opening without adding the angle iron for support. However, once again, don't use a heavy hammer or you will crack the entire wall.

The vent itself is easy to install. With the right size hole cut through your wall, rough frame the opening between studs with 2 X 4's. A piece of 1 X 2 is then installed all the way around to act as the backing to which the weatherstripping is attached so you can get a good seal when you close the vent flap. The flap itself is constructed from 2 X 2 exterior plywood and styrofoam insulation as outlined in Figure 11-7. Then hinge the flap in place and drill holes just large enough to admit some nylon rope through the wall above and below. The flap is tied open or tied closed, using this rope and some inexpensive cleats you can buy at the hardware

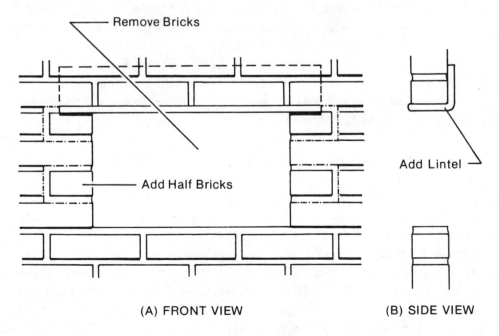

Remove Bricks

Add Half Bricks

Add Lintel

(A) FRONT VIEW   (B) SIDE VIEW

FIGURE 11-8

store. Make sure to caulk around the area where the vent frame meets the exterior siding to prevent rain from getting in.

STORM WINDOWS

Once you've rearranged the windows in your home, you should definitely consider adding storm windows. These help slow down the cold winter air which seeps in around windows and also cut in half the heat which escapes by conduction through the glass. Sealed double-pane "insulating" glass is the easiest to maintain and gives fewer moisture problems than windows with storms added, but insulating glass is also more expensive. If you use storm sashes, make sure to weatherstrip the inner sash thoroughly. This should prevent moisture from condensing on the inside of the storm sash. If you have moisture condensing on the inside of the main sash, the storm sash is allowing too much air to enter this space and you should weatherstrip it also. (Not a bad idea in any case.) Folded newspaper or felt-hair weatherstripping placed around the storm sash before it is set in place should take care of this problem.

Adding storm sashes to an old window not only cuts down on heat losses and drafts, but it also helps to keep the humidity in the house at a healthy level. Water condenses onto a glass surface which is cooler than the dew point of warm room air, lowering the humidity of this air. Thus, one of the biggest arguments for double glazing is that it allows an inside relative humidity of 30 to 40 percent, while with single panes a relative humidity above 15 percent is impractical in cold climates.

New storm windows will cost about $1.50 to $2 per square foot of window area. This represents about a 7-year investment in areas with an average winter climate (New York City-New Jersey area). In warmer areas it will take a bit longer to get your investment back, and a shorter time in colder areas.

Aluminum storms can be assembled at home for a fraction of the cost of ready-made ones. Detailed instructions on how to make aluminum storms, install windows, and add weatherstripping is in the *Reader's Digest Complete Do-It-Yourself Manual*, pages 122-130 (see Bibliography, Chapter 9). As another way to save money, you can pick up an old set of storm windows at a local garage sale; but before you go, be sure to measure your window sizes carefully.

Although their life is limited, plastic storm windows work fine and cost very little. Six-mil polyethylene (avoid using less than 4-mil) is available at just about any building supply outlet both in rolls and in "storm window kits." Plastic in the kits costs several times more than buying a roll, however. The plastic should be cut to a size slightly larger than the window frame. The edges are then rolled in a thin strip of cardboard and tacked or stapled to the window frame. See Figure 11-9.

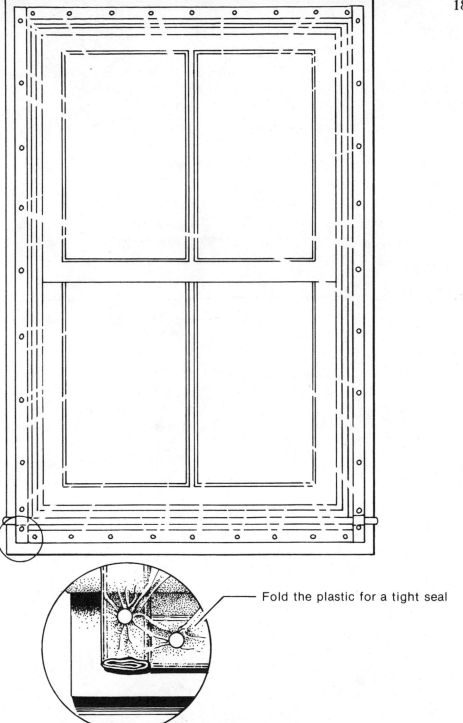

Fold the plastic for a tight seal

FIGURE 11-9: PLASTIC STORM WINDOW

Frames for plastic storms can be made from wood furring strips, attached at the corners with small angle irons (Figure 11-10). The storm frame should be made to just fit into the window frame. After plastic is stapled to it, the frame is attached to the window opening with screws. (See *Save Energy: Save Money,* available free from CSA, Washington, DC.) The plastic will last longer when used on these frames because you don't have to rip it off in the spring. Just unscrew the frame and put it in the cellar or some other shady place till the next winter. If you paint the storm window frame, you can also use it over and over, even when the plastic wears out after a couple of seasons.

Should you put the storm windows on the inside or outside of your home? The plastic storms last longer on the inside, but they are more pleasant on the outside where they won't smell and detract from the appearance of your interior. In hard-to-reach areas such as the upper floors of a home, the window panels will be easier

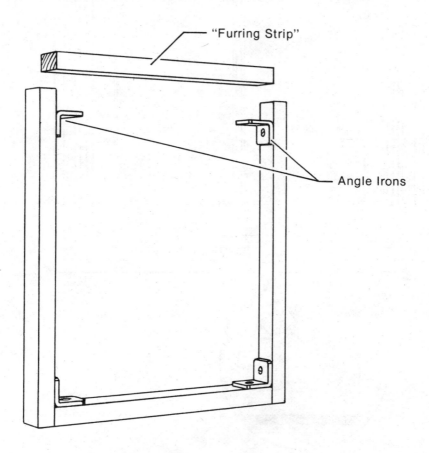

FIGURE 11-10: PLASTIC STORM FRAME

to install on the inside. If placed on the inside, the storm's edges should be lined with felt or newspaper to make a tight seal.

A last minute way to winterize with plastic is to attach it with duct tape. Duct tape costs about $6 per 100 yards and adheres well to just about any surface, but it will probably take off some paint when it is removed in the spring.

HEAT TRANSFER, THERMAL CURTAINS, AND SHUTTERS

If you are serious about lowering your heating bill, more has to be done to contain your home's heat than simply adding some kind of storm window. In Figure 11-11 a comparison is made between the insulating effect provided by window glass and a stud wall. The units of "total resistance" are used to measure the relative insulating value of combinations of materials to heat flow. The higher the total resistance of a combination of materials, the better the insulating value (more on this in the next chapter). As you can see, even insulating windows still lose heat eight or nine times as fast as the same area of wall does. While you benefit from the view, and the natural light and heat from the windows during the day, something has to be done at night to help retain this heat. Once the sun sets, about half the heat you lose from your home escapes around and through the windows. Depending on the climate, there are several approaches to movable insulation to cover the windows at night. As you think in terms of movable insulation, you'll find that you need not be bound to an energy-conserving strategy based on small window areas that deny you essential outdoor light and vision.

There are as many types of thermal shutters and curtains as there are inventive minds to dream them up, and we won't try to describe them all here. If you are an innovative type of person and have an idea you want to try out, don't feel limited to what we suggest here. Any way that you can move a shield of thermally resistant material to seal your windows at night and hang back out of the way during the day will be effective in saving energy.

Thermal screens or shutters have been successfully employed on both the inside and outside of the window. Even styrofoam beads blown between two panes of glass have worked well. Many prefer to work with screens on the inside where you can readily move this shield and where it is protected from the weather. If you want to get started protecting your window openings from heat loss, the best place to begin is on your already existing curtain tracks.

Curtains

Even ordinary curtains make a room warmer. The surface temperature of a curtain is significantly higher than that of its accompanying window glass; and,

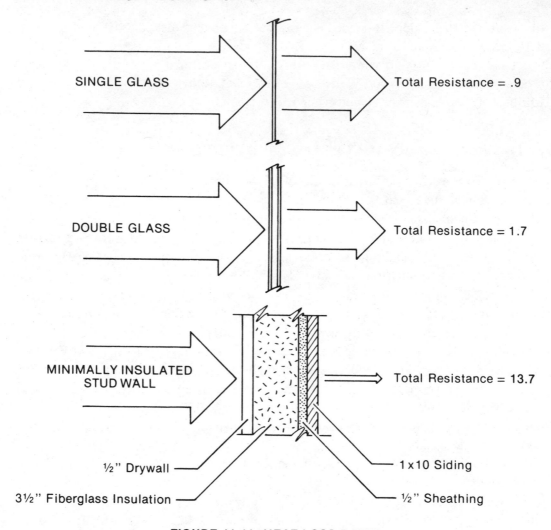

SINGLE GLASS   Total Resistance = .9

DOUBLE GLASS   Total Resistance = 1.7

MINIMALLY INSULATED
STUD WALL   Total Resistance = 13.7

½" Drywall

3½" Fiberglass Insulation

1 x 10 Siding

½" Sheathing

FIGURE 11-11: HEAT LOSS RATES

surprisingly enough, the temperature of a room's surfaces is more important to your comfort than the room's air temperature.

To make a curtain more effective we must stop the air in a room from moving down across the cool glass and onto the floor (Figure 11-12). One way to cut down on this convective flow is to put a plywood box over the top of the curtain (Figure 11-13). This prevents warm ceiling air from moving down across the glass, but a local convective air flow still develops. To eliminate this flow we need to seal the curtain to the window frame on the bottom and sides, thereby making a pocket where the cool air is trapped.

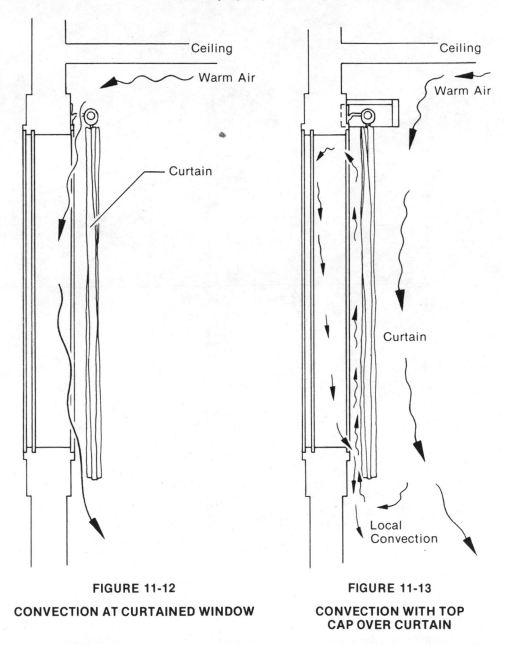

FIGURE 11-12

CONVECTION AT CURTAINED WINDOW

FIGURE 11-13

CONVECTION WITH TOP CAP OVER CURTAIN

A simple but crude way of fixing the curtain to the frame is outlined in Figure 11-14. Use your own tools and imagination to improve on this system and check the local hardware and yardgoods stores for hooks, snaps, velcro (a material which sticks together like a cocklebur on a wool sock), or any other mechanical fasteners which can be used with your curtains.

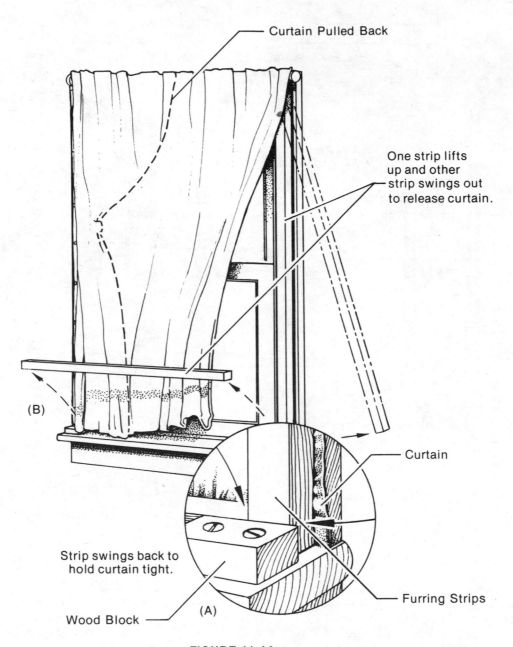

Curtain Pulled Back

One strip lifts up and other strip swings out to release curtain.

(B)

Curtain

Strip swings back to hold curtain tight.

Wood Block

(A)

Furring Strips

FIGURE 11-14

How can the curtain itself be made more effective? The curtain need not be heavy, but the thicker it is, the more thermally resistant it will be and the warmer the inside room surface in your home will be. It is important that a layer of the curtain material be of a tight enough weave that air doesn't penetrate it. Curtain liners, which are sold in some department stores (or through the Sears Roebuck catalogue) consist of about ½-inch-thick cotton batting. Hooking these to the back of your curtains will certainly help. Matching squares of velcro are sewn to the curtain liner so that by pressing the curtain against the side frame, the liner will adhere to this trim piece. The liner can be made to seal to the inside sill at the bottom of the liner, again with squares of velcro 12 inches on center or by building a small wooden slot to tuck the bottom of the liner into. The visible portion of the curtain hangs down in front of this slot. The curtain will be simplest as a single piece of material, drawn to one side during the daytime. If you want the curtain to draw in from both sides of the window, make two sections of liner which overlap by about 2 inches and sew squares of velcro to each section near the edge so that they can be joined together. If you use the thicker batts (3 or 4 inches), you can attach curtain ties to help compress the bulk of this curtain out of the way during the day. Again, try to devise a way to seal this curtain to the window frame so that it makes a cool air pocket.

Curtain Liners

If you want your windows to insulate as well as your walls, you can make a curtain liner which, although thicker and bulkier, is no heavier than the cotton ones commercially sold. A light-weight filler material called "fiberfill" is now available in a variety of thicknesses and densities under a number of trade names. This material is thermally effective in coats and sleeping bags because of its ability to puff up and create dead air space. Two brands, Fiberfill II (made by DuPont) and Polarguard (made by Celanese) are both high-performance fiberfills and are very lightweight and compactable. Some kinds of fiberfill are probably available at your local fabric store. The Stearns and Foster Co., Lockland, Cincinnati, Ohio, is a good mail-order outlet which sells fiberfill batts in 1, 2, 3, and 4-inch thicknesses.

Fiberfill is made in both loose stuffing and blanket (batt) types. The blanket kind is easier to work with, and your curtain will insulate well with the 2-inch batting. the 1-inch batting is a little thin on insulation. The 3- or 4-inch batting will hold in heat better but will be somewhat awkward and bulky to draw aside. The kinds of fiberfill available locally can be good for a curtain liner, but do try to get one which is lightweight.

To make this liner, the fiberfill blanket (batting) should be covered with a thin, low-cost material such as muslin on both sides. The muslin should be sewn together all the way around the edge of the batting and tacked through the batting every 12 inches in both directions, vertically and horizontally. This liner can be hung behind your curtain from the same hooks to which the curtain is attached. Your curtain can then be replaced or washed independently of the liner. A detail of this curtain with the thermal liner is shown in Figure 11-15. One-inch squares of velcro are glued to the trim pieces on each side of the window at 12-inch intervals.

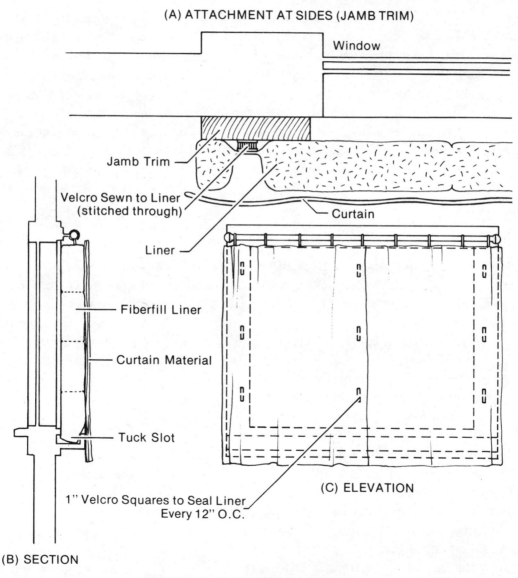

FIGURE 11-15: THERMAL CURTAIN

Interior Shutters

For some of your windows you may want to construct interior shutters. Figures 11-16 and 11-17 show details on how these can be made. In Figure 11-16, the panels contain 1½ inches of styrofoam sandwiched between two layers of masonite or plywood. By adding a 2 X 2 strip to your window frame and adding a thinner strip to the front of your inside sill, you can make a contact surface all the way around your window. The strip which is attached to the sill should be cut or planed so that when installed it is even with the other contact surfaces. If the inside trim piece around your window is tapered in toward the window, rip the 2 X 2 (1½ X 1½) pieces out of a 2 X 4 at a slight angle with a circular saw. Angle it so that, when mounted on your window frame, the face of these surfaces will be parallel to the wall. By setting this 2 X 2 strip right at the inside edge of the trim piece, you gain the extra ¼ inch you need to use a full 1½ inches of insulation in the panels.

The panels are assembled with a 1 X 2 around the styrofoam, attaching two layers of ¼-inch plywood or masonite. Hinges are bolted to the panel and recess holes are bored into the frame to allow clearance for these bolts. The panels are then hinged to the outside edge of the frame and black compressible weatherstripping is applied at the locations shown in Figure 11-16a. Even with your well-caulked and weatherstripped windows (and this should be done anyway), some infiltration will probably still occur, so it is important that the weatherstripping of this hinged panel be applied carefully.

Unlike a thermal curtain which is entirely dependent on good window caulking and weatherstripping to be effective, this type of shutter makes an additional barrier to air infiltration. For this reason, and to stop the seepage of warm house air between the panel and window, it's important to latch these shutters firmly closed. Check your local hardware store and find a fastener you like. Avoid the straight hook-and-eye type, for these are too loose and won't maintain the compression seal. However, the flat-type hook fasteners do quite well (Figure 11-16b). Whatever fastener you use, be sure that it will maintain a firm pressure on your weatherstripping. Note that you will probably need to add small blocks on which to mount these fasteners.

In places where the width of a window is too great for a single pair of shutters without creating a serious clearance problem within a room, panels can also be made in multiple-accordian fashion (Figure 11-16c). By applying hinges alternately on the inside and outside of the panels, this multiple shutter will fold aside easily. Compression-type weatherstripping applied to the sides at each joint will ensure a good seal. This multiple shutter should also be latched top and bottom at each joint.

By adding a 2 X 2 frame around a window you create a space which allows for the thick thermal shutter in Figure 11-16. All windows found in homes today will

198

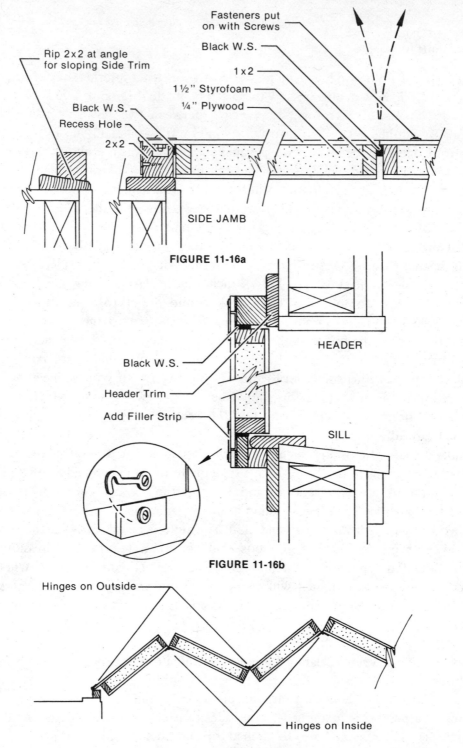

Fasteners put
on with Screws

Black W.S.

1 x 2

1½" Styrofoam

¼" Plywood

Rip 2x2 at angle
for sloping Side Trim

Black W.S.

Recess Hole

2 x 2

SIDE JAMB

FIGURE 11-16a

Black W.S.

Header Trim

Add Filler Strip

HEADER

SILL

FIGURE 11-16b

Hinges on Outside

Hinges on Inside

MULTIPLE-ACCORDIAN TYPE

FIGURE 11-16c

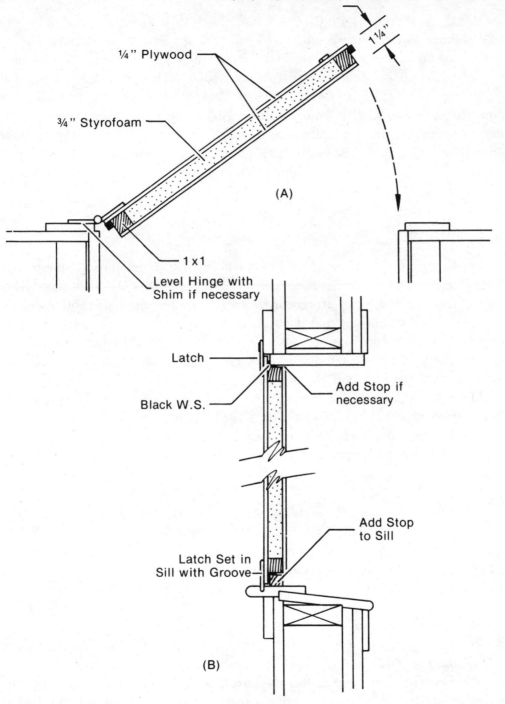

¼" Plywood

¾" Styrofoam

1 ¼"

(A)

1 x 1

Level Hinge with
Shim if necessary

Latch

Black W.S.

Add Stop if
necessary

Add Stop
to Sill

Latch Set in
Sill with Groove

(B)

FIGURE 11-17

accept this thermal shutter with the 2 X 2 frame added. However, you may prefer to install a thinner panel, which is mounted flush with the existing trim, although this panel will not insulate as well. Figure 11-17 shows one such panel made from ¾-inch styrofoam, again sandwiched between two layers of ¼-inch plywood. For this panel to work you will need a full 1¼ inches of clearance between the face of your window trim and the face of your window sash. These panels use 1 X 1 (¾ X ¾) lumber, whereas the thicker ones used 1 X 2's. By nailing a 1 X 1 stop to your windowsill, a compression seal can be added here. The bottom latch will then fasten to a groove in the sill. At the top, a simple flat hook will work well.

CLOSING OFF DOORS

If you have more than one door in your home, try to get through the winter using only one. With the other(s) thoroughly weatherstripped and sealed, the savings and added comfort make the effort worthwhile. In winter, close off those doors on the windy side of the house, and if possible don't use a door on the leeward side either. The positive pressure on the windy side forces air into the house, and the low pressure on the leeward side will suck air out. Thus, fully sealing doors on these sides of your home will have the best effect.

Sealing up the door can be easy. First, weatherstrip all the way around, and don't forget to block air which comes in where the door meets the floor. If you want to really stop cold air infiltration, then tack over the frame a plastic sheet like the ones mentioned before for storm windows. A curtain or drape can be hung in front of the door and makes a pleasant appearance.

Air-Lock Foyer

Once you've closed off the doors you don't need for the winter, the next project you may want to consider is an entrance foyer. These foyers act as air-locks, preventing the cold air from coming into the house each time the door is opened or closed. Another heat-saving feature of the entrance foyer is that the entrance door to the house faces a still air space. Hence, the air infiltration that normally occurs around the door will be eliminated.

An air-lock foyer is easy to build. Since the purpose is to create a still-air space and minimize the cold air that comes in the house when the door is opened or closed, the best entrance foyer is also the least expensive—a small one! It's not even necessary to insulate the walls: just build a simple construction with the outer door weatherstripped to keep the wind out. The foyer can be placed just outside your

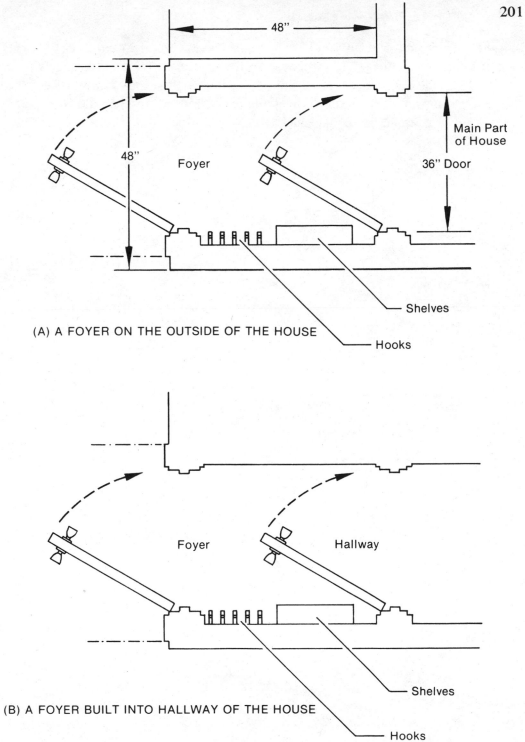

(A) A FOYER ON THE OUTSIDE OF THE HOUSE

(B) A FOYER BUILT INTO HALLWAY OF THE HOUSE

FIGURE 11-18

present entrance door; or, if you have a long hallway, you might consider building it on the inside of the house. In either case, a few shelves and hooks added to the side wall will provide an area for wet or muddy winter clothes, thus saving some "clean-up" energy as well. The outer door of the foyer, by the way, should open *out* to ensure that in a fire no one could block the entrance door. See Figure 11-18.

chapter 12

INSULATING YOUR HOME FOR WINTER AND SUMMER COMFORT

by EUGENE ECCLI

This chapter will explore why and how you should insulate your home. Insulating is one of the most important design considerations when building or renovating a house; the amount of insulation you use and where it's placed affects not only your winter and summer fuel bills, but your comfort as well.

Insulating your home is a good investment. As we saw with the Minnesota test house (see Chapter 10), about one-third of the heat in the average home is lost through walls, ceilings, and floors that are already moderately insulated. Thus, by increasing the amount of insulation in your home, you can make real savings. The money and effort spent to heavily insulate your home will be repaid in just a few short years through lower heating and cooling bills. On a cost-benefit basis, insulation is on a par with the winterizing techniques of weatherstripping and caulking mentioned in the last chapter. If you really want to save fuel and money, it's imperative that *both* winterizing and insulation get top priority.

Since the amount of heat that gets through the walls, ceilings, and floors of your home is reduced by heavy insulation, you'll be more comfortable too. The reason is that the temperature of these surfaces of your home will be higher in winter, and a big part of our *feeling* of comfort comes from the radiant balance between our skin temperature and that of our surroundings. Thus, insulating your home not only saves money, but will also give you that pleasant sense of well-being. Moreover, this is true in summer as well as winter; so even if you live in a warm climate, insulation will be just as beneficial because it helps maintain summer comfort levels at a low cost.

There are other benefits that this winterizing-insulating strategy brings. Not only will your fuel bills be lower, but the initial cost of heating and cooling equipment will be less because you can use smaller units. In addition, a whole host of opportunities for creating heating and cooling comfort in conjunction with your local environment open up once these needs have been reduced. There are the possibilities of using wood heating, natural ventilation for summer cooling, solar house heating, and greenhouses that can provide not only food but warmth in winter—all these elegant and vital relationships with our surroundings come into reach once we learn how to reduce our heating and cooling needs.

LEARNING THE LANGUAGE OF INSULATION

As we all know, some materials resist the flow of heat better than others. An aluminum cup filled with a cool drink, for example, feels colder to the touch than a glass cup. This process of different materials passing heat through themselves is called *conduction,* and it can be exactly measured. In our homes, we want to use those materials that do the best job of retaining heat in the winter and keeping heat out in summer. These materials are called *insulators.*

Most insulators for home construction work because they trap air in many small pockets within the material. The tiny hairs in fiberglass, for instance, create thousands of small "dead air spaces" in the material. Another type of insulator, spray plastic foam, creates thousands of tiny closed cells filled with nonconductive gases. The movement of heat through these materials is slow because the trapped air spaces or closed cells are so small that the heat cannot easily move from the warm side to the cold side by convection. See Figure 12-1.

Insulators are compared to each other by using the term *resistance.* The higher the resistance of a material, the better it insulates. For example, a material with a resistance of "5" per inch of thickness would retain heat five times better than a material with a resistance of "1" per inch of thickness. Note that the term *resistance* not only applies to a particular material but also must be specified for a certain *thickness* of that material. The reason for this is that the resistance of materials put in contact with each other is additive. If I have two 1-inch-thick pieces of material in contact with each other, each with a resistance of 5 per inch, then the *total resistance* to heat flow is 5 + 5 = 10 (see Figure 12-2).

This idea of total resistance is used to evaluate the effectiveness of different kinds of building construction. The higher the total resistance of a floor, wall, or ceiling, let's say, the better that part of the building will be in saving energy winter and summer. From these comparisons, certain standards have been set up by the building industries that indicate a range of values for the total resistance of building

FIGURE 12-1

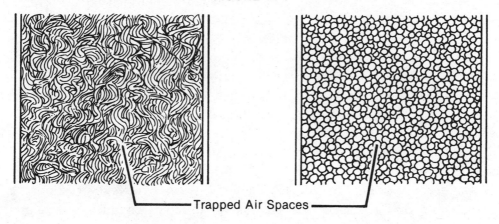

Trapped Air Spaces

Fiberglass-Type Insulation   Closed Cell-Type Insulation

elements. For your home construction to be at least as good as these standards makes economic sense. You may have heard of standards like R-11 or R-19. What these mean is that the total resistance to heat flow should add up to be 11 or 19 when all the insulation in a wall, floor, or ceiling is taken into account. As time has gone on and fuel prices have increased, the standards have increased from the older recommendation of a total resistance of R-11 to recommending a minimum total resistance of R-19. In other words, it makes economic sense to insulate more heavily today than it did in the past.

How do heating engineers come up with these standards? They compare the cost of the insulation with the cost of heating a house. As you increase the amount of

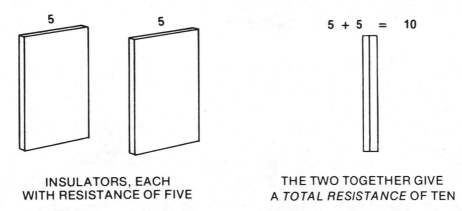

INSULATORS, EACH
WITH RESISTANCE OF FIVE

THE TWO TOGETHER GIVE
A *TOTAL RESISTANCE* OF TEN

FIGURE 12-2

insulation, the size of your heating bill goes down, but the initial expense of buying the insulating materials goes up. At some point, there is a balance between initial cost of insulation and cost of fuel bills, which adds up to a minimum total cost. Right now, the amount of insulation with the total resistance that gives the lowest total cost for heating and cooling (including both initial costs for insulation and fuel costs) has a resistance of about 17. However, since the cost of fuels is still rapidly increasing, a more realistic figure is a total resistance of *at least 20*. Going higher, to a resistance of 25 or 30, still makes sense, particularly for the ceiling areas. It's just a bit more of a long-term investment but one that will pay real dividends in the years ahead.

To put the cost-saving benefits of insulation to work requires several things. You have to choose the right kind of insulation for different circumstances and then choose a type of construction for your home that will provide enough space for this insulation so that its total resistance is up to at least 20.

TYPES OF INSULATION

There are several features that you'll want to compare as you look over the insulators that are commercially available: cost, resistance factor per inch of thickness, fire rating, and durability. Before we examine the various qualities that different insulators have, let's look at the forms of insulation that are available:

Blankets. Blankets come in rolls, usually from 16 to 24 inches wide, and are unwound and cut to fit in place between wall studs or ceiling and floor joists.

Batts. Batts are similar in size to blankets but are precut to specified lengths.

Boards. Boards are precut to standard sizes and are usually applied with a mastic as perimeter insulation around the foundation or over existing walls for renovation work.

Loose-Fill. This type of insulation comes as a loose granular material in bags and can be poured into place between joists or behind walls.

Blown Insulation. This is a spray-foam type of insulation which is installed under pressure behind walls and between joists in ceilings and floors by professionals.

Now, let's look at the qualities that the various commercial insulators have:

Fiberglass. Fiberglass is perhaps the most common of all insulators today; it is available in rolls or batts. The resistance of fiberglass is about 3.5 per inch of thickness. It's fire and vermin proof, moderate in cost, and easy to install. However, over some years it will tend to mat down or settle on itself. The result is that the material is compressed a bit and loses some of its insulation value.

Rock Wool. Rock wool is much like fiberglass, except that the material is made from melted rock instead of glass. The resistance value is a bit less than fiberglass—about 3 per inch—and the problems with matting are even worse.

Insulating Sheathing. This is a material made from wood fibers and is usually applied on exterior walls as a backing for siding. It comes in various thicknesses, usually between a ½ and 1 inch. The resistance is about 2 per inch of thickness; and though it's water repellent, extended exposure to moisture will cause damage through mildew and rot.

Plastic Insulation. The most common type of plastic insulation is polystyrene, which comes either as a blue "closed-cell" foam board or as white granule boards where the granules have been heat and pressure treated so they adhere together. The resistance of these materials is about 5.5 for the blue closed-cell type and 4 for the white granule type. Polyurethane, on the other hand, comes as closed-cell foam boards or can be professionally blown into place inside walls, ceilings, or floors as a foam. In either case, its resistance value is about 6.5 per inch. Polystyrene and polyurethane tend to be expensive. Moreover, while both are good insulators and do not attract moisture or vermin, there are *extreme* fire hazards involved. Both will burn very rapidly and give off toxic gases, particularly when placed in a vertical position. These materials should therefore *never* be installed where they will be exposed. The best applications are for perimeter insulation on the outside of foundation walls.

Loose-Granule Mineral-Based Insulators. The most common of these insulators are vermiculite and perlite, available as loose-fill insulation at moderate costs. The resistance is about 2.5 per inch. These materials are fireproof and will not deteriorate from moisture, rot, or insect problems. However, settling out can be a problem, particularly when they are placed in a vertical position between wall studs.

OTHER OPTIONS

What follows is a description of some relatively new insulators that are available at moderate cost.

Cellulose Fiber. These fibers are often made from recycled paper that has been chemically treated to resist fire, rot, and vermin. In addition, this kind of insulation has good sound-deadening properties. It comes as a loose-fill insulation, though this variety has the settling out problems of the fiberglass, rock wool, or loose-granule insulations just described. The cellulose fiber material can, however, be blown into place in walls, in ceilings, and under floors at densities that are nonsettling. The resistance of these materials is about 3.5 per inch. There are many formulations of cellulose fiber insulation. Some have fire- and insect-resistant properties that are better than others, so make *certain* that either the product literature or the bag of insulation specifies one of the following standards:

— The N.C.I.M.A. (National Cellulose Insulation Manufacturers Association) seal;

— or the insulation meets federal specification FF-515C;

— or the insulation meets A.S.T.M. specification C 739.

Urea-Formaldehyde. This is a moderately priced pneumatically blown insulation, sometimes referred to as "Arctic Foam." It is fire-resistant, chemically stable, non-toxic; it will not attract moisture or insects. This material shrinks between 1.8% and 3.0% when it sets, creating small spaces between the studs. This is not a problem, except perhaps where exterior sheathing is extremely loose so that air can blow into the wall. As with all blown insulations, you will need the services of a competent installer to get your money's worth. The resistance of this material (after shrinkage) is about 5.0 per inch of thickness.

WHICH TYPE OF INSULATION IS BEST?

The best type of insulation for your use will depend upon the type of construction you have or build with and how much money you want to spend. Pick up product and installation literature and do comparison shopping at your local building supplier, or with different installers in the case of blown insulation. In addition, there are several inexpensive government publications that contain specific instructions about the cost-benefits and application of various insulation techniques. These are particularly valuable if you have an older home and can't always insulate as much as we suggest in this chapter:

Making the Most of Your Energy Dollars
U.S. Government Printing Office
Washington, D.C. 20402
#C13.53:8 16 pp., 70¢

In the Bank . . . Or Up The Chimney
U.S. Government Printing Office
Washington, D.C. 20402
#023-000-00297-3 69 pp., $1.70

The basic criteria for insulating in these times of high fuel prices is to use as much as possible. Remember, it's often hard to add more insulation once the house is built, so put in as much as you can at the *beginning*.

   Let's look at how the heat escapes through a typical building element and see what kind of construction works best. As an example, we'll use the standard 2 X 4 stud wall. First, let's see what the resistance is for such a wall *without* insulation. There are six resistances that must be added to find the total resistance of this building element (see Figure 12-3):

1. The resistance of a thin film of
   air on the outside of the wall,
   usually taken to be equal to .2   .2
2. A wood siding material   .8
3. Insulating sheathing, ½ inch thick   1.0
4. An air space between the studs   1.0
5. Gypsum wallboard, ½ inch thick   .5
6. The resistance of a thin film of
   air on the inside surface of the wall   .7

               Total resistance is only 4.2

   From this example, we can see that ordinary building materials provide nowhere nearly enough insulation by themselves. We want an insulation that will give a total resistance of *at least 20*. Thus, an uninsulated wall provides only about 20 percent of the insulating qualities we need. Next, let's fill the 3½-inch air space between the studs with different insulators. In this case the value of 1 for 3½ inches of air in item #4 of the previous example is replaced with the resistance value of 3½ inches of insulation. The total resistance for fiberglass would be:

Item #4 (3.5 inches X a resistance
   of 3.0 per inch of fiberglass)   10.5

   If we do the same calculations for all the insulators mentioned before, we find that a standard 2 X 4 stud wall *filled* with 3½ inches of insulation has the following resistance:

Fiberglass	10.5
Mineral wool	10.5
Polystyrene (white)	14.0
Vermiculite or Perlite	11.9
Urethane	22.9
Cellulose fiber	12.3
Urea-formaldehyde	17.5

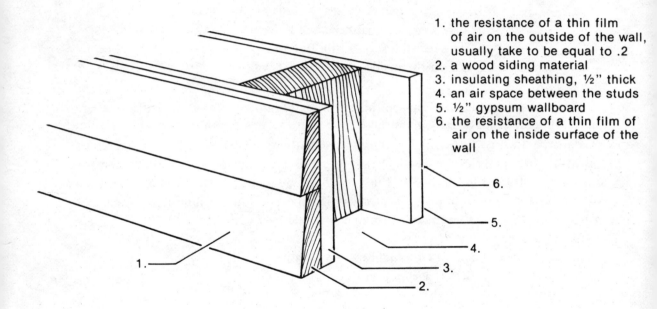

1. the resistance of a thin film
   of air on the outside of the wall,
   usually take to be equal to .2
2. a wood siding material
3. insulating sheathing, ½" thick
4. an air space between the studs
5. ½" gypsum wallboard
6. the resistance of a thin film of
   air on the inside surface of the
   wall

FIGURE 12-3

As you can see from the figures, the only insulations which surpass, or even come close to, a total resistance of 20 are plastics, cellulose fiber, and urea-formaldehyde. The plastics, however, are a great fire hazard; so only two of the insulators mentioned really do the job properly.

As we suggested before, use a total resistance of 20 as a *minimum* figure. You can increase the total resistance by putting even more insulation in your walls. As suggested in Chapter 6, this can be done by using 2 X 6 studs in the walls spaced 24 inches apart (instead of the usual 16 inches). Since the actual depth of these 2 X 6 studs is 5½ inches, you have 2 inches more space to use for insulation than you did with 2 X 4 studs, which had only 3½ inches of space. The total resistance for the various insulations using 2 X 6 studding would be as follows:

Fiberglass	16.5
Rock wool	16.5
Polystyrene (white)	22.0
Vermiculite or Perlite	13.8
Urethane	35.8
Cellulose fiber	19.3
Urea-formaldehyde	27.5

As you can see, with 2 X 6 studs, your range of insulators expands. It's just as easy to install the insulation, too, as rolls and batts can be purchased in the 6-inch thickness. The only one on the low side is the vermiculite or perlite material. While the insulation value of the plastic polystyrene and polyurethane is good, their being fire hazards makes them ill-suited to this kind of construction (though, as mentioned before, they are excellent for perimeter insulations where fire danger is small).

You gain another plus factor with 2 X 6 studs. Since wood studs conduct more heat than the insulation which is placed in the space between them, the entire wall will lose less heat with the thicker 2 X 6 studs spread out on 24-inch centers than it would with the 2 X 4 studs on 16-inch centers. Wood conducts heat so much faster than the insulations we've mentioned that for the well-insulated houses we are talking about, the difference in heat loss due *only* to the thickness and more spread out arrangement of the 2 X 6 studs is 5 percent less than for the 2 X 4 stud arrangement! The extra 2 inches of space that the 2 X 6 provides for insulation reduces the heat loss even more, so that the total saving is on the order of 35 percent!

Won't the 2 X 6 stud arrangement cost a lot more? The answer is that it doesn't have to. Since the 2 X 6's are spread out to 24-inch centers, the cost here is little greater than it would be for 2 X 4's placed on 16-inch centers. The extra cost normally comes in because the plates above and below the studs which hold them in place are then 2 X 6's instead of 2 X 4's. However, if the post and beam construction method for horizontal studs is used (as outlined in Chapter 6), this is an expense which can be avoided. Also, roughsawn 2 X 6 lumber (again, see Chapter 6) is little more expensive than roughsawn 2 X 4's. Thus, the material cost of using the 2 X 6 scheme need not really be much greater.

Another question people ask is whether the interior sheathing bows more when there is 24 inches between supports than when there is 16 inches. While this is true for some of the thin wall panelings, the sheetrock recommended in Chapter 8 should present no problems.

INSTALLING INSULATION IN YOUR HOME

Before we examine where you should place insulation in your home, let's take a look at some of the more important installation procedures which you should keep in mind:

• As mentioned in Chapter 10, a vapor barrier is very important for many reasons. Before you install your insulation, read over the recommendations in that chapter.

• Some insulations, like fiberglass and rock wool, have a polished reflective surface on one side of the insulation. When you install this type of insulation, make certain that the reflective material faces *into* the house. Also, this type of insulation comes with a ½-inch flap at each side so that the insulation can be stapled to the sides of the studs. Be sure you staple often enough to get a vapor-tight seal and also so that there is a ½ inch of air space between the face of the insulation and the inside sheathing of the wall, ceiling, or floor. The reason for the ½-inch space is that the shiny material on these insulators reflects back any radiant heat trying to enter, and an air space is required to do this. Since a ½-inch air space is also a fairly good insulator against conduction, the heat must then cross this space by convection. See Figure 12-4. Thus, the best application of these reflective-faced insulators is under the floor, since heat will have the greatest difficulty convecting across an air space in the downward direction. The next best application is in walls, where the ½-inch space then insulates about as well as the same thickness of fiberglass would. The least effective application is in the ceiling, because the heat easily convects across the air space in this case.

• On fiberglass or rock wool insulators, *do not* compress the material when installing. Buy the proper thickness and then let the material fluff-up with air before you staple it in place. Compressing only reduces the insulation value, as there are then fewer trapped dead air spaces in the material.

• If you use the polystyrene or polyurethane insulations for perimeter insula-

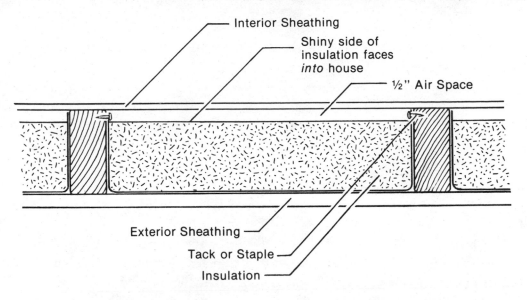

Interior Sheathing

Shiny side of insulation faces *into* house

½" Air Space

Exterior Sheathing

Tack or Staple

Insulation

**FIGURE 12-4: LOOKING DOWN ON STUD WALL FROM TOP—
USING REFLECTIVE-FACED INSULATION**

tion, as we suggest, make sure to use an adhesive or mastic that will not melt these plastics.

There are many areas to consider when placing insulation in your home. What follows is an outline of those parts of the house which should be considered. All these areas should receive attention if you want to lower your heating and cooling bills and provide the most comfort in your home.

The Attic. The attic is the most important area of your home to insulate because you lose the most heat in winter here, and gain the most in summer. If you have a flat or shed-type roof above your ceiling or an unoccupied attic space, then insulation with a total resistance of *at least 25* should be installed between the joists. Practically speaking, this means you want to completely fill the space between joists with insulation. If you already have insulation, it's very easy to add enough of either the batt, roll, or loose-fill type directly on top of what already is there to bring the total resistance up to 25 or more. Sometimes an extra inch or two of the fiberglass, mineral, or loose-fill insulators are added to offset the effect of settling that occurs with these materials. If the attic is used, the insulation should then be placed between the rafters under the roof and in the exterior walls of any attic rooms.

Where there is an unused attic, it's very important to provide ventilation. This will help dissipate the moisture that gets up through the ceiling in winter (keeping your insulation and attic joists and rafters dry) and also help in summer to prevent the build-up of attic heat. Usually, a 1-foot-square vent is placed on opposite sides of the attic wall to do this.

Insulating Walls.   As we pointed out earlier, a total resistance of *at least 20* is a good idea for all walls. You can install the batt or roll types of insulation in wood stud walls with the installation procedures already mentioned. Since it's more difficult to add insulation to walls than it is to attics, try to add as much here as you can at the beginning.

In masonry construction, the cores of concrete blocks can be filled with the loose-fill or foam types of insulation. Note that this is usually possible only before the roof or wall plates are put in place.

The Basement.   The insulation procedures for the basement will depend upon whether you have a crawl space, use the basement for storage, or have a full live-in basement. For the crawl space, first make sure that the foundation all around the house is windtight except for two 1-foot-square vent openings. Then place enough insulation between the floor joists to give a total resistance of at least 20. As mentioned before, the reflective-faced insulations with a ½-inch air space will be very effective here. To hold the insulation in place, buy some wire mesh such as screen material and tack this in place underneath the insulation.

For a storage or as yet unfinished basement, the same procedures would apply as with a crawl space, except that ventilation is necessary only where moisture has leaked into the basement through the foundation or slab.

If, as suggested in Chapters 5 and 6, you make use of the basement as a living space, then some different installation procedures apply. In this case, your best bet will be to use the perimeter insulation techniques suggested in the next section. Perimeter insulation works best here because, when placed on the *outside* of the foundation wall and under the slab, it not only insulates in winter but allows the mass of the slab and masonry walls to cool down on summer nights, helping to keep the house comfortable in the daytime. This same system of having the insulation on the outside of the walls will work well with solar heating. See Chapter 16.

Perimeter Insulation.   The term *perimeter insulation* applies to insulation placed along foundation walls and under concrete floor slabs. Several situations exist which need attention, such as a concrete slab on level ground or a slab in a basement where one side opens to the outside. For the slab on level ground, you will want to insulate around the outside of the foundation wall. In cases like this where fire isn't a hazard, the high insulation value of the plastic insulations like polyurethane can work to your advantage. Normally only an inch or so of insulation is placed around the foundation. However, studies have shown that more heat escapes in northern areas than might be expected, so a 2- to 3-inch-thick sheet placed *all the way down* to the base of the foundation is recommended. See Figure 12-5a. In cases where perimeter heating is accomplished via ducts placed within the floor slab, a 1-inch layer of insulation stretching from the edge of the slab about 3 feet inward will ensure that heat moves into the room and not down into the earth below the slab. See Figure 12-5b. A 1-inch insulation strip is also used to separate the slab from the foundation wall, and in this case the insulation is placed inside that part of the foundation walls that is below the floor slab. In extreme northern areas, check for alternate designs.

When polyurethane is installed, no protection from moisture on the outside of the insulation is necessary, as this plastic material is also a vapor barrier. However, moisture can leak through at those points where these insulation boards meet. Hence, it's necessary to install a vapor barrier to the outside of the foundation wall *before* you put on the insulation boards. One of the cement-based hydraulic compounds mentioned in Chapter 5 will work well. For those parts of the insulation which are exposed between the ground and the beginning of the side wall, run a piece of flashing from behind the wall and down over the front of the insulation to keep moisture out. See Figure 12-5a.

In the case of a basement with one side open to the outside, the perimeter insulation need be applied to the foundation wall only on the exposed side. However, contrary to what many people believe, the three other partially or fully submerged basement walls will lose large amounts of heat unless insulated. If the wall is fully submerged, you can run insulation down its entire length and use flashing for exposed areas at the surface, as outlined earlier. For a partially buried wall on sloped ground, the same basic procedure applies except that more care must be taken for the larger exposed wall area in this case. Here, unless the polyurethane is protected from the sun, it will break down to a powder which is blown away by the wind. To protect this insulation and for appearance sake, apply a cement-stucco plaster to the surface at any exposed point (see Figure 12-6).

OVERCOMING THE SPECIAL PROBLEMS OF THE OLDER HOUSE

With our recommendations for heavy insulation, what can be done with the older house that often requires a complete overhaul to be fully insulated? In this case, it's best to look at priorities and leave the difficult jobs till last.

Most older houses without insulation also suffer from problems of air infiltration. The window sashes and casings may be loose, doors may not close tightly, and loose foundation walls may be allowing the cold winter wind to suck away heat through the floor. Contrary to what people often believe, insulation is *not* the most

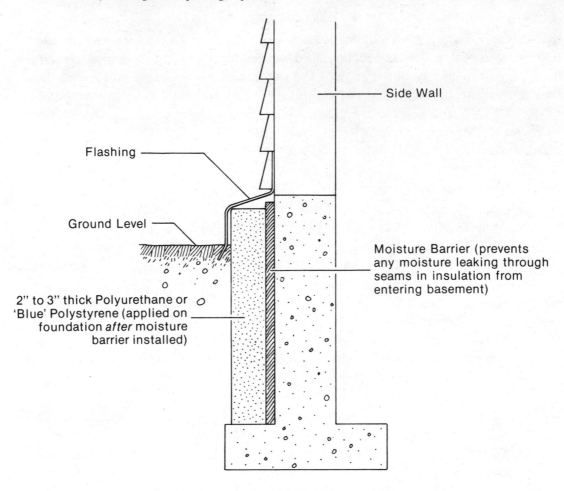

Side Wall

Flashing

Ground Level

Moisture Barrier (prevents any moisture leaking through seams in insulation from entering basement)

2" to 3" thick Polyurethane or 'Blue' Polystyrene (applied on foundation *after* moisture barrier installed)

FIGURE 12-5a

important priority in these cases. Far more money will be saved at a far lower cost and with less effort by attending to the infiltration problems first. The information in Chapter 11 tells you how you can overcome these problems.

Once the house is tight, your next biggest source of heat loss is the windows. Adding storm windows and insulation panels as described in Chapter 11 is easy to do and just as important as insulation. After this is done, you should insulate. Try first to insulate the attic. If there is not enough head space in the attic to crawl around and install insulation, you might be able to push the batt or roll insulation in place with a rake. If there are areas you still can't get to, you can leave these until the next time you have your roof redone. Then, the roofer can place insulation on the old roof and put a new layer of roofing material over it. One other option you might want to consider is the "house-within-a-house" concept outlined in Chapter 7. Basi-

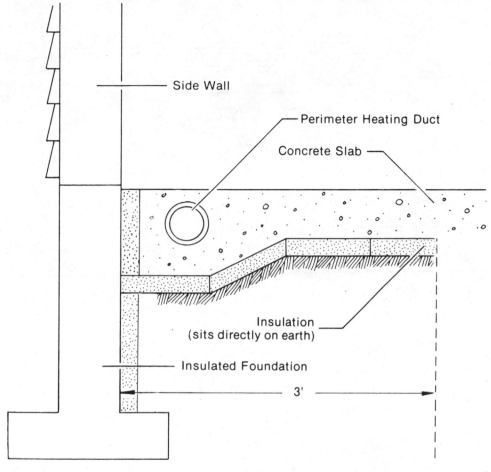

FIGURE 12-5b

cally, you insulate a certain portion of the house and use this as your main living quarters over the winter months. An example might be closing off the upstairs of a two-story, uninsulated house. Insulation placed on the ceiling and walls downstairs would then be all you would need.

By the time you have tightened up the house, added storm windows, and installed insulation in the attic, much of the value of insulating will already have been achieved. You may want to consider more insulation for your walls and floor at this point, but let this evolve within your entire plan for renovating the house. Don't renovate simply in order to insulate walls and floors. It's not that important. However, if your renovation involves removing interior finishes, then it certainly will pay to add insulation when these areas are exposed.

Once the roof is finished, the next area you should consider insulating is the

FIGURE 12-6

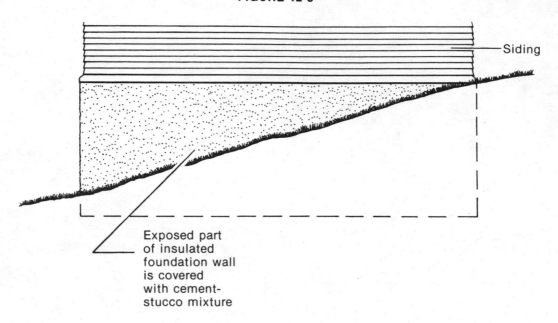

Siding

Exposed part
of insulated
foundation wall
is covered
with cement-
stucco mixture

walls. If the walls are of masonry construction, they already provide a bit of insulation value. However, if the plate at the top of the wall (for the ceiling rafters) doesn't completely cover the cores of concrete blocks (or the cavity in cavity-wall construction), you can use loose-fill insulation or have one of the foam insulations like urea-formaldehyde blown into these spaces. If the walls have a stud type of construction, it's easier to apply insulation. The simplest way is to apply it when you remove the interior wall surfaces in renovation work, as suggested earlier. In cases where you don't feel that this kind of refinishing is necessary, small holes can be drilled near the top of the wall from the outside and one of the foam insulations can be blown in. The hole is sealed up and painted afterwards. It's important to use a moisture-resistant foam in this case, as otherwise water vapor will enter the walls from the inside and be trapped there. This is a common problem with loose-fill insulations, so it would be best to check with the installer and follow his recommendations for small outside air vents if necessary.

The final area of insulation for an older house is the floor. Two common situations arise for houses with or without a basement. In the cases where you don't have a basement and there is room in the crawl space to get under, use the reflective foil-faced insulation we mentioned before. If there isn't enough room to get under the house, your first step should be to seal the foundation around the house. You should make an airtight seal, leaving only a few small openings for ventilation.

Often, stopping the cold winds before they get under the house is all that is necessary.

What about situations where there is almost no room underneath the house and where the floor is old and air leaks exist? One simple solution is to cover it over with a carpet, which will stop the influx of cold air and provide a bit of insulation as well. If you feel that the floor is so far gone that you want to install a new one, it's then easy to put in more insulation when you rip the old floor out. Another way to do this is to nail 2 X 2's directly over the old floor just above the floor joists. Then put a 1½-inch-thick piece of one of the foam-board insulators into the space between the 2 X 2's and cover your new floor with a 1-inch-thick piece of plywood.

For the older house with a basement, there are several options for insulation. If you install drainage pipes around the house in cases where leakage is a problem (see Chapters 5 and 19), it's very easy to add insulation on the outside of the foundation while it's exposed. In cases where it's too difficult to dig around the foundation, you could install either 2 X 2 or 2 X 4 studs against your wall and insulate between them, or use a mastic and one of the plastic foam-board insulations. However, if you choose to install the plastic foam boards on the basement walls, a covering of 5/8-inch fire-rated sheetrock is recommended because of the fire hazards involved.

CONCLUSION

Let's take a look at the heating and cooling load for several versions of the same house and see what the effects will be of the winterizing-insulating techniques we've described. Table 12-1 shows a comparison of the heating demand for a two-floor, 800-square-foot house. The chart shows the quantities of heat lost by infiltration; by conduction through walls, ceilings, and floors; and by conduction through the glass areas in the house. Three versions of the same house have been used:

A. Uninsulated version, with loose-fitting doors and windows (nonweatherstripped) and single-paned glass.

B. Same house, but with conventional construction: 3½ inches of insulation in walls and 6 inches in attic. Little underfloor insulation. Windows tighter fitting but nonweatherstripped. Glass is single pane.

C. Same house, but with a vapor barrier used throughout and 6 inches of insulation between 2 X 6's in walls, 9 inches of insulation in ceilings and under floors. All windows are double pane or have storm windows added; half are weatherstripped and caulked, while the other half use fixed glass with an operating vent below. In addition, all glass areas have either insulated curtains or insulation panels which are used at night, and the house door used in winter has an entrance foyer.

Table 12-1: *Comparison of Heating Demands*

	House A	House B	House C
Conduction through walls, ceiling, and floors	18	3	1.5
Conduction through windows	3	3	1.0
Infiltration	8	4	2.0
	29	10	4.5

As you can see, house C, using all the winterizing-insulating techniques, *will use less than 1/6 as much fuel as an uninsulated house, and less than half as much as the conventional house!* House C will thus not only drastically reduce heating and cooling expenses, but will be far more comfortable to live in. This kind of construction already has the natural ventilation techniques of Chapter 10 built into it, so summer comfort is easier to achieve. Moreover, once your winter heating needs have been reduced as with house C, then the sun on the south side of your house can bring warmth through either south-facing windows or a built-in greenhouse as described in Chapters 16 and 17.

chapter 13

WAYS TO SAVE WHEN USING HEATING AND COOLING EQUIPMENT

by DAVID L. HARTMAN

There are many important reasons for energy conservation in homes. A typical family directly spends $700 and $1000 per year for energy, a sizeable portion of the household budget. Conservation can reduce this burden now, and will become even more important as energy prices increase.

On a wider social level, homes account for 20 percent of all the energy consumed in the United States, and the rapid growth in this part of the nation's fuel budget is causing many strains on energy distribution networks. See Figure 13-1. The problem of maintaining adequate and economical supplies of energy involves domestic as well as international affairs. The fact that oil and gas are limited resources with approximately 30 to 70 years of availability gives strong motivation for wise energy usage. Pollution from energy extraction, transport, and the combustion of these fuels is well known. All of these problems can be reduced to a more manageable level by energy conservation.

Within a home, energy is used for the following basic functions: heating and cooling, appliances and lights, and water heating. Although the quantity and type of energy consumed for these functions varies widely throughout the country and even from house to house, the techniques of conservation can be applied everywhere.

Learning how to save energy can be fun. Spending time choosing an efficient refrigerator can easily substitute for spending time on choosing one with the perfect color. Weekends can center on constructive projects such as insulating a house or replacing inefficient lighting. The savings possible, some $300 to $500 for the aver-

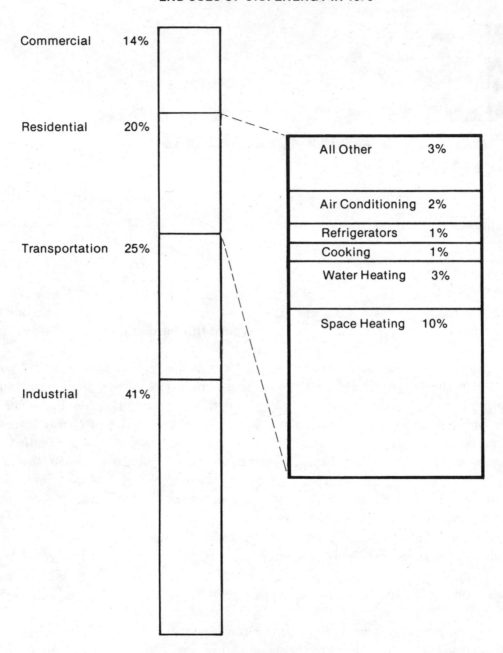

FIGURE 13-1

age home each year, will relax some of the pressures on the family's budget, too. In this chapter we'll look at the savings possible with heating and cooling systems, and in the next two chapters the ways to save with appliances, lights, and water consumption will be explored.

THE ENERGY COSTS OF HEATING AND COOLING YOUR HOME

In most homes, energy consumption for heating and cooling accounts for about half of the total annual energy for all uses. It is approximately equal to the amount of fuel a midsized automobile uses in a year–1,100 gallons of gasoline. In the northern and cold parts of the country, the energy consumption for home heating exceeds the energy consumption for cooling; in the southern and warm areas, the opposite is true. On a national basis, the energy for home heating exceeds that for cooling by 11 to one. Thus, efforts aimed at lowering the heating bill will have the greatest impact on your family budget.

The typical energy consumption for heating in a northern U.S. home is about 1 gallon of oil (or the gas or electric equivalent) per square foot of floor area per year. Domestic hot water is additional. The energy consumption for cooling ranges from less than 1 kilowatt-hour (KWH) per square foot per year in cool climates to over 5 KWH in hot areas. Since there are about 80 million homes in the United States, each with its own construction, occupancy, and operating characteristics, the actual consumption of course varies widely. The corresponding costs for heating and cooling also vary widely, but an average figure is about $600 per year for a 1,100-square-foot home in New York (as of 1975).

SELECTION OF MORE EFFICIENT EQUIPMENT

Even if your house is built by a contractor and the heating and cooling equipment selected by him, it will be up to you to check that the most energy-efficient equipment is chosen. From an energy conservation point of view, the criteria for selection differs from past practices. First, to select on initial cost is not satisfactory in itself. Instead, equipment should be selected for maximum efficiency. Saving $100 on a home boiler that is 5 percent less efficient than a better boiler would end up wasting 1,000 gallons of oil (costing at today's prices $450) over a 20-year period. Second, equipment should be selected according to real comfort needs. Air conditioning, for example, really is not necessary for many homes in the northern U.S. The savings for eliminating a typical central A/C system can exceed 50,000 KWH of electricity, or $2,500 over a 20-year period, plus first costs of the equipment.

There are many major energy conservation features to look for when buying new heating and cooling equipment. The prospective buyer must discern the facts from the many claims and should use this chapter as a guideline. An introductory book on heating systems (e.g., *The Home Comfort Handbook*—see Bibliography) or magazines such as *Popular Mechanics* or *Popular Science* can also be helpful.

Before selecting new equipment, it is always necessary to make sure the heating and cooling loads are minimized. Building modifications such as heavy insulation, double glazing, and other load-reducing methods are discussed in the previous chapter.

HEATING

Whether your home is old or new, there are several ways in which you can save by choosing the right equipment. Let's start with the selection of a new heating system.

Do not use oversize boilers or furnaces. "Oversizing" is commonly practiced because heating contractors, engineers, and everyone else who sizes heating systems use a large "factor of safety" to account for unknown variables. If more care is given to making accurate calculations of heat loss, most heating systems could be reduced by 10 to 20 percent. This, in turn, can save several hundred dollars in the cost of the heating system.

Standard calculation procedures are given in two readily available publications:

"Insulation Manual: Homes-Apartments"
NAHB Research Foundation
P. O. Box 1627
Rockville, Maryland 20850
($4.00)

"Standard for Application of Year-round Residential Air Conditioning—#230-62"
Air Conditioning and Refrigeration Institute
1815 North Ft. Myer Drive
Arlington, Virginia 22209
($2.75)

The maximum heat loss for a well-insulated and weatherstripped house as calculated from these manuals can be as low as 20 BTU's per hour per square foot of floor area for houses in the northern part of the U.S. Although few houses are insulated well enough to achieve this figure, it should be used as a realistic target. Compared to the common values of 40 to 50 BTU's per square foot, the savings in the first cost of the heating system, and in its operating expenses, will be substantial.

The importance of selecting the right size heating system is based on the principle that oversized systems cycle more often than those designed to just meet the load. The more often the system cycles, the greater the loss of heat up the chimney. For a typical house, an oversized system can waste 5 to 10 percent of the annual heating fuel. Translated into money, this would be $25 to $50 per year.

One problem with a small, well-insulated house is that it is difficult to find a furnace or boiler that is not too large. The minimum heat output for gas-fired equipment is about 50,000 BTU per hour, while oil-fired equipment only goes down to 85,000 BTU per hour. Since the houses described in this book may only require a 25,000 BTU per hour input after the insulating and winterizing procedures have been put into effect, there are few alternatives but to oversize. However, manufacturers should get the message and probably will eventually produce smaller equipment. In the meantime, the following alternatives should be considered:

— If the house has an open plan and few doors, a small gas-fired space heater will serve nicely. They are cheap and don't require any duct work.

— heat pumps, which are described later in this chapter, are also available in small sizes.

— in cooperative projects, try to build small houses next to each other. This will allow a central heating system to serve several houses and will also save on construction costs.

Further, circulating pumps and blowers should be selected without oversizing. A 1/6- to 1/3-horsepower pump or blower is suitable for most homes. Also, chimneys, draft controls, and start-stop controls must be properly selected to ensure efficient operation of your heating system. A heating contractor normally takes care of this, but the homeowner should make sure installation guidelines furnished with the equipment are followed.

Here's another option for saving with your oil or gas heating system: some 20 to 40 percent of the heat from these units is unused and goes up the flue. Part of this heat can be reclaimed by installing what's called a "flue heat exchanger." Placed in the flue, the heat exchanger extracts a good part of the otherwise wasted heat from the flue gases. This heat can be used to heat your cellar, or can be ducted to any part of the house you like. These heat exchangers cost about $100 to $140 each and can save about 10 percent of your heating bill every year. Units are available from:

Dolin Metal Products, Inc.   Isothermics, Inc.
475 President St.   P. O. Box 86
Brooklyn, New York 11215   Augusta, New Jersey 07822

As an alternative to the oil or gas burner, you may want to consider wood heating. Wood is a renewable source of fuel, and the price of cord wood makes it competitive with gas and oil. Moreover, if you have a woodlot and cut your own, it's free.

You can heat your home with wood in a variety of ways, too. If you have an old fireplace in a home you wish to renovate, the addition of a damper control and an underfloor air inlet to the fireplace (see Figure 13-2) may be all you need to provide many pleasant evenings in front of the fire. Several special C-shaped grates that capture some of the heat which would ordinarily go up the flue can be purchased. See Figure 13-3. These grates will help eliminate a part of the cost or work of obtaining wood, because you end up using less. Grates are offered for sale by:

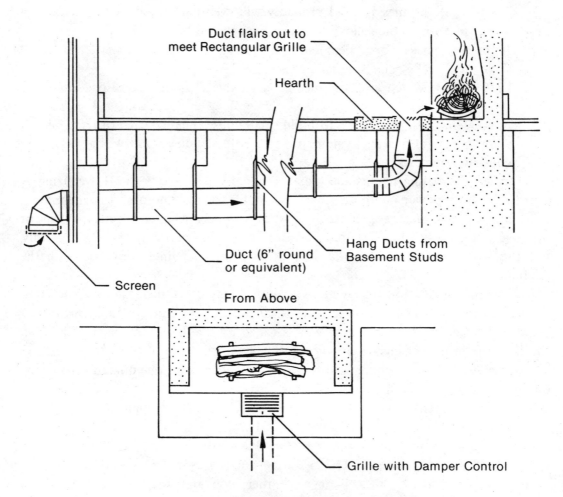

UNDERFLOOR AIR INLET TO FIREPLACE (DRAWS AIR FROM
OUTSIDE HOUSE RATHER THAN THROUGH ROOM)

FIGURE 13-2

Radiant Grate, Inc.   Thermograte Enterprises
31 Morgan Park   51 Iona Lane
Clinton, Connecticut 06413   St. Paul, Minnesota 55117

Much more efficient than the open fireplace are the enclosed chamber wood space heating units. Though they cost more (in the $200 to $500 range) than ordinary wood heaters, if you seriously want to obtain warmth from wood, this is the way to do it. The chimney or flue heat reclaimers mentioned earlier can be used here, too, making an extra saving. Several of the better-known, highly efficient wood space heaters can be obtained from:

Ashley Automatic Heater, Inc.   Riteway Manufacturing Co.
P. O. Box 730   P. O. Box 6
Sheffield, Alabama 35660   Harrisonburg, Virginia 22801

For an explanation of the features of wood stoves and the chimney heat exchangers mentioned before, as well as an extensive listing of companies and products, get a copy of one of the following:

Spectrum: An Alternate Technology   *Burning Quarterly*
Equipment Directory   80009 34th Ave. S.
Alternative Sources of Energy   Minneapolis, Minnesota 55420
Route 2, Box 90A
Milaca, Minnesota 56353   *The Wood Heat Journal*
($2.00)   Box 2301
                                        Norwalk, Connecticut 06851

or Society for the Protection of New Hampshire Forests, 5 S. State St., Concord, NH 05301

If you want the comfort and convenience of a modern central heating system, there are wood furnaces that can do this, too. Several companies now make wood furnaces with remote thermostats, forced hot-air distribution systems, and oil and gas backup units. The operation is fully automatic; and when you're away, the oil or gas fuel takes over. In a well-insulated house, all you do is fill the unit with wood about once each day.

A key point to remember about heating systems is to avoid the use of electrical resistance heating. There is no question that resistance heating uses over twice as much raw source energy (fuel burned at the power plant) as oil or gas systems. Even if the power plant is nuclear or hydroelectric, the waste is needless.

Electrically heated homes are often sold on the claim that the heating bills will be close to those of fuel-burning systems. The main reason that this could be possible is that the electrically heated homes tested were better insulated than the others. The comparison should be made on identically constructed homes to show the truth: electric resistance heat *is* expensive.

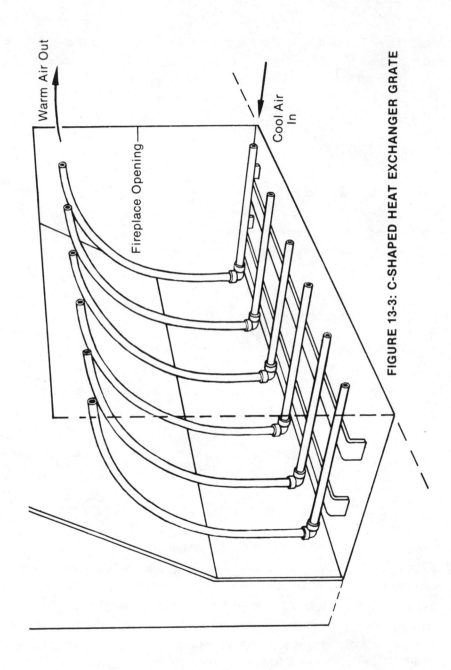

Warm Air Out

Fireplace Opening

Cool Air In

FIGURE 13-3: C-SHAPED HEAT EXCHANGER GRATE

A substantial improvement over electric resistance heat is the heat pump. See Figure 13-4. This unit operates like an air conditioner in reverse, ejecting warm air into the house and cold air outdoors. Its overall consumption of raw source energy is about one-half to one-third as much as with resistance heating, and is roughly equal to oil or gas heat. Moreover, the unit can be used for summer cooling, saving the extra initial cost of a separate air conditioner.

Although most heat pumps have been used in the south and southwestern United States, the newer designs can be used in northern areas as well. When selecting any heat pump, it is important to pick a unit that *does not* shut off when the outdoor temperature falls below 32°F. Instead, the unit should automatically defrost the outdoor coil, and should engage the auxiliary resistance heaters only as required. Units should also be selected for high efficiency. They should have an *energy efficiency ratio (EER)* of *at least* 10 at standard test conditions. Water-source heat pumps should be considered where ground water is available.

Further information on heat pumps, including performance data, can be obtained from:

*Electric Space Conditioning in
  Residential Structures*
Electric Energy Association
90 Park Avenue
New York, New York 10016

*Directory of Certified Unitary Air
  Conditioners and Unitary Heat Pumps*
Air Conditioning and Refrigeration
  Institute
1815 North Ft. Myer Drive
Arlington, Virginia 22209

For further information on fuel-burning heating systems, contact:

National Fuel Oil Institute
60 E. 42nd Street
New York, New York 10017

Gas Appliance Manufacturers Assoc.
1901 North Ft. Myer Drive
Arlington, Virginia 22209

There are other ways you can save, too. If you have a central heating system, setting back thermostats at night is a well-known energy-saving technique you should consider. Timer control thermostats are available to do the job automatically, or they can be assembled from a standard thermostat and a time clock. (For a low-cost homemade design, see *Popular Mechanics,* October 1974.) The benefits versus cost of the device are good, and they can be used on both old and new heating systems. A typical store-bought unit costing $50 to $75 will save 10 to 20 percent of your annual heating bill, worth $30 to $80. They are available from:

Honeywell, Inc.
2701 4th Ave. South
Minneapolis, Minnesota 55408

Penn Controls, Inc.
Goshen, Indiana 46526

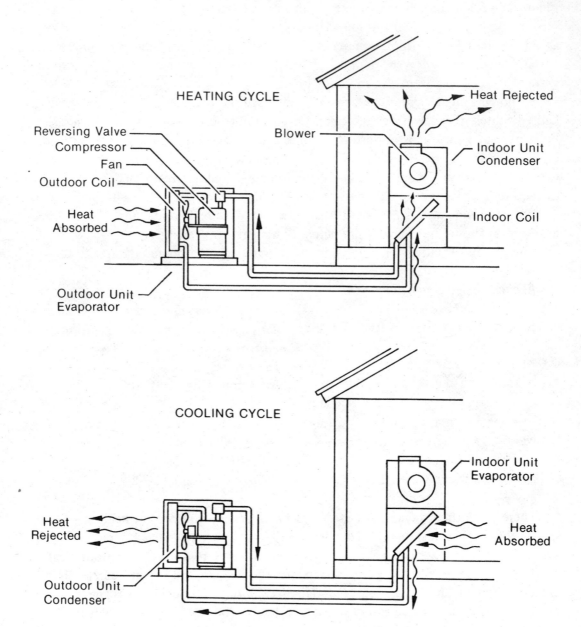

HEATING CYCLE

Reversing Valve
Compressor
Fan
Outdoor Coil
Heat Absorbed
Outdoor Unit Evaporator

Blower
Heat Rejected
Indoor Unit Condenser
Indoor Coil

COOLING CYCLE

Heat Rejected
Outdoor Unit Condenser

Indoor Unit Evaporator
Heat Absorbed

FIGURE 13-4: HEAT PUMP SYSTEM

Here's another energy and dollar saver: in large houses where there is a distinct separation between bedroom and living areas, zone control dampers (for air heating systems) or valves (for hot water systems) can be used to lower or shut off the heat when certain rooms are unoccupied. For example, most two-story houses have up-stairs bedrooms which can be shut off during the daytime. This can be done auto-matically with a zone control system operated by a thermostat on each floor. If you have a contractor build your new home, make absolutely certain that he installs these controls. This is a point contractors often overlook. (Further information on install-ing zone controls in both new and old houses can be obtained in *Popular Mechanics,* October 1974.)

COOLING

The first step in selecting better cooling equipment is to ask yourself if it is really needed. Many homes that normally have large central cooling systems can still have comfortable bedrooms at night with just a single room air conditioner, and many homes in cool climates or with good shading can get along without any air conditioning at all. (See Chapter 10.)

If air conditioning is required, then the unit should be sized just to meet the load, not exceed it. Air conditioners, like heating systems, are usually oversized, resulting in excessive cycling. This shortens the life of the unit, makes the tempera-ture control uneven, and increases energy consumption.

Procedures for sizing air conditioners are given in the same booklets mentioned earlier for sizing heating systems. Commonly, the square footage of the air-condi-tioned area is totaled and multiplied by a heat gain factor of 30 to 40 BTU's per hour per square foot. This is a shortcut *which should be replaced* by the more rigorous and accurate calculations that take into account wall and roof insulation, window area, and exposure.

If the house is designed to reduce summer heat gains, the cooling load can be as low as 20 BTU's per hour per square foot. Thus, insulation and other energy conser-vation features will help you save on both the first costs and the operating cost of your cooling system as well as your heating system. One very useful booklet on selecting and operating room air conditioners which you might want to get is:

"How to Choose the Room Air Conditioner Best Suited for You"
Association of Home Appliance Manufacturers
20 North Wacker Drive
Chicago, Illinois 60606

After the air conditioner has been sized, a unit with high efficiency should be chosen. Since EER's (energy efficiency ratio—equal to the BTU per hour output divided by the watts input) range from 6 to 12, it is important to inspect the manufacturer's literature carefully. The EER is based on a standard test, much the same as the Environmental Protection Agency's automobile mileage test.

Most sizes, though, for both central and window air conditioners are available with EER's of nearly 10. Although units with a high efficiency cost more, the dollar savings for improved efficiency will offset the added costs in the first two to four years.

MORE EFFICIENT OPERATION OF EQUIPMENT

The major ways to improve the operation of your heating and cooling system are:

- change your thermostat settings
- balance air or water flows
- tune up and service your equipment regularly
- turn off equipment when the house is not occupied

The do-it-yourselfer can probably handle all these without a repairman, and this will minimize dollar outlays for maximum energy conservation. Other homeowners will still be able to show worthwhile energy savings for their dollars invested if they have a good serviceman.

Change Thermostat Settings

One of the best ways to save some of the money you now spend to heat and cool your home is to lower the thermostat settings. Thermostats set at 68°F in the winter and 78°F for summer air conditioning should maintain a comfortable house for most people. Adjusting your body to these temperatures involves dressing for the season—warm clothes in winter and light airy clothes in summer—and realizing that people can exist at these temperatures without harming their health. In England most homes are heated to 65°F or less, and many people even claim the lower temperatures are more healthful. For the greatest savings, thermostats should also be lowered to about 60°F for winter nights. You can use an accurate thermometer if you want to check the thermostat.

As the winter heating bill is the most costly, try to keep the heat as low as you can all winter. The best way to do this is to get used to a lower temperature slowly, as winter comes on. Most people do fine at 65°F or less, but be careful if there are small babies crawling on cold floors, or older folks who might catch colds. Go slowly, a little bit at a time, and let the whole family get used to lower temperatures.

Living with less heat can be easy. The best way is to cover up. Make sure clothes are loose fitting. This traps body heat and helps keep you warm. Also, use wool and cotton clothing, as these are much warmer than manmade fabrics like nylon or rayon. Wear several layers of light clothing. This traps warm air and makes you feel more comfortable than one heavy layer would. Long underwear can be a big help too. Wear sweaters for an extra layer of protection. Finally, when you sit and work or watch T.V., use a quilt or blanket over your legs.

Use quilts and comforters on beds to keep you warm at night. Just as with clothes, several light blankets trap warm air and keep you more comfortable than one heavy cover. Socks and longjohns are great, too.

Balance Air or Water Flow

Adjusting your heating system so you get enough heat to those areas of the house that need it is also very important. Dampers or balancing valves on central heating systems should be adjusted to make sure no part of the house is too hot or too cold. If the bedrooms or other areas away from the thermostat are too hot, for example, then the heating system will have to work longer. In this case, balancing a home heating system basically is a matter of lowering air or water flow, starting with the overheated areas farthest from the heating plant. If the problem is underheating, then you make adjustments to provide more circulation to the colder rooms.

At times you may still find that some rooms are cold. This is particularly true if you have a wood or other space heater, as the heat will then tend to get trapped in certain rooms. Also, if you use the "house-within-a-house" concept to renovate that was outlined in Chapter 7, you'll have to reestablish the heat flow patterns in the house for optimum winter comfort. If certain areas are cold, try some of the following suggestions:

First, check to make sure that nothing is blocking the heat in the cold room. Are radiator valves open, and is the area around the radiator clear of furniture which might trap heat? If you have a hot air system, be sure the register is not blocked and that the furnace filters are clean. Another idea is to move furniture away from the outside wall or windows in the room. It may be more comfortable to sit or work near an inside wall, where it's warmer. If the room still seems cold, use it as a storeroom. You can turn a cold spot to good advantage this way.

On the other hand, if you need the area for work, try one of these ways of heating it:

Wall Grille.   Cut a hole between the door and ceiling and install a grille or vent. This lets the hot air trapped on the ceiling of the warm room into the cold one.

Fan.   If you still have a problem, use a fan to move air from room to room. Place the fan on the floor in the doorway between the hot and cold room. Blow air from the cold room *into* the warm room. See Figure 13-5. The heavy cold air is easier to move and will force warm air through the door to heat the cold room quickly. Be very careful when using the fan, though, as small children may want to play with it on the floor.

For homes without central heating, it's often hard to get heat to the upper floor. The best way to overcome this is to put vents in the upstairs floors. Different sizes are available at hardware stores. Cut holes in the upstairs floor and downstairs ceiling. Place the vents in. See Figure 13-6. Close them each morning to keep heat from going up to the bedrooms, and open them each night to allow heat to rise into the upstairs part of the house.

There are a few other things you can do to make this daytime segregation between the upstairs and downstairs more effective. First, close the doors to the bedrooms. This way, only the heat conducted up through the ceiling gets upstairs. If

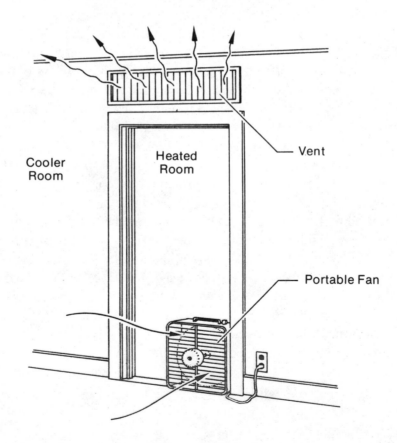

Cooler Room

Heated Room

Vent

Portable Fan

FIGURE 13-5: FAN BLOWS COOL AIR INTO WARM ROOM

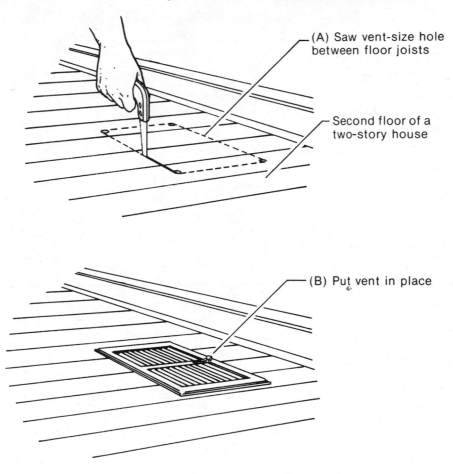

(A) Saw vent-size hole between floor joists

Second floor of a two-story house

(B) Put vent in place

FIGURE 13-6: INSTALLATION OF A FLOOR VENT

you don't have doors, you can add them; or try to enclose the stairway with wood paneling and put a door at the bottom. The beauty of this system is that it not only cuts your heat bills during the day, but opening the vents and door at night lets the heat rise. Thus, the unused downstairs tends to cool off at night more than the bedrooms, and you can save even more fuel.

Tune Up Your Equipment

Once you've turned down the thermostat and put the heat right where you need it, tune up your heating system to peak efficiency. Oil, gas, and electric heating systems are normally serviced once a year, preferably in the fall. Servicing should include cleaning of heat transfer surfaces, adjusting the firing rate, checking combustion air, and oiling motors. Air filters on forced-air systems should be changed according to the instruction manual.

A typical tune-up costing $30 to $40 can save that amount or more in fuel bills. Further savings are possible if a homeowner does the servicing himself or herself. Although few people are presently equipped to do the job, it is basically no more difficult than an auto tune-up. Details of servicing procedures are given in such magazines as *Fuel Oil and Oil Heat* and *Fuel Oil News,* which are available in most large libraries. Servicing manuals can be purchased from:

Petroleum Marketing Education
  Foundation
P.O. Box 11187
Columbia, South Carolina 29211

National Fuel Oil Institute
60 E. 42nd Street
New York, New York 10017

Test instruments can be obtained from:

Dwyer Instruments
P. O. Box 373
Michigan City, Indiana 46360

Bacharach Instruments
625 Alpha Drive, RIDC Industrial Park
Pittsburgh, Pennsylvania 15238

A complete kit should cost from $60 to $80, including instruction manuals, and will pay for itself in two years.

On coal and wood heating systems, you should certainly clean the flue at least once each season. A good booklet that explains how to do this is:

Save Energy: Save Money (OEO Pamphlet 6143-5)
Community Services Agency
Washington, D.C. 20506
(It's free.)

Turn Off Equipment When It's Not Used

The last energy-saving operating procedure is to shut off equipment when the house is unoccupied, or when the equipment is not needed. There is no significant reason why air conditioners, for example, should be left on while a family is away for a weekend, or even for an evening. Also, air conditioners should be shut off when outdoor temperatures are below 78°F, and the windows should be opened instead.

Turning down winter heating, on the other hand, is a little different. When you're away, even for a few hours, turn the heat down. If you're away for several days, lower your heat to about 50°F. When you return, it will take a while to warm up the house again. Try to schedule unpacking or housework for these times, so you don't have to turn the heat up past its usual setting. Let the house reheat *slowly*. Quick heating wastes some of the fuel and money you saved while away.

When a separate water heater exists, boiler pilot lights or oil burners should be shut off during the summer. This will cut the stand-by heat losses considerably and save both the fuel to operate the pilot light and the energy to remove the boiler's waste heat from the house.

Every home can save heating and cooling energy by the selection of better equipment and improved operation. A goal of 25 percent savings is reasonable and achievable, and it can cut your bills by $125 per year. Additional savings of up to another $75 per year can easily be obtained by better insulation and other construction improvements. The best way to prove the results of a conservation program is to keep track of the fuel bills. If energy consumption drops, then your program has been effective. A nice way to check on your progress is to plot consumption each month for two or three years, as this will help show the long-term gains.

chapter 14

SAVING MONEY WITH APPLIANCES AND LIGHTS

by DAVID L. HARTMAN

In most homes, appliances and lights account for about one-third of the total energy consumed. Many people feel that it is hard to save here; but actually the opportunities are at least as good as for heating, cooling, and hot water. The key is to determine what your needs are and how to satisfy them at the lowest cost in energy and dollars.

The quantity of energy consumed for these purposes depends on many factors, but the average is about 7,500 kilowatt hours (KWH) of electricity for a four-person family. Of this total, the refrigerator consumes about 22 percent; the range, 14 percent; dishwasher, 4 percent; clothes washer, 2 percent; dryer, 12 percent; TV, stereo, and miscellaneous, 23 percent; and lights, 23 percent. See Figure 14-1.

The corresponding costs for appliance and lighting energy vary throughout the country. In New York, at 9¢ per KWH, the annual bill would be over $675! At a more average rate of 4¢, the bill would still be about $300.

SELECTING MORE EFFICIENT EQUIPMENT

When buying appliances and lighting for your home, make sure they are efficient. A variety of efficient equipment is now available, and manufacturers will soon be producing even better models to meet the public demand. Before buying, it is very important to make decisions based on factual energy consumption data. This will take some effort, but the rewards are real energy savings as well as lower operating costs.

The Association of Home Appliance Manufacturer's (AHAM) is working on

238

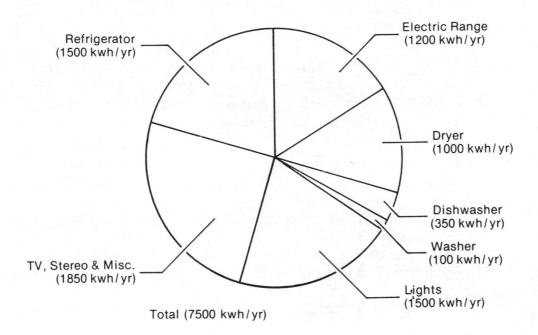

Refrigerator
(1500 kwh / yr)

Electric Range
(1200 kwh / yr)

Dryer
(1000 kwh / yr)

Dishwasher
(350 kwh / yr)

Washer
(100 kwh / yr)

Lights
(1500 kwh / yr)

TV, Stereo & Misc.
(1850 kwh / yr)

Total (7500 kwh / yr)

**FIGURE 14-1: APPLIANCE AND LIGHTING ENERGY FOR A
TYPICAL FOUR-PERSON FAMILY**

performance standards and testing procedures for major appliances, and the government is considering regulations requiring better labeling that gives energy consumption information. So, check the labels and see how much it's going to cost you to operate the equipment. Another way to check for energy-saving features is by looking through magazines like *Consumer's Report* and *Consumer's Bulletin* at the local library. They have articles that will increasingly use the AHAM and government information, and they also report on features like the quality and durability of equipment.

Refrigerators

The most important energy-saving feature to look for is more insulation. Conventional refrigerators have only about 1 inch or so of fiberglass insulation in the sides, top, bottom, and door. There are no technical reasons why 2 to 3 inches of foam insulation, which is superior to fiberglass, cannot be used instead. The only problem is that the refrigerator will be a little larger on the outside for a constant inside volume, and it will cost more. Generally, though, the extra first cost will be saved by lower operating costs in 3 to 4 years.

A number of manufacturers now advertise energy-saving refrigerator models with many good features The energy savings for these models are between 30 and 50 percent! Data on all major refrigerator models are given in a useful booklet called "Directory of Certified Home Refrigerators and Freezers," published by the Association of Home Appliance Manufacturers, 20 N. Wacker Dr., Chicago, Illinois 60606. It contains standardized data on refrigerated volume and monthly electric consumption, which, together with current prices, can be used to help you choose the best refrigerator for your needs.

In addition to more insulation, a number of other key features are worth checking out. Refrigerators should have a separate freezer-section door so that you don't have to open up and, consequently, warm the refrigerator to get at the freezer. Manual defrost units, when properly cared for, use 20 to 40 percent less energy than self-defrosting models. Options like ice makers and water coolers consume energy and are hard to service. Casters and removable grills make cleaning the condenser easier, which is important to keep the unit at top efficiency. Doors should always have tight-fitting gaskets to prevent air leakage. Finally, stay away from refrigerators with electric door heaters. See Figure 14-2 for features you should look for when selecting a refrigerator.

Ranges

Since energy consumption for ranges is fairly evenly distributed between the cook top and oven, conservation efforts must be aimed at both for maximum

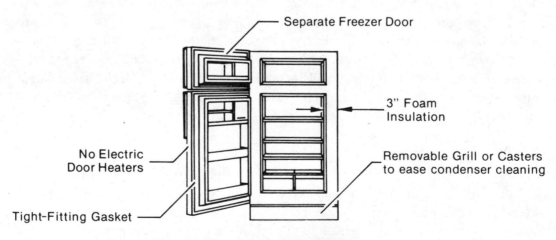

Separate Freezer Door

3" Foam Insulation

No Electric Door Heaters

Removable Grill or Casters to ease condenser cleaning

Tight-Fitting Gasket

CONSTRUCTION FEATURES
OF AN ENERGY SAVING REFRIGERATOR

FIGURE 14-2

savings. Electric range cook tops should have "calrod" surface elements for good thermal contact with the bottom of pots and pans. Ceramic cook tops suffer in this regard and should not be considered. *All ovens should be well insulated.* Manufacturers have not got the message yet and provide high insulation only on certain types of self-cleaning ovens, the "pyrolytic" or high-temperature type. The difference in insulation is about 50 percent; conventional ovens have about 1½ inches of fiberglass, while the "pyrolytic" units have about 3 inches. The best bet is to choose a "pyrolytic" oven and then use the self-cleaning feature sparingly until well-insulated, manual clean ovens are available.

Gas ranges have a similar problem with lack of insulation—so, again, get one with a self-cleaning feature and use it wisely. Gas cook tops, however, all have about the same burner efficiency, so there is little to choose from. The best opportunity for energy conservation is to buy a range with an electronic igniter. This feature substitutes for the pilot lights (there are three of them in most ranges) and can save about 30 percent of the annual gas consumption. "Caloric" gas ranges are the only ones with this feature right now, but other manufacturers should also be offering it soon.

Other ways to save cooking energy are to use efficient cooking utensils and special-purpose appliances. A pressure cooker, for example, can cut cooking time for boiled foods in half. Electric fry pans and toaster ovens can be used for small meals instead of heating up the main oven. Microwave ovens certainly offer large energy savings, but they *should not be used by people with heart pacemakers,* as this can be dangerous.

Dishwashers

The main factor in dishwasher energy consumption is the amount of hot water used. The standard water consumption is about 15 gallons per wash. A "short wash" cycle using about 10 gallons is a highly desirable feature available from most manufacturers. The short wash can be used for all but very dirty dishes.

Another feature to look for is an "air dry" cycle. This lets you throw a switch to cut out the drying heaters. It can save about one-fourth of the dishwasher energy.

Clothes Washers

The vast majority (90 percent) of the energy used by clothes washers goes into the heating of water. Therefore, this is the key area where you can save. New units should have wash water temperature controls and a variable water level. A "sud-saver" feature, which allows wash water to be reused, is very desirable. When used in conjunction with a laundry tub, this will save 20 of the normal 35 gallons of water used.

Only one manufacturer at this time, Maytag, has all these features on their washers; the washers are also noted for their durability. The company's address is:

> The Maytag Co.
> Newton, Iowa 50208

Laundry tubs are available from:
> E. L. Mustee and Sons
> 6911 Lorrain Ave.
> Cleveland, Ohio 44102

Clothes Dryers

The best way to dry clothes is to do it outdoors. A simple rope clothes line costs about $2 and can save $10 per year, as well as give your wash that fresh-air smell. When a dryer must be purchased, it should have a moisture-sensing shut-off control and an adjustable thermostat. With these features, the heater is not on full blast all the time; heat is supplied only when the clothes are damp. The savings are on the order of 10 to 15 percent.

Although dryers are normally vented to the outdoors, a simple bypass damper plus an extra lint filter will allow the warm dryer exhaust to help heat the house in the winter. If this is added, the pressure drop for the damper and filter should not exceed the maximum recommended for the dryer. If it does, damage to the heating element could occur.

Other Appliances

An energy-conscious home can certainly get along without appliances such as hairdryers, portable electric heaters, and electric can openers. By not buying them, it is possible to save on first costs as well as on operating costs. Television sets should be solid-state and should not have an "instant-on" feature. The key is to think about what your needs really are. A small compromise in convenience can make a *big* difference in energy consumption.

Lights

The greatest lighting energy savings will come by switching to fluorescent fixtures. These use about half the energy of incandescent fixtures, and the bulbs last two to three times longer. A good booklet to read on lighting is:

"Design Criteria for Lighting Interior
  Living Spaces"
Illuminating Engineering Society
345 E. 47th Street
New York, New York 10017
(Free)

A couple of common misconceptions exist regarding fluorescent lights: first that the color rendition is poor and second that they are expensive. When "warm white deluxe" bulbs are used, people look far more natural than under typical office lighting. The color rendition for "warm white" bulbs is very close to the color rendition of incandescent. With these bulbs, fluorescent lights can be used in nearly every room of the house, except perhaps dining areas. The actual initial costs for fluorescent lighting vary depending on the fixture type, but costs of equivalent fluorescent illumination are generally no more expensive than quality table lamps. Typical installation costs are about 50¢ per watt for either one. Since fluorescent lights are twice as efficient, half as many watts are needed. Therefore, it is possible for the installation cost to be even lower than for conventional systems.

When choosing a lighting system, it is important to consider how to use it effectively. Switches should be located where people won't forget to turn them off. "Task lights," such as desk lamps, can often be used to provide adequate light with less energy than ceiling fixtures. Also look for new energy saving incandescent bulbs and fluorescent tubes.

MORE EFFICIENT OPERATION

Whether your appliances are new or old, you can learn to operate them in ways that will keep the operating costs down. The savings here can be just as important as with the newer and more efficient equipment. Another benefit is that appliances which are used efficiently last longer and give more trouble-free operation.

The first step is to *use appliances only when you have to.* Choose only the equipment that is really a help in getting work done, and then learn how to use it sparingly. The best way to learn how to do this is to keep a record of how much energy you are using. Look at the bills each month, and keep track of the number of kilowatt-hours of electricity or cubic feet of gas used. A good goal is to try to cut down the amount used by about a third. As the following ideas are put to work, keep checking your progress. Further information is available in the references for this chapter.

Refrigerators

Make sure that refrigerator doors close and seal properly. You can check this by closing the door on a piece of newspaper. If the paper pulls out easily, you need a new gasket. Also, keep the condenser on the back of the refrigerator clean, so it can operate efficiently.

Dishwashers

Use the dishwasher sparingly. Most four-person families can get along on one wash a day by waiting for a full load. Washing by hand can save energy as long as the water is not allowed to run continuously. Short wash cycles (sometimes called "crystal" or "plastic" cycles because they are designed for the washing of delicate dishes) and air dry cycles should be used if the machine has them.

Clothes Washers and Dryers

Use cold water whenever possible. It will clean as well as hot water when the right type of detergent is used. The only exception is for diapers and oily work clothes.

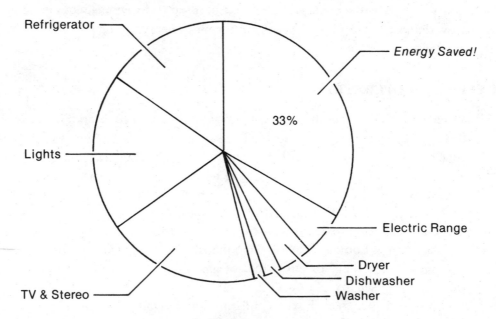

APPLIANCE AND LIGHTING ENERGY CONSUMPTION FOR AN ENERGY CONSERVING HOME (FAMILY OF FOUR)

FIGURE 14-3

Clothes dryers should be run just long enough to dry the clothes. Often they run 10 to 15 minutes longer than necessary. When weather permits, dry clothes outdoors.

Lights

Turn off all lights that are not being used and whenever you are away from home. If security is a problem, use a burglar alarm. Often, high-wattage bulbs can be replaced with lower ones. Try cutting the total connected wattage by 30 percent. Substituting a 50-watt bulb for a 100-watt bulb will cost about 30¢, but it will save over a dollar per year. If you have a gas yard-light, replace it with a switch-operated fluorescent fixture.

Every home can save a substantial part of the money spent on appliance and lighting energy. It is a matter of selecting efficient equipment and learning how to use it properly. A reasonable goal for both new and existing homes is a one-third reduction in the amount of energy used and what the owner has to pay for the operation of this equipment. See Figure 14-3. The resultant yearly savings for the typical four-person home mentioned at the beginning of the chapter is about 2,500 KWH per year, worth $100 at 4¢ per KWH.

chapter 15

WATER HEATING AND WATER CONSERVATION

by DAVID L. HARTMAN

As the prices for the energy to clean, pump, treat, and heat water increase, more people will find that saving water is a worthwhile thing to do. Even at today's costs for water and hot water heating, a small investment in more efficient equipment that helps reduce the amount of water used can save $70 or more each year. There are other reasons for saving water, too. By using less water, many people can save on their initial and maintenance costs for a septic system. Saving on water also helps reduce pollution where the treatment of waste water is inadequate.

Water conservation need not cause any hardship. It is a matter of choosing equipment that uses less water to perform the same tasks. Many families are already doing this for numerous reasons. In localities where water is scarce, such as Arizona, water conservation is an accepted habit. In other localities, such as Washington, D.C.'s Suburban Sanitary Commission area, plumbing codes have been amended to *require* water-conserving plumbing fixtures; private citizens initiated the amendment, and the public response has been favorable.

The water we use adds up quickly. A typical four-person home consumes about 300 gallons each day or *110,000 gallons per year* for indoor uses. These uses include toilets, lavatories, bathing, laundry, utility sinks, and kitchen dishwashing and food preparation. Outdoor uses, such as car washing and lawn sprinkling, consume additional water; but it normally doesn't drain into a sewer system. Hot water consumption is about 75 gallons per day, requiring 180 gallons of oil or 6,000 KWH of electricity to do the heating. This energy represents between a tenth and a quarter of the total energy consumed by an average home. (House heating and cooling is about one-half of the total; appliances and lights about one-third.)

246

A breakdown of water consumption (see Figure 15-1) is as follows: toilets consume 40 percent; lavatories, 3 percent; bathing, 33 percent; laundry, 12 percent; utility sink, 2 percent; and kitchen, 10 percent. The actual water consumption for a home varies widely, depending on the living habits of the users as well as the efficiency of the fixtures and appliances.

Although the cost of water is often overlooked, it is far from insignificant. For a municipality providing sewer and water, the cost for both services is usually combined on one bill. Typically, this would be about $75 per year at a rate of 50¢ per 100 cubic feet. For a home with a well and septic tank, the costs are about the same when the initial costs are averaged over the equipment's lifetime. The cost for water heating would be about $120 per year for oil or gas and over $200 per year for electricity. Thus, the total costs can be $200 to $300 per year.

SELECTION OF MORE EFFICIENT EQUIPMENT

Many water-conserving fixtures that can help save money are available right now. They haven't been widely publicized and therefore few people have used them. A

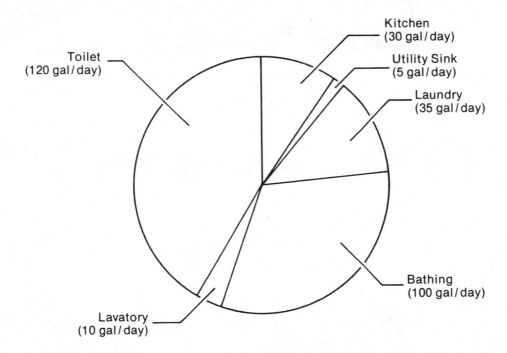

FIGURE 15-1: WATER CONSUMPTION IN A TYPICAL HOME (FAMILY OF FOUR)

different situation prevails in England, where water costs are higher than in the U.S. There, low-flush water closets, for example, have been available for many years and are widely used. In this country, with the costs of supplying our water needs increasing, we can begin now to help ourselves. Listed below are the major fixtures we use and the water conserving features to watch for.

Toilets

Tank-type toilets normally consume between 5 and 6 gallons per flush. Water-saving toilets available in the U.S. consume 3 to 3½ gallons. Foreign toilets consume as low as 2 gallons per flush but are not readily available. The practical minimum has never been determined, but 3 gallon units are compatible with standard American plumbing.

The best water-conserving toilets are specially designed to give full flushing action with a reduced amount of water. Figure 15-2 shows one technique used commercially. Bricks in the tank and retrofitted flush valves (as opposed to readjusting the tank float, which does not work well) will generally work adequately in existing toilet tanks. See Figure 15-3. In some cases, though, two flushes are necessary; and if this occurs, it may often cancel the benefits of the water-saving features. Suitable toilets are available from many manufacturers, including American Standard, Kohler, Crane, and Briggs. Your plumbing supply house can order them if they

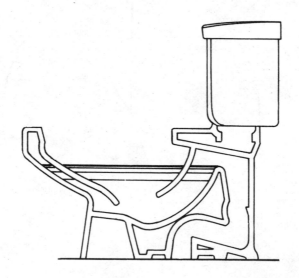

FIGURE 15-2: THE SMALLER TANK AND REDESIGN OF THE BOWL ALLOW THIS TOILET TO GIVE PERFECT FLUSHING ACTION WITH ONLY 3.5 GALLONS.

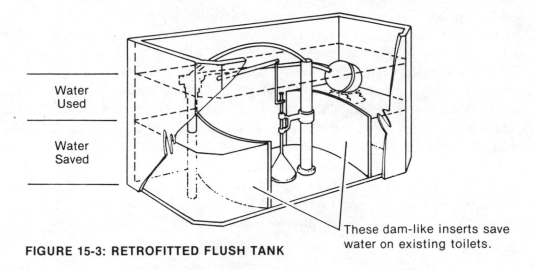

Water
Used

Water
Saved

These dam-like inserts save
water on existing toilets.

FIGURE 15-3: RETROFITTED FLUSH TANK

are not locally in stock. For existing toilets, tank inserts and special flush valves are available from:

Metropolitan Water Savers	Water Wizard
8001 Forbes St.	Box 184
Springfield, Virginia 22151	Croydon, Pennsylvania 19020

In addition to water-saving toilets, a number of waterless units are also available. The best known is the Clivus Multrum which composts toilet wastes as well as kitchen scraps. This unit, costing about $1500 complete, is allowed by the codes in some areas to be used as a substitute for as much as 40% of the normal size of the septic system. In cases where percolation is bad and hence a septic system costs upwards of $5000, the investment in the Clivus system can be a worthwhile one. More information on the Clivus can be obtained from:

Clivus Multrum, U.S.A.

14 A Eliot St.

Cambridge, Massachusetts 02138

Other water-saving units include chemical recirculating toilets, incineration toilets, and grey water systems for large housing projects. Information on these can be obtained from *Spectrum* magazine mentioned in Chapter 14.

Lavatory Faucets

The key to effective water-conserving lavatory faucets is a low flow rate combined with a wide spray pattern. The best designed faucets have a flow restricter located in the fixture body or in the spray head. Existing faucets can also be modified to save water by installing flow control valves in hot and cold supply lines. In either case, the ideal flow rate is about a ½ gallon per minute, compared with normal flow rates of 2 to 3 gallons per minute.

Low-flow faucets can be obtained from plumbing supply houses, as can other fixtures. Two styles to consider are "Unatap" model faucets, manufactured by the Richard Fife Company, and "Colortemp" model faucets, manufactured by the Speakman Company. Both models are high quality and fit all sinks. In-line flow control valves (see Figure 15-4) for existing faucets are available from:

Control Division
Eaton Corporation
191 E. North Avenue
Carol Stream, Illinois 60187

Showers

Water-conserving shower heads save energy as well as water. A 10-minute shower with a conventional 3 gallon per minute flow-rate head will consume 30 gallons. Some heads are even more wasteful with flow rates as high as 6 gallons per minute.

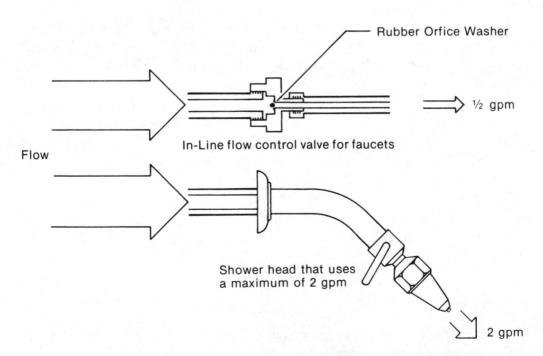

Rubber Orfice Washer

½ gpm

In-Line flow control valve for faucets

Flow

Shower head that uses a maximum of 2 gpm

2 gpm

FIGURE 15-4: FLOW CONTROLS FOR FAUCETS AND SHOWERS

If the shower is used once a day, 365 days a year, it will take about 100 gallons of oil per year to do the heating, at a cost of about $40. A water-conserving head should have a flow rate of only 2 gallons per minute. This will cut water consumption and water-heating energy in half. A typical low-flow head costing $5 can, therefore, save $20 or more per year, making it a very good investment.

In addition to a low flow rate, the shower head should have a good spray pattern since many people don't enjoy their shower unless it feels invigorating. A variety of other units are available at most hardware stores as well, including the European "telephone" shower head which lets you turn the water on and off to soak, soap, and rinse, resulting in very low water consumption.

Kitchen Faucets

Many kitchen sink operations (such as rinsing dishes and cleaning vegetables) require running water. Commonly, the water is on full blast when, actually, a lower flow rate will do. Low-flow faucets may require a little more patience during the filling of pots, but 15 seconds for 2 quarts should not strain many people.

A good flow rate for most faucets is 2 gallons per minute. This can be obtained by installing in-line flow control valves in the supply lines. Water-conserving faucets are also available from some plumbing suppliers. One note, though: flow controls should *not* be used on faucets which supply a dishwasher because they will prevent full washing action.

Appliances

Water conservation measures for appliances are discussed in detail in Chapter 14. Since dishwashers and clothes washers contribute greatly to the hot water heating load, they should not be overlooked in an effort to conserve water.

Outdoor Water Uses

A water conservation ethic does not mean that you will have to stop watering your lawn. The environmental impact of outdoor uses of water is generally lower than indoor uses because the water does not have to be heated and it does not flow into a sewage treatment system. However, many homes use a great deal of water outdoors, over 200,000 gallons on lawns where automatic sprinklers exist. Therefore, the many opportunities to save should be especially considered in homes where extensive lawn sprinkling is practiced.

One way to save water and work is to plant "groundcovers" (such as ivy or ice-plant) in place of most of the lawn grass. Any small patch of lawn left can be mowed by hand, saving the fuel and upkeep of a gas or electric lawn mower. The time saved from lawn-mowing chores can be better spent saving energy instead of wasting it. Information on groundcovers can be obtained from many landscaping stores. A useful booklet is:

"Growing Groundcovers" GPO #1969 O-358-548
Superintendent of Documents
U.S. Government Printing Office
Washington, D.C. 20402
(15¢)

Other ways to save include locating sprinklers so that they don't spray the house or sidewalks. For car washing, try using a self-closing nozzle. They let you shut off the water while you soap up the car and can save *several hundred gallons* per wash.

Hot Water Heaters

After the demand for hot water has been minimized, consideration should be given to improving the efficiency of your hot water heater. In a typical house, heat losses from the tank account for about a quarter of the energy consumed for hot water heating. Additional losses occur from uninsulated pipes. Gas or oil burners also play a large role in energy loss since they operate at efficiencies ranging from only 50 to 80 percent. (Electric heaters are even worse. They are 100 percent efficient *in the home,* but are only about 33 percent efficient when the losses from power generation are included.)

The best way to improve water-heating efficiency is to use a solar water heater and to size it to carry nearly the full heating load. Then, the conventional system will be needed only as a back-up. Refer to Chapter 18 on solar water heating for details.

To cut heat losses in any water heater (even a solar type), consider wrapping 3 inches of fiberglass insulation around the entire tank. Batt insulation can be conveniently installed with the foil or paper surface *away* from the tank. Tank insulation kits are available from Johns-Manville insulation dealers for about $20. Pipes should be insulated with a ½ inch of foam or fiberglass insulation. Together, these measures can cut water heating energy consumption by 15 to 20 percent, or a savings of $15 to $20 per year.

MORE EFFICIENT USE

The best way to encourage water conservation is to check your water consumption with a meter. By determining the consumption each month and keeping track of it on a chart, you will have a factual indication of the progress made by water conservation. Many homes already have these meters where required by the local water utility company. Otherwise, they can be installed by most plumbers or handymen. Suitable units can be obtained for about $50 from:

<table>
<tr><td>Badger Meter, Inc.</td><td>Hersey Products</td></tr>
<tr><td>4545 W. Brown Deer Rd.</td><td>250 Elm St.</td></tr>
<tr><td>Milwaukee, Wisconsin 53223</td><td>Dedham, Massachusetts 02026</td></tr>
</table>

A number of good booklets on techniques of water conservation, including lists of water-conserving fixtures and appliances, can be obtained from:

Washington Suburban Sanitary Commission

4017 Hamilton St.

Hyattsville, Maryland 20781

Changing habits to conserve requires a little discipline but should otherwise hardly influence your lifestyle. Simply turning off a faucet after filling a sink is an example of a more general rule: *don't leave the water running when it is not needed.* Similarly, turn off the hose when washing a car. Just wet it down and then use the water to rinse.

Showers can be a good energy and water saver compared to baths, when a low-flow shower head is used. A 10-minute shower with a 1½-gallon-per-minute head

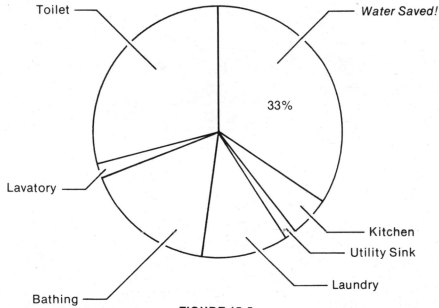

FIGURE 15-5
WATER CONSUMPTION IN A WATER-SAVING HOME (FAMILY OF FOUR)

will use 15 gallons of water, while a bath in a tub (2 X 5 feet with 5 inches of water) will use 30 gallons. In money terms, it is a 6¢ shower versus a 12¢ bath, when water and hot water costs are included.

Running hot or cold water until the desired temperature is reached is a real waste. It might take a minute of time and 3 gallons of water down the drain just to get a small glass of cold water. A better idea is to store drinking water in the refrigerator. Also, correct a leaky faucet by replacing the washer. A small trickle can add up to several hundred gallons in a month.

Some other guidelines include trying to use the dishwasher no more than once a day for a family of four, using cold water for washing clothes and turning down the water heater thermostat to 120°F, adjusting lawn sprinklers so they don't water the house or the street, and keeping oil or gas water heaters in tune.

The results of a home water-conservation program will show that a lot of water can be saved by simple, common-sense methods without any sacrifice in real water needs. A home with water-conserving appliances and plumbing fixtures can easily save a third for indoor uses and as much or more outdoors. See Figure 15-5. For a typical four-person family this would be about 30 gallons of hot water per day and 115 gallons of total water. The corresponding dollar savings would be $30 a year for water and $40 for oil or gas to heat the water.

Section IV

WORKING WITH NATURE

chapter 16

DESIGNING THE SOLAR-TEMPERED HOME

by EUGENE ECCLI

Using solar energy to provide part of the heating for your home can be a great way to work with nature. The sun showers us with clean, safe, and renewable energy that can be brought into our homes on cold winter days to provide warmth and relaxation. Moreover, if we design our homes properly, we can have this comfort at a moderate cost.

The sun has been used to great advantage by people of different cultures throughout history. Really, the only period during which the sun's importance has been neglected has occurred during the past hundred years in countries where fossil fuels have been very inexpensive. Now that this age of abundant fuel is drawing to a close, people are beginning to relearn and expand upon traditional ways of using the sun.

While there is much that can be done to make use of the sun to heat our homes during winter, there are several factors you should be aware of. First, the cost of putting the sun to work in your home varies, depending upon the approach you use. Some methods are fairly inexpensive, while others are very expensive. In all cases, though, the cost-benefits of the winterizing-insulating techniques explained in previous chapters are *far* greater than any solar system yet designed. Thus, before any worthwhile effects can be gained from solar design, it's *absolutely* essential that you reduce your heating (and cooling) needs as much as possible with the energy conservation techniques outlined in those chapters. To many people, this may seem like hard advice to follow. After all, who actually sees that extra insulation in the walls or the heat saved via weatherstripping and caulking? And doesn't the beckoning force of a relaxing and comfortable relationship with the sun seem almost to demand the greater part of our attention?

257

While it's certainly easier to make friends with the sun than with a roll of insulation or weatherstripping, let's remember that these energy-conserving techniques do two terrifically important things for you: they make your home much more comfortable all year round and they allow for this comfort at a cost *low enough* that you'll have the extra investment necessary to bring the sun into your life.

Reducing your heating and cooling needs will not only save you money that can be invested in solar heating, but since your winter comfort needs are now smaller, the size and investment in a solar heating system will be substantially reduced as well. This is a very important point because, even in a well-sealed and well-insulated home, some solar systems can be expensive. A full system—one that could provide 50 to 70 percent of your heating needs in most areas—could cost about $5,000, perhaps more. Since many people don't have this kind of money and these systems are barely economical at the present time, we have concentrated in this chapter on what's known as "solar tempering." This is a method of making use of windows, shading devices, and masonry elements on the south side of your home so that the sun can provide from 15 to 40 percent of your heating needs. An advantage of this system is its low cost—in the $300 to $1,300 range if you do the work yourself. Now, let's take a look at how we can design for a comfortable, cost-effective, and happy relationship with the sun.

DESIGNING THE SOLAR-TEMPERED HOME

The position of the sun and the amount of light striking our homes varies constantly over the day and the year. Hence, designing with the sun in mind is largely a matter of learning how to take advantage of the benefits and avoid the problems that these changes cause. Throughout this chapter we'll be concentrating on using windows on the south side of a house to bring in light and heat from the winter sun. Several factors have to be considered at this point:

1. While the sun provides heat during the day through your solar windows, all that is gained will exit through the glass again at night unless insulating curtains or panels are placed over the windows at dusk. This is true even if you use storm windows, so the nighttime insulation barriers are essential. The designs in Chapter 11 show you how to construct them.

2. The next consideration is the size of the windows. Too small a window area on the south side of the house will not take full advantage of the winter sun, but too great a window area will tend to overheat the house on sunny days, even in winter.

3. There are both daily and seasonal changes that have to be accommodated. The daily changes occur because of the changing position of the sun in the sky; i.e.,

there is more sun available at noon (usually the hottest part of the winter day) than there is during either early morning or late afternoon. The seasonal changes occur because, although the position of the sun in the sky is about the same in February as it is in October, daytime temperatures are very different. Hence, if the areas of the windows bring in enough heat in October, there will be underheating in February because it's colder then. These daily and seasonal imbalances have to be at least partially rectified if your solar windows are to provide an even and efficient form of heat from the sun.

4. Keeping your home cool is important, too. Though we will gain useful heat from south windows in the winter, these same windows will overheat the house in summer. Hence, some kind of shading device that allows the winter sun to enter but blocks the summer sun must be used.

Since the basic methods of constructing insulating curtains and panels have been outlined in Chapter 11, we'll now consider how to size solar windows to take as full advantage of the sun's heat as possible while minimizing overheating problems.

SIZE OF SOUTH-FACING GLASS AREAS

Overheating occurs because of what we call the "cloudiness factor." In each area, part of the sunlight is blocked by clouds and, at times, overcast weather. Hence, what happens is that if you design the size of your solar windows to just bring the house up to 70°F at noon on an average day during the winter months when, because of a partial overcast, you are not getting full sunlight, then in clear weather the house will overheat in the middle of the day. While this overheating isn't severe, your design must try to minimize this possibility while taking as full advantage of the sun as possible. Choosing the proper area for your solar windows requires several considerations. The first concern is the size and shape of the house. As a basic example, the area of windows on the south side of a square, two-story house with 800 square feet of floor space should be 60 square feet to make use of solar tempering. We can call this the "open-space house." (See Chapter 6.) Can we derive a formula to use with other house areas? Yes, we can. If your house is not the same size as the open-space house, you can still calculate the number of solar windows by comparing the heat-loss area of your own house with the heat-loss area of the open-space house. This will give you the square footage of solar windows you'll need. To find this figure, first divide the exterior square footage of your house by the 2,200 exterior square feet for the open-space house and multiply the result by 60. The 2,200 square feet for the open-space house was derived by measuring the square footage of the four outer walls (including the areas of windows and doors) and adding the area of the floor and roof. If your house has a floor of 20 X 40 feet (like the rectangular house in Chapter 6), its exterior area would then be 2,700 square feet. In this case, you would divide 2,200 into 2,700; this gives you about 1.2. Multiply the answer here by 60 to get a *rough estimate* of the square footage of solar windows for your house—in this case 1.2 × 60 = 72 square feet. Let's call the result of these calculations *alpha* (α) and set it aside for a moment.

Another factor that affects the area of windows is the degree to which the house has been winterized and insulated. The base number of 60 square feet of window area pertains to the open-space house where a full program of energy conservation has been put into effect. For this house, we derive the next factor by multiplying α by 1. This type of house would correspond to house C in Table 12-1 of Chapter 12. If, however, your house is of a more conventional construction (i.e., less well-insulated) or a partially insulated renovated house (like house B in Table 12-1), then multiply α—the base number for your house—by a factor of 1.5 to 2, depending upon the degree of tightness and insulation. Call this factor *beta* (β).

The final factors that affect the number of windows include the climate. Two variables come into play here: the amount of sunshine in an area and how cold the winters are. The first factor, sunshine, is measured by your weather station as the percentage of possible hours of sunshine that actually occur each month of the year. The range in most parts of North America is between 30 and 70 percent over the colder winter months of December, January, and February. To design your solar window area, contact your local weather station and ask for these "percentage of possible sunshine" figures for the winter months of December, January, and February. Then average them and use the multiplier below that applies to your area:

- If percentage of possible sunshine is between .3 and .45, multiply (β) by 1.3.
- If percentage of possible sunshine is between .45 and .55, multiply (β) by 1.
- If percentage of possible sunshine is between .55 and .7, multiply (β) by .65.

Call this factor *gamma* (γ).

The other climate factor to consider is how cold it is in your area during the winter. The people at the weather station measure this with a unit called *degree-days*. In most parts of North America, the total number of degree-days over the entire heating season ranges from 2,000 to 8,000. The composite base number (γ) applies only to an area with 5,000 degree-days. Call your local utility or oil company and ask how many degree-days there are during the average winter in your area. If these are more or less than 5,000, divide the degree-days for your area by 5,000 and multiply by (γ). This will give you a *final number* which we can call *delta* (δ). *It represents the area of glass you should have on the south side of your home.* Below I have summarized these calculations and have given an example:

1. Start with the base number of 60 square feet of solar window for a square, two-story, 800-square-foot house. The exterior surface area for this house is 2,200 square feet for outer walls (including window and door areas), floor, and roof. If your house is of a different size, then calculate the exterior surface area, divide it by 2,200, and multiply the result by 60.

   Call the result (α).

2. Then include the degree of winterizing-insulating:
   A. If all the energy conservation features we recommend in this book have been put to work in your house, multiply (α) by 1.
   B. If the house is more conventionally winterized-insulated (see Chapter 12), then multiply (α) by 1.5 to 2, depending upon the degree of tightness and insulation.

      Call the result of these second calculations (β).

3. Now add the effect of cloudiness in your area:
   A. If the percentage of possible sunshine over the winter months is between .3 and .45, multiply (β) by 1.3.
   B. If the percentage of possible sunshine over the winter months is between .45 and .55, multiply (β) by 1.
   C. If the percentage of possible sunshine over the winter months is between .55 and .7, multiply (β) by .65.

      Call the result of these calculations (γ).

4. Finally, add in the coldness factor:
   A. Find the number of degree-days for your area and divide this number by 5,000; then multiply the result by (γ).
   B. The final result, which we can call (δ), gives you the area of south-facing windows you will need.

   EXAMPLE: Suppose we have a one-story, square, 900-square-foot house that is well sealed and insulated, in an area with 70 percent of possible sunshine and 4,700 degree-days. The total exterior area of the house is 2,900 square feet. Then, (α) = 60 $\times \dfrac{2,900}{2,200}$ = 79 square feet. Since the house is well sealed and insulated, (β) = $\alpha \times 1$, and for 70 percent possible sunshine, (γ) = $\beta \times .65$ = 51 square feet. With 4,700 degree-days in the area, (δ) = $\gamma \times \dfrac{4,700}{5,000}$ = 48 square feet. Hence, the optimal area of south-facing windows for this house would be 48 square feet. If for construction purposes it would be easier to install another window area, say in the 45 to 55 square-foot range, this would still work well.

SHADING

Later in this chapter we'll explore the general characteristics of shading devices for the east and west sides of your home. Right now, let's concentrate on those particular shading devices which will work best with *south* windows. In this case, the best type of shading device is called an "overhang." The overhang is merely a horizontal projection which protrudes out from the house a certain distance. See Figure 16-1. The length of the overhang is important. Basically, it ought to be long enough to cast a shadow over the windows from 8 A.M. to 4 P.M. from May through August. The overhang of 4 feet marked in Figure 16-1 will cast a shadow long enough to keep any windows up to 6 feet beneath the overhang in shade about 85 to 90 percent of the time during the overheated summer season at a latitude of 40° N (New York City-New Jersey area). If you live farther north, say, latitude 48° (southern Canada), then the overhang can be longer, 4½ feet. On the other hand, in the deep South and Mexico, latitude 32°, the length of the overhang can be shortened to 3½ feet. If you live *in between* the latitudes mentioned, you can interpolate between these figures to find the length of the overhang for your area. For example, if you live at 44° N, then the length of the overhang is the basic 4 feet plus half of the

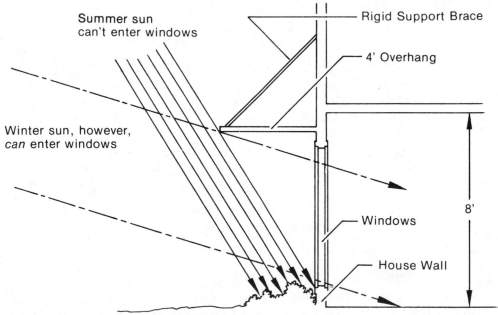

Summer sun
can't enter windows

Rigid Support Brace

4' Overhang

Winter sun, however,
can enter windows

8'

Windows

House Wall

Basic 4' overhang will shade windows up
to 6' below it at 40° N latitude

FIGURE 16-1

extra 6 inches that would be added to this by moving up to the 48° N latitude, or 4 feet 3 inches.

There are several other considerations about the design of the overhang that we should understand. Normally, an overhang used for shading is fixed in place, but as you will see later, an *adjustable* overhang has some very fine advantages. Figure 16-2 shows two ways of changing the effective length of the overhang by either sliding an outer section under the inner one or by flipping this outer portion back over the inner part. Another thing to think about is how the wind, rain, snow, and summer heat will affect the design of your shade overhang. Figure 16-3 shows the overhang sloped at about 3° for rain runoff. Note that a painted 2 X 4 frame with 3/8-inch exterior grade plywood has been used. The 2 X 4 frame will give adequate support if the span between rigid supports to the house wall is no more than 8 feet. In cases where there is not enough room for these rigid supports above the overhang (as may be the case with some one-story homes), use 4 X 4's running from the ground to the overhang as support. The rigid supports attached to the house should be able to hold 60 pounds per square foot from the overhang. In very heavy snow areas, a stronger support structure may be required, so check your codes.

TAKING ADVANTAGE OF DAILY AND SEASONAL CHANGES

Your south-facing windows are to be used in conjunction with a 4-foot overhang that has an adjustable outer section. We've discussed using the overhang to keep out unwanted summer sun. Now let's look at how we can use it for another purpose—to balance out the sun through the winter day. Most of the sun's heat comes in during the middle of the day. The reason for this is that the sun is usually most intense at zenith and the outside air has warmed up a bit. On the other hand, the early morning and late afternoon sun is less intense; it's generally colder outside at these times. Hence, we get a pattern of little sun in the morning and late afternoon when we need it most.

To complicate this imbalanced daily pattern, the amount of sun relative to the daytime heating needs varies over the winter. As mentioned above, the position of the sun in February is about the same as it is in October, but the average daytime heating needs in February can be 300 to 400 percent of what they were in October. The reason is that the cycle of the sun's position centers on December 21 when it's lowest in the sky, but the low point in outdoor temperature comes in late January because it takes the earth time to cool off in early winter and to heat up in early spring.

As intertwined as the above cycles are, it *is* possible to more or less balance out the daily and seasonal changes in the amount of sun coming through south-facing

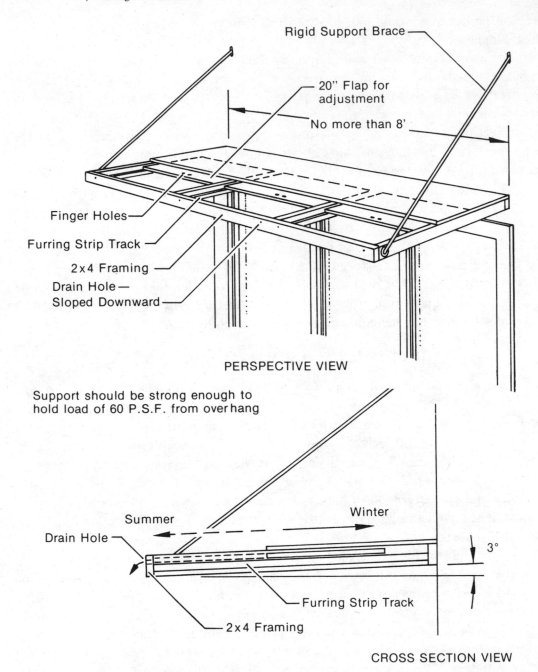

Rigid Support Brace

20" Flap for adjustment

No more than 8'

Finger Holes

Furring Strip Track

2 x 4 Framing

Drain Hole—
Sloped Downward

PERSPECTIVE VIEW

Support should be strong enough to
hold load of 60 P.S.F. from overhang

Summer   Winter

Drain Hole

3°

Furring Strip Track

2 x 4 Framing

CROSS SECTION VIEW

FIGURE 16-2a: ADJUSTABLE OVERHANG — SLIDING TRACK TYPE

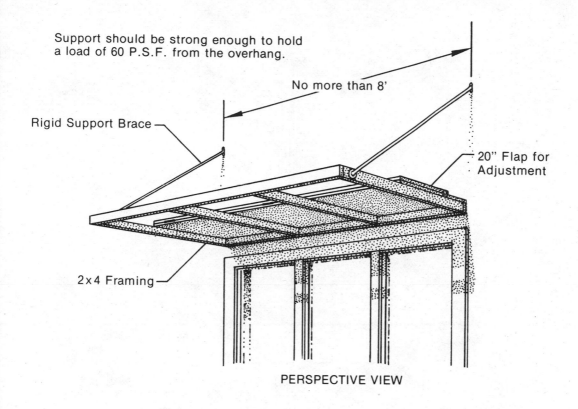

Support should be strong enough to hold a load of 60 P.S.F. from the overhang.

No more than 8'

Rigid Support Brace

20" Flap for Adjustment

2 x 4 Framing

PERSPECTIVE VIEW

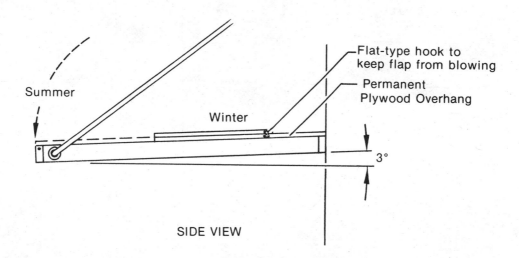

Summer

Winter

Flat-type hook to keep flap from blowing

Permanent Plywood Overhang

3°

SIDE VIEW

FIGURE 16-2b: ADJUSTABLE OVERHANG — HINGED FLAP TYPE

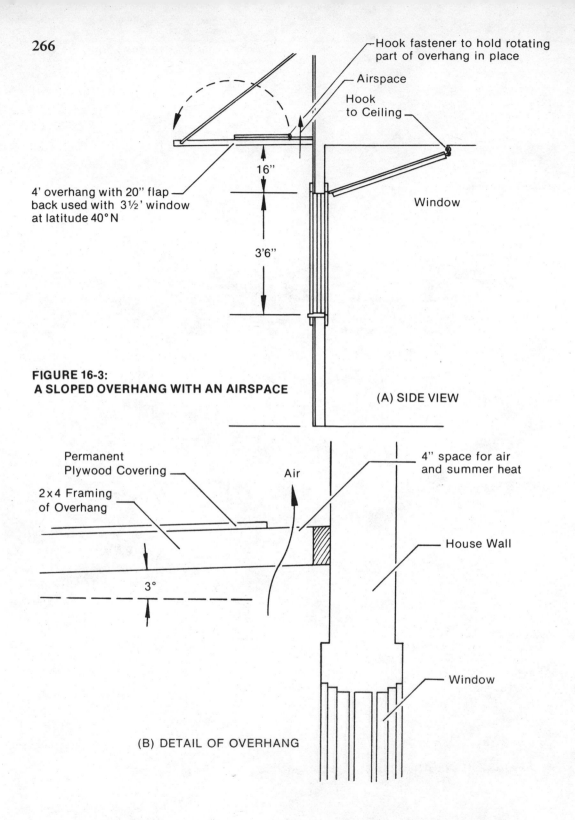

Hook fastener to hold rotating part of overhang in place

Airspace

Hook to Ceiling

Window

4' overhang with 20" flap back used with 3½' window at latitude 40° N

16"

3'6"

FIGURE 16-3:
A SLOPED OVERHANG WITH AN AIRSPACE

(A) SIDE VIEW

Permanent Plywood Covering

2 x 4 Framing of Overhang

Air

4" space for air and summer heat

House Wall

3°

Window

(B) DETAIL OF OVERHANG

windows. To do this involves adjusting the overhang during the winter season and positioning the windows on the south wall in the most advantageous way possible. Let's look at the overhang first.

Make one of the adjustable outer flaps mentioned in Figure 16-2 20 inches long. This value is based on designing a 4-foot overhang at a latitude of 40° N. That is, the flap is 40 percent of the length of the overhang. If the size of the overhang changes in different latitudes to 3½ or 4½ feet, the flap is, again, *40 percent of these lengths*, or 17 and 22 inches respectively. The idea is to move this adjustable part of the overhang back during the cold part of the winter to get more sun in the windows. Doing so will help to balance out the amount of sun so that it will roughly correspond to the heating needs during each part of the winter. When those needs are small, as on the average day in October, having the full overhang in place shades part of the windows so there is just enough heating at midday. With part of the overhang flipped or slid back in late winter, more sun is available during the colder February-March late winter season. All you have to do is move the overhang twice each year—once to open it in early January and once to close it for the summer in early April. As you gain some experience with your own climate, by all means go ahead and make these adjustments when they seem best to you.

The next area of balancing involves the location of the windows on the wall. Basically, you want to spread out the amount of heat coming through the windows so you get a more even heating pattern over the entire day. This is particularly important in early winter and early spring. At these times, the morning and often the late afternoon temperature is cold, while at noon it is warmer outside. This difference can be partially overcome by locating the windows so that in relation to the 4-foot overhang, there will be some shading at noon. Then the full sun hits the windows in early morning and late afternoon when needed, but is partially blocked out at midday. Hence, the amount of heat is more nearly balanced with the need for it.

At 40° N, place the tops of the windows about 16 inches under the overhang, and run the windows down 3½ feet below this starting point. See Figure 16-3. Since the length of the overhang varies between 3½ and 4½ feet, depending on the latitude, a 3½-foot window length would start 10 inches below the overhang at 48° N, and 21 inches below the overhang at 32° N. Again, if you live at a latitude that falls in between these numbers, interpolate to find the distance below the overhang where it is best to start the windows in your area. Note that starting these windows up near the ceiling accomplishes several other things. First, the windows can be used for a view to the outside, and the shade patterns from a 4-foot overhang (or the equivalent at your latitude) with the windows placed this high will keep the sun out of the house all summer. Also, since the windows are on the upper part of the wall, the insulating curtains or panels will be less likely to interfere with any furniture.

All the windows on the south side needn't be shaped exactly as indicated. Some windows could extend more than 3½ feet below the starting points outlined above. Figure 16-4 shows two different arrangements for solar windows on a 20-foot-long south wall. In Figure 16-4a you will see 60 square feet of windows placed to create more balanced heating over the day as described earlier. Note how the insulation panels easily hinge and can be raised up to the ceiling during the day. Figure 16-4b shows a solution with two windows placed up high to balance the daytime heating, and two others placed vertically to give the wall a somewhat different appearance. In this case, there won't be quite as even a balance of heat, but then the 4-foot overhang will still shade the vertical windows most of the summer. Note how a bookcase divider provides a backup for the attractive long open panels during the day.

You will have other south-side windows than your solar windows if your house is of more than one story. The overhang described above will shade only first-floor windows. The second-floor windows in this case would be best served with an awning. However, deciduous trees that will not substantially interfere with the winter sunlight can be used here also if you wish. Details can be found in Chapter 5.

CONSTRUCTING AND USING YOUR SOLAR WINDOWS

Constructing solar windows can be easy. For the average house to achieve 60 square feet of south windows only means adding or building in an extra 30 to 40 square feet of window space on this south side, as you'll probably have 20 to 30 square feet of windows in that area anyway. Put the extra glass on the first floor in a two-story house, as this will make the construction and seasonal adjustment of the overhang simpler. Make sure you use two sheets of glass on the solar windows, as it will cut the daytime heat loss in half. Also, the easiest and least expensive way to install the glass will be to fix it in place and use operating vents below, if ventilation is necessary. The information in Chapter 11 explains how this can be done and where you can get materials.

There are a couple of other things you might note about this method of solar window design. As was pointed out in Chapter 9, a window area of between 10 and 20 percent of the floor area in your home is optimal. This window area keeps heat losses in winter to a minimum and yet allows for a view to the outside and the use of natural light in your home. As was also pointed out in Chapter 9, few windows should be placed on the north side of your home, as little useful light is gained here. Hence, in our example of the "Open House" with 800 square feet of floor space,

Decorative Insulation Panels

PERSPECTIVE VIEW FROM HOME'S INTERIOR

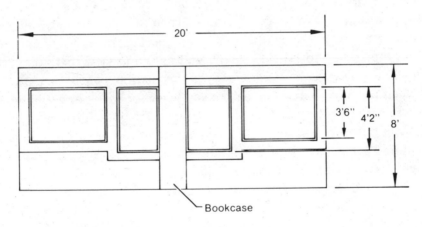

20'

3'6" 4'2" 8'

Bookcase

(B) Two horizontal and two vertical windows
broken up by center bookcase.

FRONT VIEW

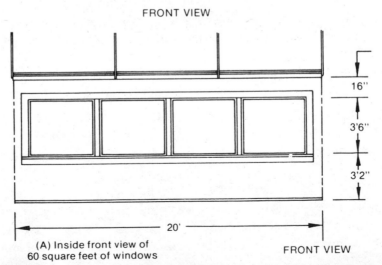

16"

3'6"

3'2"

20'

(A) Inside front view of
60 square feet of windows

FRONT VIEW

FIGURE 16-4: TWO ARRANGEMENTS FOR SOLAR WINDOWS

having a window area equal to 15 percent of this floor area gives 120 square feet of windows. Solar tempering then means reducing the 30 square feet of this 120 total that would normally go on the north wall and moving it to the south side of the house. This brings the total on the south side up to the 60 square feet we have recommended for this house. Also, keep in mind that the calculation procedure outlined in this chapter is approximate. If you find that a small change in size or spacing of windows or overhang is easier or less expensive to work with, by all means adjust the designs a bit to suit your own circumstances.

The operation of solar windows is very easy. Just open the insulation panels or curtains in the morning, then close them when the sun sets. Using two sheets of fixed glass cuts the heat loss through the windows to such a low value that *you will gain heat on almost all winter days!* The only time you might want to keep the panels closed would be on very overcast days when the outside temperature is below 20°F. In these cases, just open enough of the insulation panels or curtains on the windows in your house to provide the natural light and view that you need. By the way, there is no need to be a slave to your insulation panels or curtains. If you forget to open them some morning before you leave for work, or get home late some evening so that the panels are open a few hours after dusk, little will be lost. Once you spend a month or so opening and closing the panels or curtains, letting the sun into your home will become a pleasant habit.

Though we've designed the solar windows to eliminate much of the overheating problem, there will still be a few very clear days each month when this overheating will occur around midday. In areas where the percentage of possible sunshine ranges between .45 and .7, this will mean that the midday temperature in the house may go as high as 75°F about five or six days each winter month. In cloudier areas, because of your extra windows, the temperature may even climb to 75° to 80°F at midday about 10 days each month. Adjusting to this is easy, and in fact many people find these slightly higher temperatures relaxing. On winter mornings, you may want to use a sweater and keep the house in the 65° to 68°F range. Then as more and more sun enters your home in the morning and the temperature goes up to 70°F, take your sweater off. If you go to work, just open all the panels or curtains, as the temporary rise in temperature at midday won't hurt anything. Make sure to install the floor-to-floor vents mentioned in Chapter 13 to ensure that the heat circulates throughout the house. Also, the open, flowing floor layouts described in Chapter 6 will work to your advantage here.

How much will these windows cost? For the average well-insulated house, an added 40 square feet of solar windows, to bring the total up to 60 square feet, should cost no more than $300 (about $7.50 per square foot), including the cost of insulation panels and overhang. This assumes that the glass is double paned and fixed and that you purchase the materials and do the construction yourself. The cost of prefitted insulating glass could double this figure. One point to note is that if your house were not tightly constructed and you needed almost double the area of solar windows, the extra cost would be some $400 more. This money would be better spent on insulating and winterizing, even if you call a professional in to do the work.

Now, how much of the fuel bill will these windows save? Using the design procedures we've outlined, the net heat gain of the solar windows over the same area of an insulated wall would save about 20 percent of your heating bill. Actually, since there will be mornings when you sleep late or days when you're away and don't open the panels or curtains, the savings will more conservatively be about 15 percent. The cost to the average homeowner being about $300, solar windows would save 15 percent of a $200 fuel bill. (If a $200 fuel bill seems low, remember we've assumed the house is *very* tightly constructed.) Thus, solar windows are about a 10-year investment. In cold, cloudy areas, the money spent would be more and in warmer climates less; but the time needed to return your investment would remain about the same. Note that I haven't included bank interest in the 10-year payback period for the solar windows. Since the price of fuel is still increasing, investment in solar windows, with the resulting lower fuel bills and greater comfort, actually offers a better return for your money than any interest you can accrue at the bank.

EXPANDING ON SOUTH WINDOW HEATING

In many cases, it's very easy to use masonry floors and walls in your home to absorb some of the sun's heat. This can be an attractive solution that not only helps spread out the heat from the sun over the day, but will allow you to increase the number of south windows in your home so that a greater percentage of your total winter heating load can be supplied from the sun.

The best way to do this is to make use of masonry structural elements that would be in the house anyway. In this case, once you install the solar windows as described, the cost of using these areas is little more than the time spent to ensure that they serve both structural and heat-absorbing functions. On concrete floors, just arrange any carpeting you may have so that the sun will strike uncarpeted areas. See Figure 16-5. The open floor plans shown in Chapter 6 will easily allow the sun to penetrate to floor level and be absorbed by a slab on grade on the first floor. In this case, the same kind of overhang and window arrangements described before will work well.

If you don't have a concrete floor but do have a masonry wall at the back of your south room, you can still store solar heat. In this case, the sun will strike the wall in the early morning and late afternoon. If you have *both* a concrete slab and a

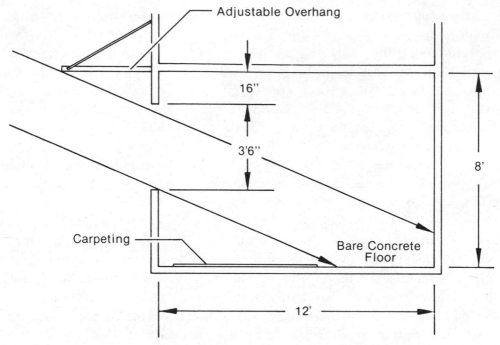

Adjust carpeting so sun can strike concrete floor.

FIGURE 16-5

stone or block wall, you can actually raise the percentage of solar heating of your home from 15 percent (for windows alone) to about 20 to 25 percent with some additional windows and the masonry floor and wall acting to store heat. When you use the block or stone wall, however, make sure to construct the usual 4-foot overhang (or equivalent for your area), but in this case with an outer 3-foot-4-inch lip (or 85 percent of the width of the overhang) that can be flipped back all winter so the sun can strike as large a portion of the masonry wall as possible. Also, make sure to raise the tops of the windows closer to the overhang. A spacing between the bottom of the overhang and the top of the windows of 6 inches would be good, for it allows more light to reach the block or stone wall at the back of the room. While changing the length of the overhang and the position of the windows would normally create an imbalance in the daytime and seasonal heating patterns as we outlined before, the extra heat absorbed by *both* the concrete floor and masonry wall will rebalance things. Note also that in this case a sliding overhang cannot be used because the fixed part of the overhang here is only the 8 inches necessary for an attachment to the house and a wind and summer heat vent. If the 3-foot-4-inch adjustable lip interferes with second-floor windows on a two-story house or the roof overhang on a one-story house when rotated up and back, just unscrew it from the overhang and store it for the winter.

With either a masonry floor in an open house layout or a masonry floor in the south room in conjunction with a masonry wall at the back of that room, you can even *expand* the number of south windows without overheating. Take the factor (δ) from the previous calculations for south windows and multiply it by 1.1 to get the total area of windows for the case where you use only the masonry floor, or multiply it by 1.2 if you use both a masonry floor and wall as heat absorbent surfaces.

There are other ways to expand even further on the use of the sun's heat and achieve esthetic effects as well. If you wish to, you can increase the area of the solar windows and build a block or stone wall within about 4 or 5 feet of the glass. With the overhang constructed so that virtually the entire basic 4-foot length can be flipped back in winter (a 3-foot-4-inch outer lip), a wall this close to the windows will have nearly its whole area receiving sun during most of the day. In this case, you would want to fill the cores of masonry blocks, if these are used for construction, to increase the heat capacity of the wall; or you could use fieldstone, or fieldstone as a veneer on a 12-inch poured concrete wall. What happens is that about 50 percent of the sun's radiant energy coming into the house is absorbed by such a thermal wall and is retained for the evening hours. The other 50 percent spreads out via floor-to-floor vents or openings in the wall to heat the house during the day. The result is that you can get about 30 to 35 percent of your heating needs from the sun!

For a wall like this, take the window factor (δ) which you originally calculated and multiply by 1.7 to get the area of windows you need. To add these windows, it would be best to arrange them running about 6 feet down the wall from 6 inches under the overhang, to ensure that the thermal wall gets full sunlight. Also, since the glass area will now be nearly twice as large as before, the best insulating material for the nighttime would be a large version of the lightweight movable curtain described in Chapter 11. Figure 16-6 shows some of these design considerations.

The construction of the wall can be a lot of fun, and decorative as well. The fieldstone discussed in Chapter 5, or 12-inch-thick concrete blocks (cores filled with cement), will work nicely, and both are exciting family projects. Make sure to check with your codes, though, to be sure you have enough extra support under masonry or wood floors. You *must* add whatever supports may be necessary so that the floor can carry the extra load.

How about the color of the masonry walls and floors that were described in this section? In the case of a heat-absorbent concrete floor or masonry wall at the back

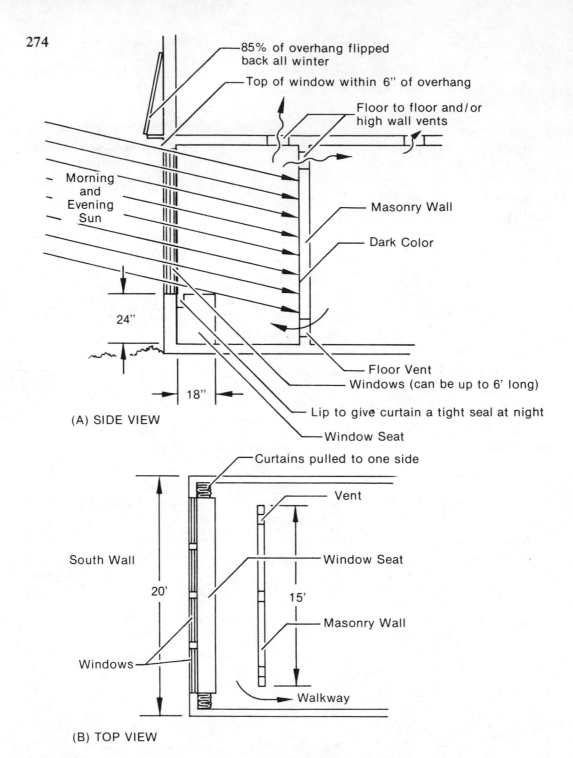

85% of overhang flipped
back all winter

Top of window within 6" of overhang

Floor to floor and/or
high wall vents

Morning
and
Evening
Sun

Masonry Wall

Dark Color

24"

18"

Floor Vent

Windows (can be up to 6' long)

Lip to give curtain a tight seal at night

Window Seat

(A) SIDE VIEW

Curtains pulled to one side

Vent

South Wall

Window Seat

20'

15'

Masonry Wall

Windows

Walkway

(B) TOP VIEW

102 sq. ft. of windows placed along a 20' wall
This is the basic 60 sq. ft. x 1.7 (multiplier for this situation).

FIGURE 16-6

of the room, it's not too important. Try to stay away from light colors, but it doesn't have to be flat black either. The medium gray of concrete blocks or the rust color of brick will absorb half the light that strikes them, and whatever is reflected will bounce around the house until most of it is absorbed. Also, adding a dark colored dye to the mortar in a block or fieldstone wall will increase the amount of light absorbed and makes a nice color combination as well. In the case of the masonry wall within 4 or 5 feet of the glass, however, it's more important that the wall absorb as much of the sunlight as possible. A flat black is actually best, but darker colors like an evergreen or maroon or a dark, nonmetallic-looking fieldstone will work nearly as well. If you do paint the wall, use a flat color and try to get a high-temperature paint. Actually, the skin temperature of the wall on the south side will never go very high (at most in the 150° to 200°F range), but a high-temperature paint will avoid any peeling and odor problems that might occur with an ordinary house paint. Check with a large paint store for the properties of different paints.

Some very exciting and decorative things can be done with the light and shadow patterns and the area around one of these thermal walls. First, you could add a strip of sheet glass or "glass blocks" at the top, about 1 foot in height, which will serve to bring natural light behind the wall. Another pleasing effect that will bring in natural light and create some patterns is the use of spaces in the wall itself. A few holes can be left open, filled with glass blocks (these come in a variety of sizes and shapes, and range from clear to nearly opaque—check with your building supplier), or you could make your own stained glass or painted glass inserts for these spaces. Just think of the endless options for light, color, and patterns that can add to the appearance of the wall from both the inside and outside of your home! The 4- or 5-foot area between the glass and the wall can also become a creative solar living area with many exciting possibilities. Plants could be placed in this area with plenty of room left over for a walkway and some chairs. In addition, you could build an 18-inch-wide by 24-inch-high continuous window seat. The area of the window seat next to the floor could serve as a bookcase, and the flat top can be used for plants or a small deck for sun bathing in winter. See Figure 16-7.

There is one other option you may want to use. If you build a home on a slope with the south-facing basement wall fully exposed to the outside as suggested in Chapters 5, 12, and 19, you can use the floor and north walls as heat reservoirs by building solar windows into the south basement wall and placing the insulation on the *outside* of the east, west, and north basement walls. In this case, the sun striking the inside lower part of the north wall will heat it. Another way to take even better advantage of this feature is to build the 4- or 5-foot creative living space we outlined above in the basement, and add to it a horizontal duct with vents at the top and just

FIGURE 16-7

in front of the south side of the thermal wall. The duct exits at the center of the wall with more ducting that goes through the wall and then down to the floor to a distribution duct. These ducts run through the floor and up the north walls to outlets. See Figure 16-8a. In this case, you would build an overhang so that virtually the whole cover could be flipped back for the entire winter, and multiply the window factor (δ) by 2. On warm days, particularly in late fall and early spring, any excess heat can be drawn from the hot area near the top of the masonry wall into the ducting system. Circulating it in the ducts through the floor and north wall will cool the air down while heating the floor slab and north wall. Moreover, the concrete in the east, west, and north walls and in the floor will cool down on summer nights and keep your house comfortable on hot days. The key here is to provide roughly 150 to 200 pounds of concrete for each square foot of solar windows. The combined mass of the thermal wall, the floor, and the north wall should be adequate to do this in most cases.

The north wall in this case can be either concrete blocks where some of the cores are aligned to serve as ducts, or it could be a poured concrete wall where metal ducts are added into the concrete forms beforehand. A good arrangement would be to use an axial fan placed in the ducting on the north side of the south masonry wall to pull air off the ceiling and push it down into the distribution header under the floor. See Figure 16-8b. Also, remember that you don't want to circulate too much air, otherwise the house will become drafty. If you hook the fan to a remote house thermostat set at 75°F (use a fan that can circulate 200 to 400 c.f.m. against the static pressures mentioned below for a house in the 600 to 1,200 square foot range), the air will circulate about three times each hour when the system is in operation, at an air temperature that is optimum for this heat storage. The pressure drop of the fan should be roughly about .1 to .3 inches of water gauge, depending on the house size and length of ducts. Check with your heating supplier for more details on equipment choice and operation. Though the cost of the fans and ducting will have to be added to the cost of the masonry wall and windows, this system can provide up to 40 percent of your heating needs.

How much do these systems cost? In the case of a slab on grade, the only cost is that of the windows. With 10 percent more windows added to the average house, which originally used 60 square feet of solar windows, the extra cost at $7.50 per square foot would be about $50 higher than the $300 stated before, or $350. With 20 percent more windows added to a house in order to use a concrete floor slab and a masonry wall on the north side of the south room, the total cost would go up to $400. For those without such heat-absorbent floors and walls, adding the wall for the creative solar living area will cost more but increases the percentage of heat obtained from the sun and provides other pleasurable benefits as well. The cost

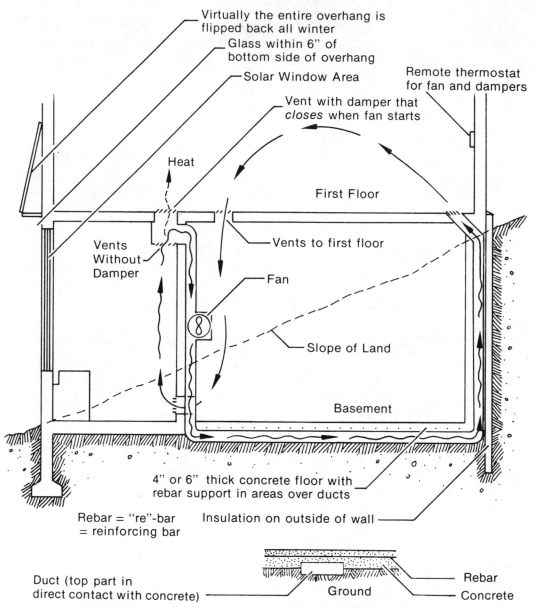

Virtually the entire overhang is flipped back all winter

Glass within 6" of bottom side of overhang

Solar Window Area

Remote thermostat for fan and dampers

Vent with damper that *closes* when fan starts

Heat

First Floor

Vents Without Damper

Vents to first floor

Fan

Slope of Land

Basement

4" or 6" thick concrete floor with rebar support in areas over ducts

Rebar = "re"-bar = reinforcing bar

Insulation on outside of wall

Duct (top part in direct contact with concrete)

Ground

Rebar

Concrete

FIGURE 16-8a: HEAT STORAGE SYSTEM FOR A HOUSE ON A SLOPE

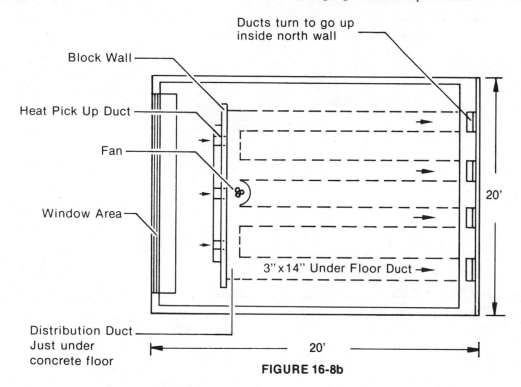

Ducts turn to go up
inside north wall

Block Wall

Heat Pick Up Duct

Fan

Window Area

20'

3"x14" Under Floor Duct →

Distribution Duct
Just under
concrete floor

20'

FIGURE 16-8b

would be about $250 for an 8-foot-high and 15-foot-long wall, including supports (and would be much cheaper if fieldstone were used). To this must be added the cost of 70 percent more windows, or a total of about $850. If your home is being built on a slope and you want to take advantage of the winter heating and summer cooling potentials of walls and floors, you would have to include the cost of doubling the number of windows, the cost of the thermal wall in the solar living area, and the cost of fans and ducting. In this case the total cost would be in the $1,300 range. Note that in the case where the solar living area is used, the payback period is now around 15 years. However, the summer cooling benefits, the relaxing presence of the winter sun in this area, and the unusual beauty of your home are additional benefits.

Is $300 to $1,300 a lot of money? It certainly can be, but think about the realities of the present-day energy situation. For the average house where the savings on the fuel bills alone can be more than $150 a year by using only the energy conservation techniques we've outlined, you'll have this extra money. Investing it in one of these low-cost and effective solar energy systems will save you even more money. With all the long-range benefits to be gained, it certainly makes sense.

Table 16-1: *Summary of Design Factors When Windows Are*
Used in Conjunction with Masonry Floors and Windows

Situation	Length and position of overhang and windows	Size of windows Multiply δ by:
(a) slab on grade	Same as with no masonry storage features	1.1
(b) north masonry wall	Use an overhang where 85% of it can be flipped back over the winter and start windows 6 inches below the overhang	1.1
(c) both a slab and a north masonry wall	Same as in (b)	1.2
(d) creative solar living area	Same as in (b) but now the length of windows increases to 5 feet or more instead of the standard 3½ feet	1.7
(e) creative solar living area with fan and duct system	Same as in (d)	2.0

USING THE SUN THROUGH NON-SOUTH-FACING WINDOWS

Houses often do not have long walls facing south. If your house faces only a little to the southeast or southwest, say 10 to 15°, the same size and arrangement of overhang and windows that have been discussed would apply. The only difference would be in the time of day when most of the sun would enter the house—earlier for the southeast orientation and later for a southwest orientation. Moving beyond 10 to 15° east or west of south becomes more difficult. While it's possible to split up the windows with an orientation of, say, 30° to even out the sunlight entering the house over the day, the overhang in these cases no longer provides adequate shade protection in summer. We'll deal with the shading problem later, but for now let's look at how other non-south-facing windows can be used to gain solar heat in the winter. Figures 16-9a and 16-9b show how the windows might be split up for a southwestern orientation or for a direct south orientation but where (because of evergreen shading, the addition of a greenhouse, or other reasons) the entire south wall can't be used for solar heat. Note how an L-shaped solar wall can be added to provide for heat storage. In these cases the same principles of construction (except for the overhang) explained before would apply. Since the overhang would not generally be used in these cases, except for that part of the L facing directly south where there are no obstructions to the sun, other shading devices for these east, west, southeast, and southwest orientations will be explained later.

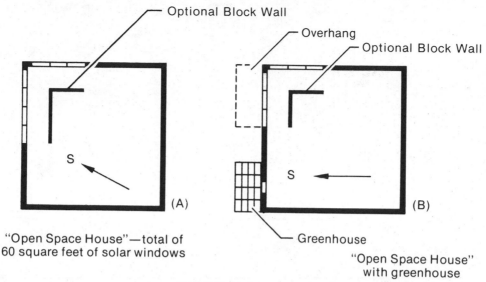

"Open Space House"—total of
60 square feet of solar windows

"Open Space House"
with greenhouse

FIGURE 16-9

Another alternative, created by Charles Marsh for his home in New Jersey, can be used either where the orientation is greater than 15° east or west of south or where concrete walls in renovation work make it difficult to install added solar windows. This solar heater operates by drawing cold air off the floor through low vents, warming it in the collector and letting it rise to flow into the house through other, high vents. There are several nice features to this design. You can build it easily and inexpensively, and you can start small and add more units later. See Figure 16-10.

To make one of these heaters, first try to get some old storm windows. This way the glass won't cost much. Once you have the windows, keep them in their frames. Plan the size of the wall heater so that the window frames will just cover the front of the heater. Now, clean that part of the house wall which the heater will cover. Screw 2 X 4's flat on the wall to make a box around this space. If you have a brick or concrete wall, attach the 2 X 4's with masonry anchors. Then paint the wall inside the 2 X 4's with flat black paint.

Next, build a box butting up against the outside of the 2 X 4's. Use 1 X 6 lumber. At the edges where it meets the house seal with caulking to close any cracks, and put some concrete blocks or bricks below the heater to help hold up its weight. Then make a box of ¼-inch plywood to fit *inside* the 2 X 4's (it should be only 4 inches deep). Place 3½ inches of fiberglass insulation between the two boxes and cover the top with 1-inch lumber. Nail this cover in place through the outer side of the large box. See Figure 16-10. Make vent holes in the house wall near the top and

DETAIL OF STORM WINDOW

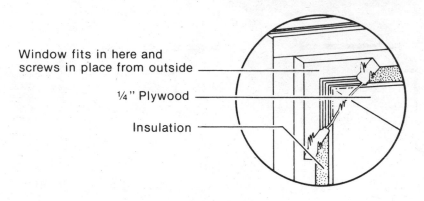

Window fits in here and screws in place from outside

¼" Plywood

Insulation

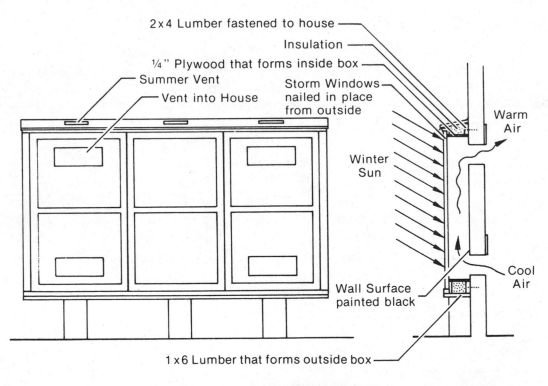

2 x 4 Lumber fastened to house

Insulation

¼" Plywood that forms inside box

Summer Vent

Storm Windows nailed in place from outside

Vent into House

Warm Air

Winter Sun

Wall Surface painted black

Cool Air

1 x 6 Lumber that forms outside box

FIGURE 16-10 : SOLAR WALL HEATER

bottom of the heater so air can come in and out. If your heater is about 3 feet wide, a hole at the top and another at the bottom, about 4 inches deep by 14 inches wide, will work fine. If it is 6 feet wide, you will need two sets of holes spaced about 3 feet apart.

Now, put the storm windows over the front of the box. If you can get enough of these, make double panes for greater efficiency. Use spacers where the storm windows meet and screw them in place. Caulk all around the edges, paint the outside of the heater with housepaint, and you're finished! You'll also want to make the vent flaps for the inside of the house, described in Chapter 11, to prevent house air from getting into the heater at night and cooling off. Also, when the snows come, if you shovel some around the front of the heater, it will reflect sun into it. When summer arrives, you can either tack an inexpensive bamboo screen over the collector to help keep out the sun (while closing off both top and bottom house vents), or you can build vent flaps to the outside on the top of the unit. These flaps needn't be large (one 4 X 10-inch flap on a 3-foot heater will work fine); and by closing the upper house vent and leaving the bottom one open, the collector will vent itself by drawing air from the house and out the top of the collector. In this way a collector which includes vent flaps on top can even serve to ventilate the house in summer.

SHADING DEVICES FOR NON-SOUTH-FACING WINDOWS

The design of the adjustable overhang that was explained earlier will work very well with a direct south orientation or with an orientation that deviates by only 10 or 15° east or west of south. However, another aspect of solar design that you'll want to be aware of is the summer shading possibilities for the windows on the other walls of your home. Also, these same design features will apply to a home where the orientation is greater than 15° east or west of south. Let's look first at windows with an orientation facing directly east.

During the summer season in North America, the sun rises over the horizon in the east-northeast compass quadrant. On warm days and between 8 A.M. and 10 A.M., there will be times when you'll want to shade these windows. There are several possibilities. First, you can use a shade tree close to the house (see Chapter 5 for more details). The easiest way to position the tree is to watch the angle the sun makes on your site on a day in late June. With a piece of cardboard held vertically and aimed at the sun so that it doesn't cast a shadow, measure the angles the sun makes between 8 A.M. and 10 A.M., and then place the tree as outlined in Chapter 5 so that the crown will throw shade on the east windows. Another very fine way to shade these east windows is by using the traditional awning that slopes down from above and the sides to create complete shade. However, in this case your view to the outside is blocked all day unless the awning can be tied up once the sun has moved away. Also, the area between the window and the upper part of the awning tends to collect heat, so leave a small air space or sew a strip of screening material onto the top part of canvas awnings.

Moving indoors, you can reduce the effect of sun coming in your east windows by keeping your insulating panels or curtains closed on these windows until the sun moves to the south side of the house. A lighter-colored curtain material facing the outside will help to reflect the sun's rays back out, or aluminum foil can be placed on the insulating panels to do the same thing. Also, opening your windows at the top slightly will allow any heat resulting from that part of the sun's radiation which was not reflected back out the window to escape outdoors again. In the case of insulating panels used with fixed glass, you may not want to seal the panel closed completely so that there will be some circulation of air between the panel and window to prevent excessive heat build-up. Also, a standard light-colored venetian blind can be used. However, in the case of fixed windows, neither the curtain nor the panel nor the venetian blind will stop more than 50 percent of the sun's heat from entering the house. In this case, the awning or shade tree solutions would be best.

On the west side of the house the overheating pattern in summer is most severe. Here, the intense early and late afternoon sun can make the interior really uncomfortable unless something can be done to prevent this light from getting into the house. This overheating will be particularly severe if a substantial portion of the west wall has windows used for solar heating. Several options exist, besides those mentioned for east windows, for creating shade on the west side of your home. The first would be to use deciduous trees, evergreens, or shrubs on the southwest side (to block the early afternoon sun) and on the west side (to block the late afternoon sun). The examples in Chapter 5 will help you do this. Note that a dense growth of evergreens or shrubs should be used *only* where a windbreak and shading are more important than westerly summer breezes or winter sun. In the case where these breezes are blocked, try to take advantage of the cool air and the enclosed private area that the shrubbery or evergreens provide by building a patio on the west side and use vents and windows in other parts of the house for summer cooling and winter heat.

It is possible to have both privacy and a summer breeze on the west side of your home if you want, and these shading solutions can be elegant as well as practical. Figure 16-11a shows a combination shading device where slanted fins near the house provide shade in the early afternoon and the open block wall shades later in the day. A two-story version of this is shown in Figure 16-11b. It uses a deck to shade the lower portion and bamboo screens to block the sun in the second-floor area.

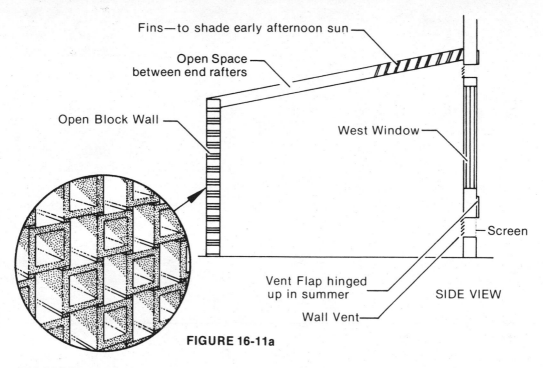

Fins—to shade early afternoon sun

Open Space between end rafters

Open Block Wall

West Window

Screen

Vent Flap hinged up in summer

Wall Vent

SIDE VIEW

FIGURE 16-11a

PERSPECTIVE VIEW

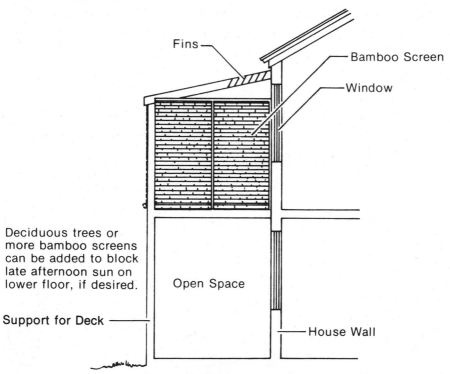

Fins

Bamboo Screen

Window

Deciduous trees or more bamboo screens can be added to block late afternoon sun on lower floor, if desired.

Open Space

Support for Deck

House Wall

FIGURE 16-11b

For orientations pointing southeast or southwest, a combination of the techniques just described will have to be used. On the southwest, for example, the shading device must provide protection not only from the early and late afternoon sun but from the midday sun as well. In the same way, on the southeast side the early and midmorning sun as well as the midday sun will have to be deflected. If you use windows on either the southeast or southwest wall for solar heating, you will want to make sure that the winter sun can reach these windows past any shading devices used on these sides of the house. Here, deciduous trees that can block the noontime summer sun will work well. Another alternative would be to use the same shading techniques just described for the west side but with an adjustable shading device that blocks the sun in the south and west directions. Figure 16-12 shows the same basic layout as before but in this case with a removable lip on the overhang that runs past the permanent finned part, and roll-up bamboo screens that can be raised or lowered at different times of day to block the sun and yet allow indirect light into a patio area. This same technique could be used to block the summer sun from solar windows or a patio area on the southeast side of the house. For individual windows where an outdoor deck or patio is not desired, an awning or deciduous trees would work best for these orientations.

A good way to learn about the effects of any design is to build a scale model, a ½ inch to 1 foot, and place it out in the sun to watch the changes in the shadows

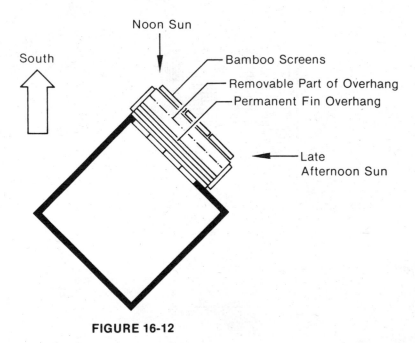

FIGURE 16-12

over a summer day. You can even simulate different types and spacings of trees; and by observing the shade patterns every few hours over a summer day, you can learn many things about just which design will work best for you.

The study of solar energy is an art as well as a fascinating science, and there are many other low-cost and effective designs that you can use to bring the benefits of the sun into your home. Once you are aware of the basic opportunities the sun can provide for winter heating, you can move on to explore other dimensions of bringing the sun into your life. The next two chapters will explain the principles of green-house construction and solar water heating. For further reading material, see the references for this and the next two chapters.

chapter 17

THE GREENHOUSE AS A SOURCE OF FOOD AND WINTER HEAT

by RON ALWARD

The greenhouse is essentially a trap for solar energy. It allows most of the radiation from the sun and sky to enter through the transparent cover, and it retains inside the structure much of the heat generated by this radiation. How it does this is fairly simple. Glass, and most transparent plastics, allow the short-wave radiation that comes from the sun and sky to pass through to the inside of the greenhouse, where it is absorbed by the plants, soil, benches, floor, and structural members and is converted to heat. These surfaces warm up and in turn re-radiate long-wave radiation, most of which cannot pass through the glass and some of the plastic coverings normally used for greenhouses. This heat, then, is trapped inside, causing the interior to warm up considerably above the ambient temperature. On sunny days, the interior temperature can become unbearably warm, even in winter. On cloudy days, when there is no direct sun and the only available energy is the diffuse radiation coming from the cloudy sky, then the greenhouse will become only slightly warmer than the outside.

The direct and diffuse radiations coming from the sun and sky play a second role in the greenhouse. They supply the light energy that is necessary for plants to grow. Visible light energy of certain wavelengths is required for plants to carry on the process known as photosynthesis. Essentially, this is a process in which light energy triggers a chemical reaction (in the presence of water and carbon dioxide) to produce carbohydrates in living plants. These carbohydrates are essential to the growth of the plant.

So, two vital ingredients are obtained from the sun (or the sky) in greenhouses—energy to provide an environment warm enough for plants to grow and energy to trigger the chemical reactions that permit growth.

288

WHY A GREENHOUSE

If you are an urban or rural dweller or fall somewhere in between, a greenhouse could assist you in these four ways:

1. It could provide a source of inexpensive, good-quality food that you grow yourself.
2. It could be a source of additional heat for your house on cold winter days.
3. It could help to moderate the humidity in your house during those cold, dry winter months.
4. It could provide a summer- or springlike environment on a year-round basis.

WHERE TO BUILD A GREENHOUSE

A greenhouse does not have to be a separate structure sitting all by itself out in a field. It can just as easily be a part of your house. In fact, it is *much cheaper* to build a greenhouse as an addition to an existing house than it is to put one up separately, say, in the backyard. If built in the right location with respect to existing windows and doors, it can be a very good heat source for the house on those cold, but sunny, winter days.

In this section, we are going to be concerned only with greenhouse structures that are extensions to existing buildings. This is not to imply that the separate units are unimportant, but the assumption is being made that most homeowners reading this would like the maximum return for only a moderate investment in time, energy, and money. We are not going to tell you where to build your greenhouse because that decision is ultimately your own. Instead, some possibilities and preferences will be indicated.

Here in North America, the sun is always to the south of us during the season when the greenhouse is needed, that is, during the fall, winter, and spring months. It is important, then, to have a good southern exposure for the greenhouse. The best location is against a south-facing wall; however, southeast- and southwest-facing walls are almost as good.

Most people will probably consider constructing a greenhouse at ground level, building in the appropriate foundations for wall and roof supports. However, it is equally as possible to convert a south-facing balcony or porch to a greenhouse. Other possibilities include flat shed or garage roofs, or even south-facing rooms of the house. See Figure 17-1.

Choose an area that is well drained and is not going to be shaded by trees or other buildings. If there are house windows or a door on the wall where you want to put the greenhouse, so much the better. These can provide immediate access to the

FIGURE 17-1a: A GREENHOUSE ATTACHED TO SOUTH SIDE OF HOUSE

FIGURE 17-1b: A GREENHOUSE AS A BALCONY ON SOUTHERN SIDE

FIGURE 17-1c: A GREENHOUSE OVER A GARAGE ROOF
(Check with codes to see if size and spacing of roof rafters
are sufficient to carry extra load.)

greenhouse from the living area. In addition, they can also be ventilation ports acting at the same time to cool the greenhouse and heat the house in winter.

HOW TO BUILD A GREENHOUSE

In this section, we will deal with some of the specific requirements for the environmental design of greenhouses before getting into the materials needed for the various components and the methods of construction.

If we look back into the history of greenhouses, we see that the basic form of today's large, open structures was originally developed and perfected for use in the lowlands of Europe. There, the relatively mild, overcast climate, particularly during the winter season, dictated a structure that would permit maximum penetration of the low intensity, diffuse sky radiation. A comparison of climatic conditions between the lowlands in Europe and most localities in North America reveals some very great differences that must be accounted for in the design of greenhouses on this continent. For one thing, most populated areas in North America, even in the colder regions, receive more direct solar radiation during the winter season than do the areas of greenhouse concentration in Europe. At the same time, however, winter temperatures, particularly in the northern states and Canada, can be much lower.

Large, transparent greenhouse areas unfortunately permit extensive heat losses during cold winter nights and overcast days. At the same time, the north transparent walls and roof of a conventional greenhouse permit very little of the total solar radiation to reach the interior. It seems somewhat obvious, then, that we should take account of all these factors in a workable structure that can reduce the amount of heat loss and yet retain a satisfactory radiation intensity to permit plant warmth and growth.

One of the objectives in greenhouse construction is to build a structure that will permit the maximum amount of solar radiation to enter and stay and also allow the maximum visible radiation to strike the plant canopy. In addition, a uniform distribution of solar radiation is desirable so that all plants can receive their fair share of sunlight.

If we take all these factors into consideration, the construction of a greenhouse off the south side of an existing house offers some considerable advantages. First of all, it eliminates most of the heat loss that normally occurs when the north wall is transparent. This wall has been replaced by one of the insulated walls of the house. Of course, this insulated wall also blocks out some of the sky radiation from the north, but it is only a small fraction of the total radiation in most regions. In order to bring light in on the north side of the plants, it is necessary only to paint or cover the section of the house wall inside the greenhouse with a good reflecting paint or other material. Glossy white or aluminum paints are very good reflectors of solar radiation. Entering through the south transparent wall and roof area, radiation will now not only fall on the southern portion of the plants but will be reflected off the north wall and strike the north side of the plant foliage as well.

In addition to its insulative value in cutting down on heat losses from the greenhouse, the house wall is an effective windbreak. Convective heat losses from the greenhouse, caused by cold winds from the northwest, north, and northeast are significantly reduced.

Another advantage of building a greenhouse against an existing house wall is, of course, the saving in cost of materials. A lean-to type of structure, which is easily supported by the south wall of the house, can be built. See Figure 17-2.

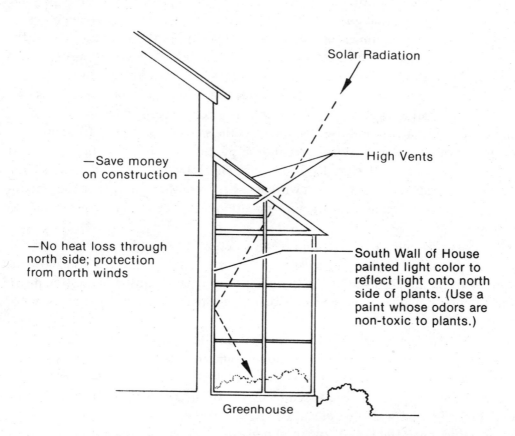

FIGURE 17-2: ADVANTAGES OF A GREENHOUSE BUILT ONTO A SOUTH SIDE WALL

COVERING MATERIALS

There are a number of options for the transparent cover of your greenhouse, depending on the amount of money you want to spend and whether a temporary or permanent greenhouse is being considered. We will go briefly through a few of the advantages and disadvantages of some of these materials.

Glass

This has been the traditional greenhouse transparent cover. It is one of the best transparent materials for solar radiation, and it allows very little long-wave heat radiation to escape. It is expensive, however, unless you can find a good second-hand supply of glass from other greenhouses or buildings being demolished. It is also a good idea to check with glass factories for any reject glass. Window and door manufacturers often have slightly broken pieces and glass with minor flaws. All of these can save you considerable amounts of money. Usually only one layer of glass is used, although two layers would eliminate a lot of convective and conductive heat losses. The structure and framing necessary have to be strong enough to support the weight of glass, which is heavy, and the wind and snow loads. The glass framing has to be fairly close since glass sizes are usually small for structural safety and cost. This causes increased shading on the inside of the greenhouse. As an indication of this shading effect in a clear-glass greenhouse, the light intensity in shadows is reduced to about 20 percent of that elsewhere. It must be remembered, though, that these shadows are constantly moving within the greenhouse as the sun's position during the day changes, so the shading effect on any one area is limited. If diffusing glass were used instead of clear glass, shaded areas would receive upwards of 75 percent of the light intensity of unshaded areas. Again, however, diffusing glass is more expensive.

Glass is easily breakable and can be a bit of a hazard, particularly in areas where children often play and throw things. Nevertheless, it is a widely available material, and most people are familiar with its use and handling. All in all, if you can afford it, glass is perhaps the best material for greenhouse covers.

Polyethylene Film

This is probably the cheapest covering material for a greenhouse. However, it is short-lived when exposed to direct sunlight since the ultraviolet radiation from the sun causes it to deteriorate. Polyethylene film that has been treated to resist ultraviolet radiation deterioration is readily available, but at a slightly higher cost. However, even with this treatment, the plastic film will rarely last more than 1 year and often has to be replaced twice a year. On the other hand, polyethylene film and all other film plastics are lightweight and do not require heavy structures to support them. Therefore, shading caused by structural members is not as much a problem as with glass. Polyethylene film is almost as good as glass in letting in solar and sky radiation. It diffuses this light as it enters, which is an advantage, but it also lets out much of the long-wave heat radiation. Therefore, it is not as good a heat trap as glass, and it is necessary to use two layers of polyethylene to provide an acceptable heat trap in the winter.

With film plastics, it is *absolutely* necessary to build the roof slope steep enough for water and snow to run off; otherwise, they collect and stretch the plastic sheet.

If you are considering a temporary greenhouse or you don't mind changing the cover every year, then polyethylene film may be your best bet. Otherwise, use glass or a rigid plastic material.

Vinyl and Other Plastic Films

There are a number of longer-lasting plastic films on the market, but often these are almost as costly as glass. They offer the advantage of requiring less structure to hold them up, but they do deteriorate with age. It is rare to find an acceptable plastic film lasting more than a few years, although fluoroplastics (Tedlar, Mylar, Solar Membrane) are reported to last as long as 20 years.

Rigid Plastics

Rigid plastics offer some of the advantages of glass—long life and good light transmission. They are usually costly, often more expensive than glass. However, they are much lighter and can be used in wider and longer panels. As a result, there will be fewer supporting bars required, making for a lighter and less costly greenhouse structure that will allow more solar radiation to reach the plants. Most of these plastics, it should be pointed out, do deteriorate with age, although not nearly as quickly as polyethylene film. As years go by, these plastics tend to yellow, become more brittle, and increasingly block out the sun. However, you can count on most rigid plastics to last for at least 10 years. Some are even guaranteed to maintain 90 percent of their original light transmission after 20 years. Rigid plastics, unlike the plastic films, have some good structural characteristics, particularly if corrugated sections are used. There is not the severe roof-slope restriction as mentioned for plastic films under snow and rain loads.

Light entering through rigid plastics tends to be dispersed so that no clear images appear. This has the benefit of scattering the light more evenly over the plant canopy thereby reducing any distinct shadows. However, it does have the disadvantage of not allowing a clear view of the outside, a factor that will have to be considered if the house wall has windows in the area enclosed by the greenhouse.

Acrylic plastics (Plexiglas, Rohglass) tend to expand and contract with increasing and decreasing temperatures, so allowances have to be made when these are being installed. Fiber-reinforced plastics, such as Filon and Kalwall, do not expand or contract appreciably under normal operating conditions.

A Brief Summary of Selection of Transparent Covering Material

If you want the most inexpensive covering material, polyethylene film is it; however, it will last only a few months to a year when directly exposed to sunlight. This material is excellent for structures of low initial cost, but be prepared to change this film plastic every so often if you are contemplating a permanent greenhouse. If you live in an area of severe hailstorms, glass coverings are *not* recommended. They will shatter too easily. Rigid plastics are best for this environment.

If the greenhouse encloses a window on the house wall and you want to be able to see beyond the greenhouse interior when looking out the window, then film and rigid plastics are not usually suitable materials. They tend to blur or obscure images of objects that are beyond them. Clear Plexiglas is one exception to this, but it can be very costly.

Make sure that all film plastics are tightly secured so that they do not flap in the wind. Also, ensure that roof slopes are adequate (usually 28° or steeper) to permit snow and rain to run off. Otherwise, any build-ups that occur will stretch the plastic.

STRUCTURE

For good light transmission and reduction of shadows, the glazing bars and supporting structure should be as narrow as possible and as widely spaced as practical, while maintaining good strength characteristics. For these reasons and also from the point of view of ease and cost of maintenance, aluminum or galvanized steel, either pipe or channel section, are sometimes considered preferable to wood. However, wood is often easier for the average homeowner to work with than metals, and it is usually less costly. In addition, aluminum is a material that requires a lot of energy to fabricate. For that reason alone, it is going to become much more costly both to the environment and the pocket-book as time goes on. Furthermore, it is easier to find suitable scrap wood than aluminum or steel.

The color and reflectivity of the structural members are also very important for the transmission of solar energy to plant level. White, silver, or aluminum painted surfaces reflect more light onto the plants than do darker surfaces. Since the humidity level inside greenhouses is usually quite high, use good quality paints that are little affected by moisture or humidity.

SHAPE AND DIMENSIONS

These are going to depend on a wide variety of factors. The space you have available, both against the south wall of the house and on the ground, is one of the limitations. Snow, wind, and ventilation are other determining factors. To eliminate snow build-up on the roof, it will have to be pitched at an angle of no less than 28° from the horizontal. The south wall will have to be high enough to allow all the snow accumulating during a heavy snowfall to slide off the roof. A taller greenhouse requires a smaller rate of ventilation than a low one for the same effect in reducing temperature and humidity. Therefore, the ventilation ports on both the sides and the roof can be somewhat smaller in a taller structure.

HEATING

Depending on where you live, you may or may not be able to get through the winter months without heating your greenhouse except, perhaps, by leaving any interconnecting doors, windows, and vents into the house open. You have a number of choices to make here. First, you may decide only to extend the growing season at either end and not add any supplemental heat to the greenhouse. This can give you 2 to 3 months of extra growing season, enough to grow two good crops of any number of vegetables. During sunny days in the late fall, winter, and early spring, you can use the heat build-up inside the greenhouse, whether there are plants in it or not, to assist in heating the living areas of your house. This can be done by simply opening the doors, windows, and vents that lead into the main house from your greenhouse.

Secondly, you may decide to expand the growing season even further, still without buying a space heater for the greenhouse. This can be done, particularly in smaller greenhouses, by placing movable insulation panels against the transparent cover just before the sun sets and while the greenhouse interior is still warm. See Chapter 11 for information on constructing insulation panels. This has the effect of trapping inside the greenhouse much of the day's radiation, which has been stored as heat in the soil, floor, tables, and benches. If the outside temperature does not fall much below 20°F in your area and you have fairly sunny winter days, this approach may be able to get you through the entire winter season. Excess daytime heat in the greenhouse can be directed into the living areas of the house. This air has the added advantage of smelling fresh, being more humid, and containing a surplus of oxygen.

Thirdly, there is the possibility that the furnace in the main house can provide adequate temperature control by simply running another heating duct into the greenhouse. You would have to check if the furnace has the extra heating capacity, which it probably does. The house thermostat can serve as the temperature control for the greenhouse, particularly if interconnecting windows and doors are left open to allow for air circulation.

Finally, you may decide to heat the greenhouse with a small furnace or space heater. Check with local manufacturers or distributors on size and location requirements. A thermostatically controlled heater with the same settings as for the interior of the main house is best. This will allow you to keep all doors, windows, and vents leading into the living areas open so that the main house furnace can assist in heating the greenhouse. Where heaters are concerned, watch out for city bylaws and permits. Also note that since glass conducts heat so well, the insulation panels mentioned earlier will reduce the extra heating bill for the greenhouse substantially.

COOLING AND VENTILATION

During the season from late spring to early fall and on sunny winter days, the excessive heat build-up inside the greenhouse will have to be reduced. In winter, this heat can be transferred to the living areas to help save on energy consumption. During the summer, however, the heat will have to be allowed to escape to the outside. Natural ventilation provides the simplest, least expensive, and most satisfactory method if sufficient ventilator openings are provided. East and west end-wall openings will allow wind to enter and escape with the excess heat. Ventilators near the peak of the roof are a real advantage in hot, still weather. It has been found that these roof vents should have an open area no less than one-sixth of the ground area of the greenhouse. With these open, in addition to several of the end-wall vents, the cooler outside air will enter and displace the more buoyant, warmer air out through the roof vents in what is commonly known as the chimney effect.

Other nonenergy-consuming cooling techniques include the use of shading devices. The simplest of these is a whitewash solution of salt and lime, which can be applied with a broom to the exterior surface of the greenhouse. When it dries, this solution reflects some of the sunlight so that the interior temperature build-up will not be so great. This solution washes off slowly with rain so that it is usually gone by late summer. If not, simply wash it off with a hose. A disadvantage of this technique is that the salt tends to be damaging for immediately adjacent lawns or gardens. Wood or bamboo slats, particularly of the roll-up type, are also very good shades when securely mounted on the outside of the greenhouse roof.

OTHER CLIMATE CONTROL

Humidity control is a problem in many areas. At low altitudes, along large bodies of water, and in the more southerly moist climates, excessive humidity in greenhouses can cause leaf mold and diseases. Adequate ventilation is the only inexpensive, realistic solution to this problem.

In dry, desert areas, lack of humidity is often the problem. This can be partially regulated by keeping the soil and/or floor moist. You can water your plants with a garden hose.

METHOD OF CONSTRUCTION

There are a number of good plans available for greenhouse construction. Look in your Yellow Pages to find commercial greenhouse firms in your area, or check with the nearest agricultural school or research station. Often one or more of these groups will have brochures available describing how to build freestanding greenhouses. These plans can be readily adapted to the construction of a lean-to greenhouse.

We will describe here a simple 2 × 4 wood frame, lean-to greenhouse using rigid plastic as the transparent covering material. The size we selected is arbitrary. In this case, the greenhouse extends out from the house a distance of 10 feet and is 24 feet long. This size allows ample ground area to grow a wide variety of crops. The roof slopes at about 29°, so the longest span from the wall is less than 12 feet. If your south wall is less than 24 feet or you would like to use part of the wall for solar windows, just scale down the plans that follow to suit your own needs. The cost of this structure is low.   NOTE: It is advisable to check on whether or not you will require a building permit before you begin construction. For details on this, see Chapter 4. Prior to construction, it is wise to consider if and how you are going to bring service lines, such as water and electricity, into the greenhouse. If these will come in underground, do it before you begin construction. Also, you might consider *bringing in topsoil, if needed, prior to construction,* as this can be a laborious task after the walls are up.

Foundations

A greenhouse is usually a light structure, so foundations can be fairly simple. Either several foundation posts can be embedded in concrete or a continuous concrete beam, say 8 inches × 8 inches, can be laid under the wall of the greenhouse. The type of foundation you use will depend on the type of soil and climate in your area. If you have a soil with very good drainage, the 8-inch concrete beam will be an adequate foundation. If you don't have good drainage and live in a cold climate, you may want to place the foundation posts deeper than the 30 inches shown in our drawing. Check locally about which method and depth of foundation is best.

If foundation posts are being considered, these should be good-grade 4 X 4-inch lumber, 8 feet long and well treated to prevent rot and termite damage. These posts should be set in concrete in 12-inch diameter post holes dug every 4 feet along the length and breadth of the proposed walls. See Figure 17-3 for details. Figure 17-4 illustrates the construction of a continuous concrete beam. Foundation bolts should be embedded in the concrete every few feet so that the bottom plates of the stud walls can be firmly bolted in place. Ground insulation is essential in most areas. It will help keep the soil temperature inside the greenhouse warm particularly along the walls. To install this insulation, there are a number of acceptable methods. One

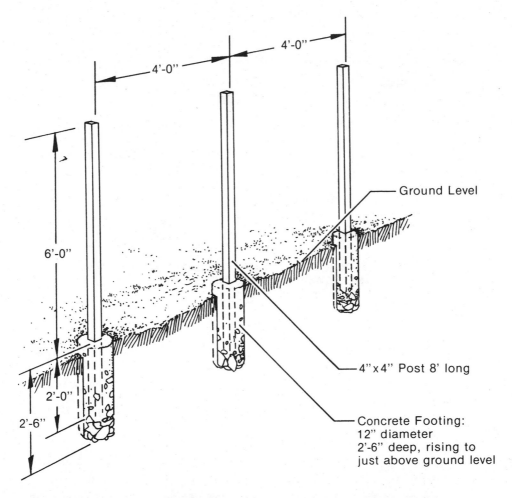

FIGURE 17-3: EXAMPLE OF FOUNDATION POST INSTALLATION

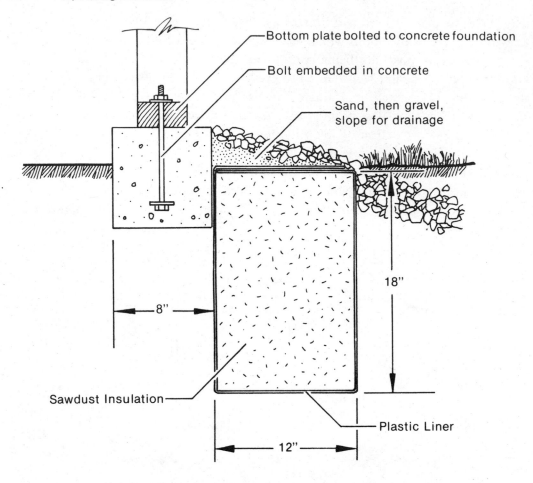

**FIGURE 17-4: ILLUSTRATION OF CONCRETE BEAM FOUNDATION
AND SAWDUST INSULATION**

is to place 4-inch-thick polystyrene sheets in the soil all around the outside of the foundation, as in Figure 17-7, and another is to use sawdust, as indicated in Figure 17-4. Sawdust will tend to pack down eventually, so increase its height by about 3 inches. Be careful to protect these sawdust insulating materials from water by wrapping them in a plastic liner.

Walls

The spacing of wall studs will depend upon the roof load and wind load characteristics and the transparent covering material used. In many areas, 48-inch stud spacing with moderate- to heavy-weight transparent plastic sheets (5 to 8 ounces per square foot) is sufficient. The plastic sheets, however, should be fiber reinforced. In other areas, where the snow or winds are more severe, a 24-inch stud spacing will be necessary, and in this case it will be possible to use a lighter weight plastic sheet (say, 4 ounces per square foot). You will have to check with your local building authorities or some other knowledgeable person about such facts as design, wind and snow loads, and the recommended stud spacing and materials necessary to meet these standards. We will briefly illustrate some plans for a 2 X 4-inch stud wall with studs every 24 inches on center. This type of structure will be overdesigned for most loading conditions encountered.

The lower portions of the east and west end walls will have to contain ventilation ports, as mentioned elsewhere, and possibly a door. The size and placement of these openings should be determined in advance and incorporated into the stud-wall construction.

Standard stud-wall techniques should be followed, using a bottom plate, which is bolted to the foundation, and a double 2 X 4-inch top plate. The vertical south wall will rise 6 feet, thus avoiding the need for any inter-stud blocking. Figure 17-5 illustrates end- and south-wall construction.

The two end walls have to be attached to the south wall of the house. In order to do this, it would be best to have lined up the foundations with existing studs in the house wall so that the greenhouse end walls can be nailed directly into the house studs. In the event that this is not possible because, for example, the house wall is masonry or the studs are not ideally located, drill several holes into the house wall where the greenhouse end-wall studs will stand. Expansion or toggle bolts or plastic plugs can then be used to attach the end-wall studs to the house wall.

Wall covering should be made of transparent, flat, fiber reinforced plastic paneling. Buy a good make that is guaranteed to last at least 10 years and that is at least 85 percent transparent to solar radiation. The width should correspond to the stud spacing. Drill holes through the plastic first before nailing, as some plastics chip. Nails should have a neoprene or rubber washer attached to their shank in order to provide a weatherproof seal when nailed tightly against the panel. All vertical ventilation ports can be covered inside and outside with this flat, rigid plastic in order to provide dead air space that will act as insulation on cold nights.

Roof

The roof is essentially the same construction as the walls except that no double 2 X 4-inch plate is required at one end. Roof vents will have to be included in the design, as illustrated in Figure 17-6. At its lower end, the roof structure will be attached to the vertical south wall of the greenhouse, using gussets or other conve-

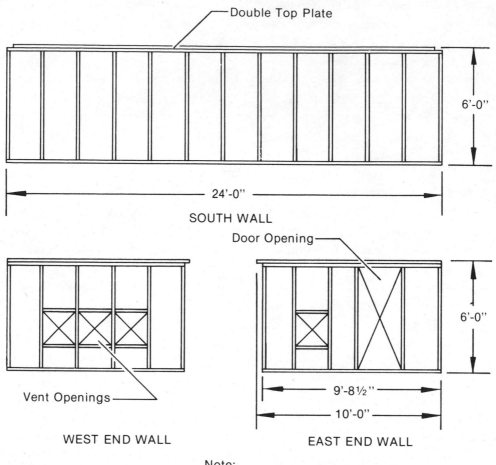

Note:
—All frame members are 2"x4" wood.
—Studs are nominal 24" on center.
—Door opening and end-wall top plate overlap
will alter this spacing somewhat.

FIGURE 17-5: STUD-WALL CONSTRUCTION

nient attachment. At its upper end, the roof will have to be attached to a 2 X 4-inch horizontal stringer, which is firmly fixed to the house wall. The triangular gap between the end walls and the sloping roof should be framed in by using pieces of 2 X 4 to continue the wall studs up to support the roof structure. Ventilation ports can also be included in these end walls near the peak of the roof, as illustrated in Figure 17-7.

Near the upper edge of the sloping roof, ventilation ports will be necessary. As

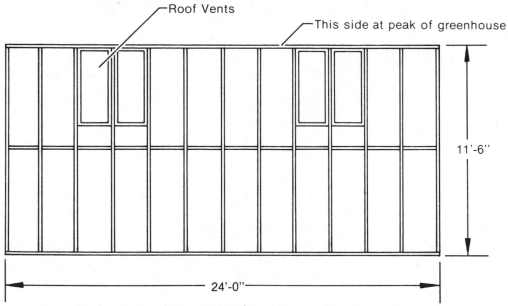

Note: All frame members, except for vents, are 2"x4" wood.

FIGURE 17-6: ROOF RAFTER CONSTRUCTION

mentioned elsewhere, the ventilation openings near or along the roof should be about one-sixth of the ground area contained by the greenhouse. In the structure described, the ground area is 24 X 10 feet, or 240 square feet. Thus, roof vents should have a total area of about 40 square feet. If end-wall vents are installed near the peak of the roof at both the east and west ends, as illustrated in Figure 17-7, the total area of roof vents can be reduced to about 32 square feet. Vent openings totaling approximately this area can be obtained by installing four hinged roof ventilators, each 4 feet long, between the roof rafters, as illustrated in Figures 17-6 and 17-8. These roof ventilators should be constructed using a 2 X 2-inch wood frame to fit easily between the rafters. Details are given in Figure 17-8. The vents should be allowed to open to a horizontal position to give maximum ventilation. A simple opening and support system for the vents is a long pole, hinged at its upper end to the underside of the lowest portion of the vent and attached to the rear wall by resting in one of the slots of a board that is attached to the lower north wall.

For the roof covering, it is best to use 4-ounce corrugated rigid plastic reinforced with fiberglass, the same quality as used for the walls. This material usually comes in 26-inch widths, which allows suitable overlapping with adjacent panels so that each piece effectively covers a 24-inch-wide strip. These panels usually can be bought in

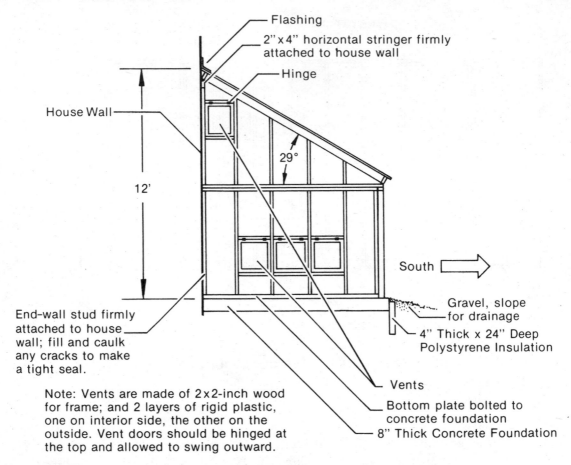

**FIGURE 17-7: WEST END WALL AND ROOF OF
LEAN-TO GREENHOUSE**

lengths up to 30 feet. Other widths, including 50½ inches for 48-inch coverage, can also be purchased.

If you are unfamiliar with the installation of the corrugated plastic sheet, ask your dealer for this information.

After you buy the transparent fiberglass panels, be careful *not* to store them stacked up outside exposed to the sun. If you do, the temperature can build up in the stack to 160°F or higher. This will cause the panels to permanently cloud over and thus reduce their transparency. All panels should be stored in a dry, shaded area with adequate ventilation.

Buy the accessories that are needed for installation from the place where you buy the corrugated panels. These include corrugated wood molding, special nails with attached neoprene or rubber washer seals, and nondrip spacers, which hold the

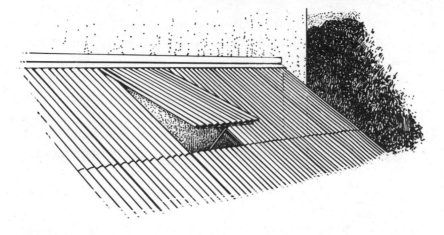

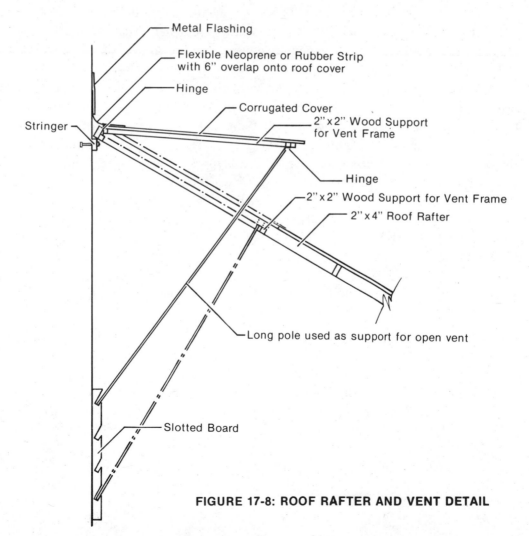

Metal Flashing

Flexible Neoprene or Rubber Strip
with 6" overlap onto roof cover

Hinge

Corrugated Cover

2" x 2" Wood Support
for Vent Frame

Stringer

Hinge

2" x 2" Wood Support for Vent Frame

2" x 4" Roof Rafter

Long pole used as support for open vent

Slotted Board

FIGURE 17-8: ROOF RAFTER AND VENT DETAIL

panels a quarter inch above purlins or blocking so that condensation can run freely to eaves or sidewalls. Buy enough corrugated wood molding to run a strip along the upper and lower edges of the roof and across the roof at the midline. Both upper and lower edges of the roof vents will also require corrugated wood molding.

If you have to overlap the ends of the corrugated panels where, for example, the lower edge of the roof vent overlaps the next panel below it, or the joints, allow at least 6 inches with the upper panel overlapping the lower one. Similarly, the sidelap should be one corrugation. Both of these overlaps prevent wind and rain leakage into the interior.

Interior Lining

It is wise to use two layers of transparent cover in areas where the winter temperature drops below freezing. If well sealed to form a dead air space between them, these two layers act like storm windows or thermopane. Two layers also reduce the amount of water vapor condensing on the underside of the roof and walls.

The vents will already be covered with a double layer of rigid plastic if the instructions for Figure 17-7 were followed. On the inside of the 2 X 4 wall and roof frames, you can install a second layer of the flat, rigid plastic sheet used for the exterior covering, only lighter in weight if possible. Or, you can completely cover the inside, excluding vents and door, with a layer of 4-mil polyethylene film sheet. The latter method is cheaper, but you will have to replace the film plastic in 2 or 3 years. This film plastic should be installed using wood lath to attach it to the 2 X 4-inch studs. Stretch the plastic as taut as possible without ripping it, and be sure that a well-sealed dead air space is obtained all around the transparent parts of the greenhouse.

COSTS

Costs are going to vary depending on local availability of materials and how much you can scrounge or obtain second hand. Using the materials mentioned for the construction of the 24 X 10-foot lean-to greenhouse, including a polyethylene interior liner, the total cost should not exceed $450, or about $2 per square foot of greenhouse area. That's really not much, considering the advantages to be gained.

chapter 18

SOLAR WATER HEATERS

by STEVE RIDENOUR

Heating water is one of the most practical uses of solar energy. Hot water is used year-round, so that the use factor is high compared to solar space heating, which is seasonal. The unit also is relatively inexpensive for the individual. An entire system will cost $300 (do-it-yourself) to $1,300 (purchased and installed), but could save a family of four about $150 per year over an electric water heater and about $75 per year over a gas water heater at present prices (though gas and oil will be at least as expensive as electric in coming years).

Solar water heaters were commonplace in Florida and California in the 1920's and '30's. Today, solar units are used extensively in the rural areas of Japan, Australia, and Israel. The annual sales for solar water heaters in Australia alone is $1,000,000.

If the solar unit is used as a supplement to a regular water heater (gas or electric) it can save up to 90 percent of the energy used by the conventional system. For certain lifestyles, though, no regular water heater is needed or desired. In this case, a solar unit can provide adequate supplies of hot water (100° to 130°F) throughout most of the United States.

PRINCIPLES OF SOLAR WATER HEATING

Solar radiation is diffuse. The average annual solar power received by a surface that faces the sun is about 1/5 horsepower per square yard. The only way that more energy can be harvested is to increase the size of the collector receiving area. Concentrators such as searchlights in reverse (special paraboloids) or an oversized magnifying glass (commonly a fresnel lens) will only raise the temperature of the absorb-

309

ing surface; no more energy will be delivered. Concentrating collectors are desirable for high temperatures such as 300° to 2,000°F; but for the 110° to 150°F range needed for domestic hot water, the flat plate collector is supreme.

The solar water heater works by passing a liquid through a device called the *collector,* which converts the sun's energy into heat and raises the temperature of the liquid flowing through the collector. See Figure 18-1. The liquid may be pure water, but in some situations an antifreeze solution consisting of water and either ethylene glycol, alcohol, or propylene glycol may be more desirable.

Figure 18-2 gives us a closer look at the solar collector. Harvesting the sun's energy and converting it into heat utilizes a process most of us understand. Black painted surfaces exposed to the sun will get hotter than any other color. A black paint which is dull (or flat) does not shine and thus does not lose energy by reflection. So, if a metal surface is painted *flat black* and water is put in contact with the back of the metal, the water will be heated efficiently. This simple device is called the *absorber.* It is usually a metal plate with integral water passages which is painted flat black on one side. Black plastics with integral water passages are also used. However, these plastics gradually degrade and must be periodically replaced.

Once the solar energy is converted to heat and absorbed by the water, the absorber must be insulated to prevent heat loss. The shaded side (back) of the absorber is insulated with the familiar fiberglass or urethane foam. Likewise, the

FIGURE 18-1: SOLAR COLLECTOR

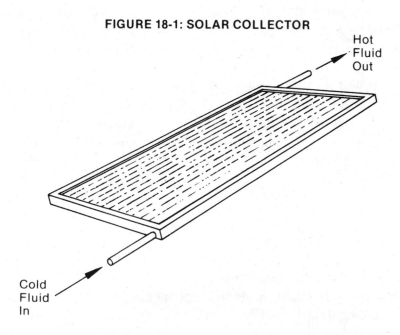

Hot
Fluid
Out

Cold
Fluid
In

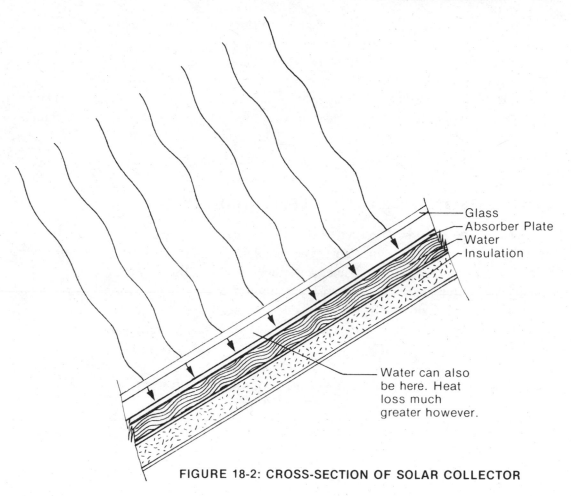

Glass
Absorber Plate
Water
Insulation

Water can also
be here. Heat
loss much
greater however.

FIGURE 18-2: CROSS-SECTION OF SOLAR COLLECTOR

sides are insulated. In the case of urethane, however, a 1-inch inner fiberglass liner is used to protect the urethane from high temperatures. Heat losses on the sunny side (front) of the absorber are reduced with a *glass cover*. The glass transmits the high-energy solar radiation (short wavelength) onto the absorber plate but retards the low-energy infrared radiation (long wavelength) emitted from the hot absorber. It also keeps air currents away from the hot absorber.

In review, then, the parts of the solar collector are (1) the absorber to convert sunlight into heat, (2) the glass cover to slow the upward heat loss, and (3) the back and side insulation to slow the downward heat loss. All of these are held together by (4) the *collector casing.*

Because of the intermittent nature of sunshine, in most solar water heating systems the hot water must be stored until needed. Thus, a *storage tank* is included, with appropriate *piping* from the collector to the storage tank. The final part of the solar water heating system is the *water circulating device.* This can be a pump with automatic controls; or, for a properly designed and constructed system, this circulation will occur automatically by thermosyphon action (also called "natural convection" or "gravity circulation"). These flow phenomena will be explained later. Most solar water heaters in industrialized countries have an *auxiliary heater* to maintain a constant water outlet temperature. All of these components for solar water heaters are shown in Figure 18-3.

THE BATCH SYSTEM—COMBINED COLLECTOR AND STORAGE

Before I go into the specifics of the most widely accepted solar water heaters, I will explain a very basic type. This type of solar heater is simple and inexpensive; it can be made portable as well. Here, water does not flow through the collector, as in most solar water heaters. Rather, a fixed amount is heated over the day in a deep absorber plate and drained off when needed. It has found wide acceptance in Japan and India where the hot climate and lifestyle make it really useful. There are several

FIGURE 18-3: HIGH TECHNOLOGY SOLAR HEATING SYSTEM WITHOUT HEAT EXCHANGER

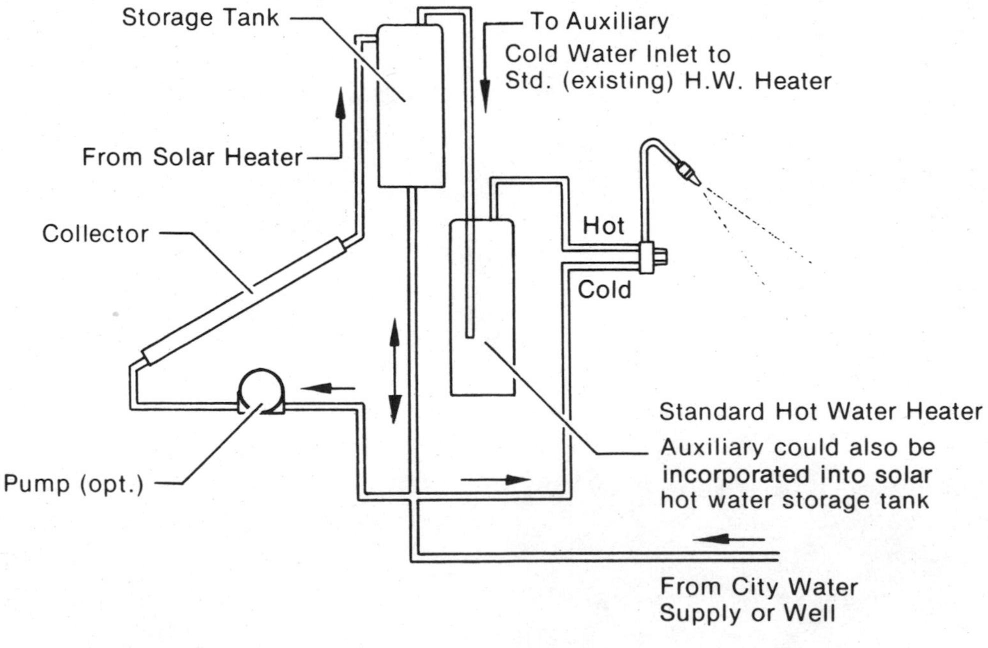

Storage Tank

To Auxiliary
Cold Water Inlet to
Std. (existing) H.W. Heater

From Solar Heater

Collector

Hot

Cold

Standard Hot Water Heater

Pump (opt.)

Auxiliary could also be
incorporated into solar
hot water storage tank

From City Water
Supply or Well

disadvantages, though. It is usually filled manually, and the quantities of hot water produced are small—on the order of 20 gallons or less. Water temperatures are lower than 130°F, especially in the winter. The water will not stay warm overnight; and, in fact, if left in the collector, it may freeze on cold, clear nights. Finally, the unit does not easily tie into a piped hot water system unless it is the gravity flow type.

Figure 18-4 depicts two of the more familiar batch solar water heaters. The tilted tank (Figure 18-4a) is simply a shallow sheet-metal tank painted black on the top side with an inlet funnel and an outlet valve. It is covered with one window. A series of tilted cylinders is shown in Photo 18-1.

The Japanese water "pillow" type is depicted in Figure 18-4b. It is basically a 50-gallon water bed made of black polyethylene. The Japanese fill the "pillow" in the morning and let it heat throughout the day. In the evening it is drained into a tub for the traditional Japanese hot tub soak. It is simple and effective, there are no freezing problems, and a transparent plastic canopy added in the winter retains heat on cold days. It sells for roughly $10 in Japan and lasts about 2 years. However, since plastics usually impart a peculiar taste to water, it is probably not used for supplying drinking or cooking water.

A firm has cashed in on the potential portability of the plastic batch solar

FIGURE 18-4: BATCH TYPE SOLAR WATER HEATERS

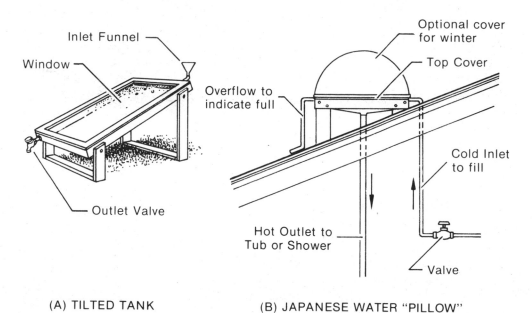

(A) TILTED TANK   (B) JAPANESE WATER "PILLOW"

Photo 18-1: A hatch-type solar water heater manufactured by a Japanese firm. The metal cylinders are filled in the morning, and the hot water is drained at night for the traditional tub soak.

heater. Basic Designs, Inc., sells a 10-quart water "pillow" shower for about $6.95. It weighs 10 ounces and was built for backpacking. They even have a heat sensor that changes colors when the water reaches 90°F. It's well designed and well built. Write to:

> Basic Designs, Inc.
> 3000 Bridgeway
> Sausalito, California 94965

CONTINUOUS-FLOW SYSTEMS—SEPARATE COLLECTOR AND STORAGE

While the batch systems just described may be quite adequate for certain life-styles, I think that most Americans would appreciate a solar heater incorporated into their pressurized water system. Let's look at how to do this.

The usual temperature for domestic hot water is 130° to 140°F. From an engineering standpoint, it is better to separate the collector from the water storage unit in order to maintain these temperatures with low thermal loss. With the collector and storage system separate, the water from the storage tank must flow into the collector to be heated and then back to storage again. There are two ways to do this: a thermosyphon system and a forced-circulation system.

Thermosyphon Systems

Remember the old gravity-flow hot water heating system? It looked something like Figure 18-5: huge pipes, lots of "water hammer," usually too hot or too cold; but it didn't use a pump to move the water around or a control to tell the pump when to start or stop. There were no contacts to corrode, no bearings to wear out, no seals to leak, no electricity to make it work. It was just a heat source. Granted, most of these units were not very efficient, but pumped systems with thermostats came along before the really efficient gravity-flow systems could make their usefulness known. Thus, they went the way of the Model "A."

The hot water was moved around in these heating systems by gravity flow. This flow can be likened to a U tube. See Figure 18-6. If heat is applied to the left side of the U tube, the water density in it will decrease slightly. The balance of the U tube is now upset and the colder, heavier water on the right side will flow in to balance out the situation. It, in turn, will be heated and rise; thus a flow is established. The flow will continue until the heat source is withdrawn or the temperature of the whole system is about the same. At this point the densities of the two U-tube "legs" are equal and the flow stops.

These thermosyphon flow principles can be used to great advantage in a solar water heater. Look at Figure 18-7. As solar radiation falls on the absorber in the collector and heats the water in it, this lighter hot water rises. Actually, it is being pushed up by the heavier, colder water in the collector inlet—tank outlet pipe (return pipe). As the colder water reaches the absorber, it in turn is heated and rises

FIGURE 18-5: GRAVITY FLOW HOT WATER SYSTEM

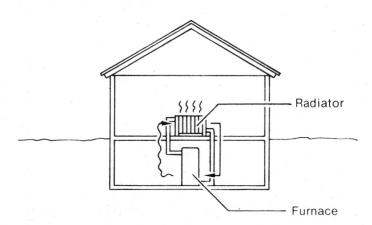

Radiator

Furnace

FIGURE 18-6: "U" TUBE & GRAVITY FLOW

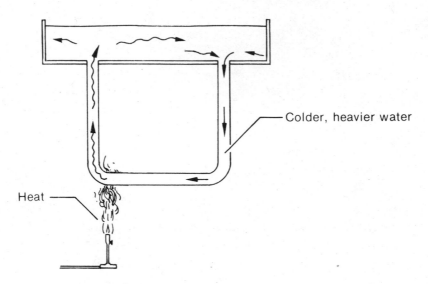

Colder, heavier water

Heat

and more cold water pushes in from the bottom of the collector. Thus, a flow is established and will continue until the water fails to gain heat from the sun, i.e., until there is insufficient heat to raise the temperature of the water as it passes through the absorber. The flow starts only when heat energy can be added to the water, and it stops when it can't. Perfection! No thermostats, no relays, no motors, no pumps, and no electricity. Natural!

FIGURE 18-7: LAYOUT FOR THERMOSYPHON FLOW

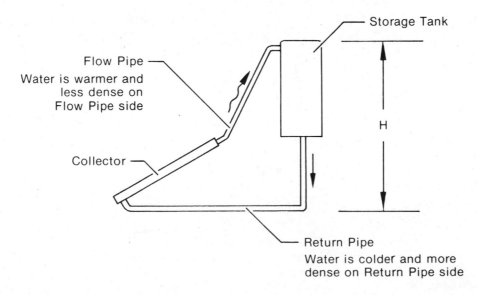

Storage Tank

Flow Pipe
Water is warmer and
less dense on
Flow Pipe side

H

Collector

Return Pipe
Water is colder and more
dense on Return Pipe side

Photo 18-2: Thermosyphon solar water heater in Mt. Isa, Queensland. This unusual mounting pedestal is possible because of low rainfall in the area.

Photo 18-2 is an unusual arrangement for a thermosyphon solar water heater. This unit serves a single family in Australia.

Forced-Circulation Systems

In certain situations thermosyphon circulation is unfortunately not possible. Because of shading, the collector may have to be placed above the storage tank in some cases. Also, large thermosyphon systems can sometimes get unwieldy to install at ground level because of large, sloping pipes (however, a flat roof may help solve this problem, as shown in Photo 18-3).

If a thermosyphon system can't be used, you have to go to a forced-circulation system. Basically, the number of components in a forced-circulation system is the same as in the thermosyphon system, but a circulation pump and some automatic controls are added. As a result of the extra pressure, a pressure vent, some one-way

Photo 18-3: Large bank of collectors for thermosyphon circulation unit on hostel building at Griffith, New South Wales, Australia.

valves, and an expansion tank are desirable. The general layout of one such system is shown in Figure 18-8; Photo 18-4 shows a large forced-circulation system.

One tricky point for pumped systems is how to get the automatic controls to duplicate the mechanism of thermosyphon flow—to wit: the circulating pump must run when heat can be added to storage and must stop when it cannot. Attempts at setting a thermostat on the collector output at, say, 100°F usually result in the pump starting off fine in the morning; but after the water reaches its peak temperature in the afternoon, the pump continues to operate, resulting in heat being lost until the collector output temperature is back down to 100°F. Thus, the best control is a differential unit. It measures the *difference* in temperature between the *top* of the collectors and the *bottom* of the storage tank. The unit is usually set at 5°F difference. Heat is then added to storage when it is available, and a few degrees lag on either side of the 5°F can be used to prevent the pump from cycling too often.

Heat Exchangers

If you live in a climate where freezing is a problem at night, you'll want to add

Photo 18-4: Large bank of collectors for a forced-circulation unit on a college dormitory in Adelaide, South Australia.

one other feature to your continuous-flow solar hot water heater—a heat exchanger. Actually, there are several ways to keep the water in continuous-flow systems from freezing in the collector at night. You can drain the collector each evening before a suspected freeze, circulate warm water through the collector, heat the collector electrically, or use the heat exchanger. However, the heat exchanger has the best advantages; for once installed, it eliminates the routine of drainage or the expense of heating the collector.

The heat exchanger is just a separate hot water loop added into the continuous-flow systems already outlined. There are several ways to do this, which will be explained later in this chapter, but basically this separate loop allows the water in the collector loop to be treated with an antifreeze. Although these antifreeze chemicals eliminate the need for nighttime drainage or collector heating, *they are poisonous.* Hence, a separate sealed container for drinking water which is in thermal contact with the hot fluid that circulates in the collector is necessary.

If you live in a climate where freezing never occurs, the heat exchanger is unnecessary. In this case, ordinary tap water can be used in the collector, though a copper-tube collector plate will be required by code for this potable (drinking) water. Lime deposits may in time slow down the circulation in thermosyphon systems with copper absorber plates. This will occur to some extent almost anywhere;

but if lime problems are really serious in your area, you might want to consider a noncopper absorber plate and use one of the heat exchangers.

PUTTING SOLAR ENERGY TO WORK FOR YOU

Sizing the Tank and Collector

In designing a solar system, our first concern is the quantity of hot water required. Our friends at the local electric and gas utility would have us believe that we use 50 gallons per person per day. This quantity *is not the average but the maximum.* It is used for sizing hot water heaters. Most modern U.S. families use 20 gallons per person per day. If your lifestyle does not include a shower or bath daily, or an automatic dishwasher or clothes washer, then naturally you require less. I think that most people would find 10 gallons per person per day to be quite adequate, and cutting down to less than 5 gallons is still entirely possible for certain lifestyles.

The size of the tank is proportional to the number of people using the hot water. Allowances must be made for tank heat losses, auxiliary heating or storage during adverse weather, and heat-exchanger losses. Generally, if no auxiliary is used, one day's storage is desirable. Tank heat losses, coupled with a modest reserve, require a 10 percent increase. A heat exchanger requires another 10 percent increase. Table 18-1 is provided for clarification.

Thus, a family of four using 10 gallons per person per day with an auxiliary heater and heat exchanger would require a tank size of 4 × 10 × 1.1 × 1.1, or about 48½ gallons. With no auxiliary they would need two times the above: a 97-gallon

---

Table 18-1: *Guide to Tank Size*

No auxiliary and no heat exchanger	Tank (gal.) size = 2 (reserve factor) × no. people × gal./person/day × 1.1 (heat loss factor)
Auxiliary but no heat exchanger	Tank (gal.) size = No. people × gal./person/day × 1.1 (heat loss factor)
Heat exchanger but no auxiliary	Tank (gal.) size = 2 (reserve factor) × gal./person/day × 1.1 (heat loss factor) × 1.1 (heat exchanger factor)
Auxiliary and heat exchanger	Tank (gal.) size = No. people × gal./person/day × 1.1 (heat loss factor) × 1.1 (heat exchanger factor)

---

tank. These figures are approximate. Tanks are usually sold in 60- and 100-gallon sizes, though smaller sizes are sometimes available for vacation cabins.

After the quantity of hot water desired is known, the collector to supply it can be sized. The calculations for collector size in the continental U.S. are given in Table 18-2. For example, a family of four, using a 48-gallon tank, would need 72 square feet of collectors in Vermont; 48 square feet in Kansas; and 24 square feet in Arizona. Again, these are approximate figures. Since collectors are usually manufactured in 15- to 30-square-foot modules, you'll have to select a set of collectors with a total area near what you've calculated for your needs if you buy a system.

The Design of System Components

In this section we'll explore some of the design considerations you will need to be aware of to put together your own solar hot water heater. If you buy a commercial unit, use this information as a guide. It's also possible to build your own units. References at the end of this chapter list booklets and sets of plans that explain various "how-to-do-it" designs in depth. Another route you might want to follow is to buy one of the kits manufactured by companies that are listed at the end of this chapter.

The primary areas of design you should look into are:
- a thermosyphon unit *or* a forced circulation unit
- a heat exchanger, if needed
- the collector
- the storage tank

Table 18-2: *Collector/Storage Sizing in North America*

Location	Collector Area Required Per Gallon of Water Stored	Number of Transparent Covers over Absorber
Northeast, Northwest, Great Lakes (e.g., Maine, Oregon, Michigan, and Canada)	1½ ft.^{2}	2
Middle States and North-Central states (e.g., Virginia, Missouri, Utah, and the Dakotas)	1 ft.^{2}	2 (1 is optional in summer if overheating occurs)
Southern States (e.g., Florida, New Mexico, Arizona, southern California, and Mexico)	½ ft.^{2}	1

Design Considerations for Thermosyphon Units

As was mentioned before, the thermosyphon system acts like a U tube that has been heated on one side. The hot side in this case is the collector, and because this warmer and less-dense fluid in the collector rises to the top of the storage tank, colder fluid from the bottom of the storage tank drops down into the collector to be heated. See Figure 18-7. Hence, this flow is due to a *pressure difference* created by the fact that the storage tank is higher than the collector (marked *H* in Figure 18-7) and by the difference in densities between the hot collector water and the colder water at the bottom of the storage tank.

Though this pressure difference is enough to circulate the water from storage to collector and back into storage again, its size is still small. Hence, any extra friction

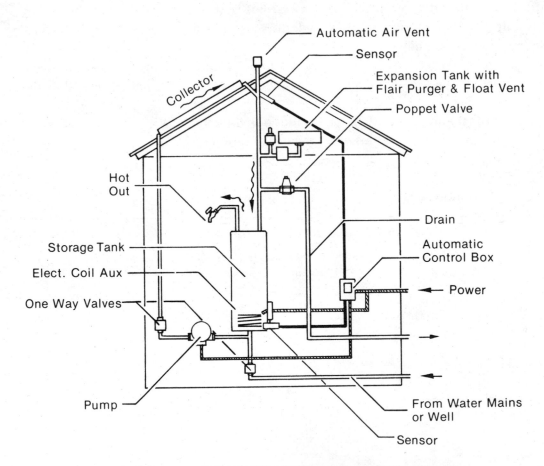

FIGURE 18-8: FORCED CIRCULATION WITH SAFETY DEVICES
(no freeze protection)

in the path of the flow will retard the natural circulation; a good design will create a situation where friction is kept to a minimum.

Let's look at an example to see just how small this pressure difference is. If the storage tank were in the attic and the collector at ground level (a good design) *H* in Figure 18-7 would be about 10 feet. If the colder water at the bottom of the storage tank were at 80°F and the hot water rising up the flow pipe were at 100°F, we'd have a 20°F temperature difference (this would occur for a well-designed unit). This difference in height and difference in temperature (and hence water density) would create a pressure difference of only .015 PSI (pounds per square inch). Compare this with the 30 PSI at the faucet in your sink when the water is coming out! It is obvious that the pressure-causing flow is very small, and the flow rate itself could be very slow. Actually, the total circulation pressure is balanced by the frictional losses of the water flowing through pipes, elbows, valves, unions, etc. Thus, one of the goals for good design in thermosyphon systems is short pipes with few obstructions. Also, relatively large-diameter pipes should be used to reduce the frictional drag.

One phenomenon that you must be aware of is reverse flow. In an improperly designed and/or installed system, the collector can lose heat at night or on cloudy days; heat is then pulled from the storage tank. See Figure 18-9. When the tank is above the collector, the hot water can seek no higher level and is stable. A cold collector will cause a slight but insignificant backflow. If the collector is raised to a level equal to the tank, the flow during heating will still be satisfactory, but now the hot water in the tank is at the same level as the collector and a cold collector will

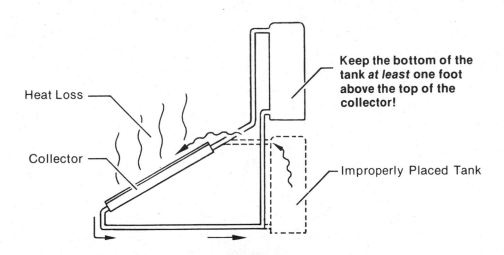

**FIGURE 18-9: BACKFLOW AND HEAT LOSS AT NIGHT
FOR THERMOSYPHON FLOW**

reverse the flow significantly. IMPORTANT: The bottom of the storage tank *must* be *at least* 1 foot above the top of the collector! Two to 10 feet is even better.

A second frequent problem for thermosyphon systems is air being trapped in the piping systems. Water at 32°F contains up to 3 percent air by volume. This air is released when the water is heated and can become trapped in elbows and high spots in pipes, where it stops the flow. Pipes to and from the tank to the collector should slope *toward* the tank at all times to avoid trapping these air bubbles. Also, an air escape vent at the highest point in the storage tank is essential to avoid air build-up.

A final consideration is the placement of the unit on the house. The attic is a good place for the storage tank, as it assures that the storage unit is above the collector. Diagrams (a) and (b) in Figure 18-10 show two possible options. If your roof is flat, the collectors can be placed at ground level and the storage tank just behind the wall and near the ceiling. See diagram (c) in Figure 18-10.

Forced-Circulation Systems

You may want to use one of these units if your house design was such that none of the options outlined in Figure 18-10 would work. For example, if the only area on your home that wasn't shaded was at the peak of the roof, there would be no way to get the storage tank above the collector. A separate system a small distance from the house might work, as was shown in Photo 18-2, but for a roof application a forced-circulation system would work best in this case.

Friction in the pipes is much less serious with these systems, as the circulation pump will overcome any resistance in the pipes. On the other hand, the forced-circulation systems are more complicated with more parts. If you buy a ready-made unit, look for the features outlined in Figure 18-8; on a homemade unit, check with your local plumbing supplier for those parts that will best fit the hot water system you have in mind. Components like the following should be placed in a forced-circulation system:

One-Way Valves. These valves accept flow in one direction only. They are normally placed on the pump outlet to keep it primed and between the water supply and the user. They are also used where convenient and sensible.

Air Vents. Air vents allow dissolved air in water to escape. This prevents gushes in the water line and stops some types of corrosion.

Pressure Release (also called a poppet valve). The purpose of this valve is to prevent excessive pressure build-up in the system under extraordinary circumstances (e.g., a thermostat does not open and the hot water is heated to steam, or the pump fails and steam is generated in the collector).

Expansion Tank. Water expands slightly as it is heated. This tank allows for expansion in a closed, pressurized system.

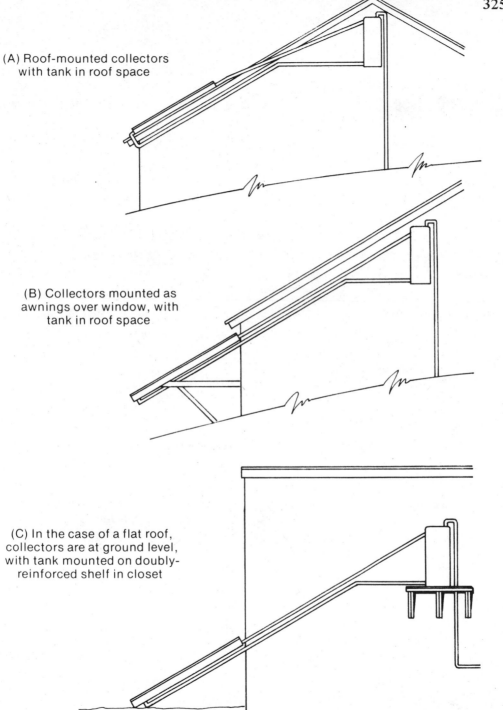

(A) Roof-mounted collectors with tank in roof space

(B) Collectors mounted as awnings over window, with tank in roof space

(C) In the case of a flat roof, collectors are at ground level, with tank mounted on doubly-reinforced shelf in closet

FIGURE 18-10: LAYOUT OPTIONS FOR THERMOSYPHON SYSTEMS

Heat Exchangers

The type of heat exchanger you use differs with the means of water circulation. For a thermosyphon system, I have seen two good ones. The first, by Steve Baer, has a foolproof, simple system of placing a small tank inside a larger barrel. See Figure 18-11. Hot water coming out of the tank can be used directly or put into an existing hot water heater as an auxiliary. Note that the solar collector and the water jacket are *not* pressurized. This makes the absorber in the collector much easier to construct, and no safety pressure vents are needed on the collector water loop.

The second heat exchanger that can be used for thermosyphon systems involves about 15 feet of ½-inch copper tubing in the upper third of a barrel. This scheme substitutes a coil of copper tubing for the tank above. (See Figure 18-12.)

For forced-circulation systems, the principles of the heat exchangers already described still apply, except that now the potable water tank or coil must be *completely* sealed within the outer collector water tank. (See Figure 18-13.) In this case you can use a commercial unit that was designed for steam heating systems. Called "Aquamate," it's available from plumbing supply houses or from

Ford Products Co.
Van Dorn Court
Valley Cottage, NY 10989

FIGURE 18-11: STEVE BAER HEAT EXCHANGER

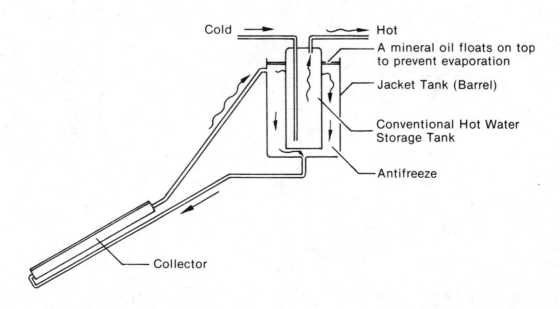

Cold →   → Hot

A mineral oil floats on top to prevent evaporation

Jacket Tank (Barrel)

Conventional Hot Water Storage Tank

Antifreeze

Collector

FIGURE 18-12: COIL HEAT EXCHANGER

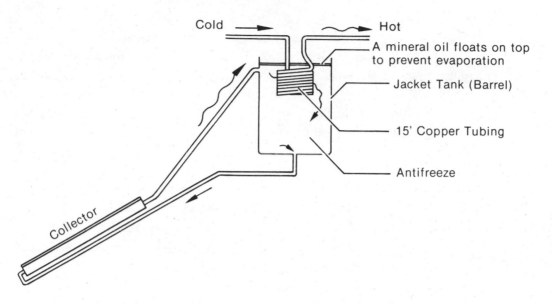

Cold →   → Hot

A mineral oil floats on top to prevent evaporation

Jacket Tank (Barrel)

15' Copper Tubing

Antifreeze

Collector

FIGURE 18-13: A FORCED CIRCULATION HEAT EXCHANGER SYSTEM

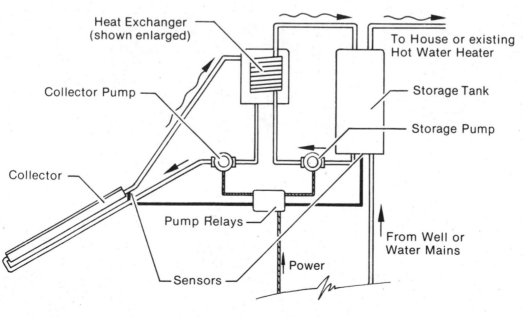

Heat Exchanger (shown enlarged)

Collector Pump

Collector

To House or existing Hot Water Heater

Storage Tank

Storage Pump

From Well or Water Mains

Pump Relays

Sensors

Power

Automatic Control System

Collectors

Producing hot water in the winter requires a double-glass cover in northern climates. Table 18-2 also lists the required number of covers for other climate zones.

The collectors should be "facing" south and inclined to maximize the incident solar energy throughout the year. The optimum direction is due south. However, ±20° east or west of south will result in an insignificant loss. Even a direct southeast or southwest orientation will be about 90 percent as effective as due south, although these extreme orientations will cause the peak water temperature to be shifted to the late morning and late afternoon respectively. Peculiar local weather conditions such as early morning fog or consistent afternoon clouds could bias orientation toward east or west of south. When in doubt, though, due south is best.

In addition to "facing" south, the collector must be tilted from the horizontal. The optimum is: tilt = latitude + 10°. An angle which is ±10° of this optimum will perform 95 percent as well. Consequently, a tilt of 45° is adequate throughout most of the U.S., and it is also easy to construct.

The basic parts of an absorber in a collector are shown in Figure 18-14. Water flows in a distribution header, goes up the risers while being heated, then flows out a

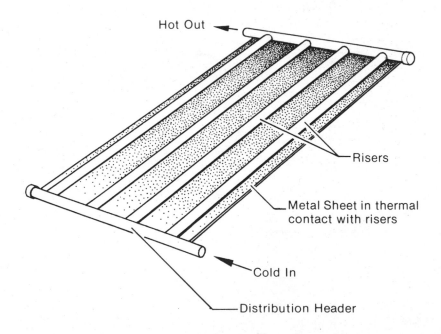

Hot Out

Risers

Metal Sheet in thermal contact with risers

Cold In

Distribution Header

FIGURE 18-14: PARTS OF ABSORBER

second header. To equalize flow in the risers, the inlet and outlet are at opposite corners of the absorber plate.

In forced-circulation systems, the risers are usually a ¼ inch in diameter and 4 to 6 inches apart. The headers are usually a ½ inch in diameter. For thermosyphon flow these passages are larger to reduce friction: risers should be a ½ inch in diameter and 3 to 5 inches apart, with headers 1 inch in diameter.

One comparison you can make between collectors is that of efficiency. The larger the "wetted" area of the collector, that part of the absorber plate which is in contact with the circulating fluid, the greater the heat-removal efficiency of the collector system. The reason for this is that the average temperature of the plate is lower and less heat can escape up through the cover or out the back of the unit. Because of this, low-pressure systems (like thermosyphon units or forced-circulation units with low-pressure pumps) that are *not* directly tied into the city or town high-pressure system offer some real advantages. In Figure 18-15, diagrams (a) and (b) show two simple designs for low-pressure absorbers. Note how the "wetted" area is greater than the others (c, d, e, f) in that figure. Moreover, the low-pressure collectors are lightweight and inexpensive, as the risers can be made with a thinner-walled metal. Maintenance of these low-pressure units is also less of a problem, as fewer leaks occur.

Whether you buy a unit or make one, there are a few other features you should look for. On all units, and particularly high-pressure units, make certain that the bond between the risers and the two headers in the collector is firm. Otherwise, leaks will occur and this kind of maintenance is difficult to deal with. Also, be certain that in any case where the risers are soldered to the absorber plate that *similar* metals are used. Otherwise, the different rates of thermal expansion and contraction between the metals will stress and, within a few years, disintegrate the solder bond. Though the collector might still work, the efficiency will drop off as the thermal contact between riser and plate will have decreased.

High-temperature, flat black paints can be purchased at auto supply shops and hardware stores. They are used on auto exhaust systems, stovepipes, and barbecue grills; and they hold up well at high temperatures. The best pigment for these paints is carbon black. The best flat black paints are sold by the major paint companies, and include 3M's Black Velvet.

Insulation behind the absorber should be of high temperature resistance—good for 350°F. A thickness of 3 to 5 inches is adequate for foams and glass fiber.

The cover glasses should be separated from each other and the absorber by a ½ to 1 inch. The clearest, cheapest, double-strength glass is best; it is usually 1/8 inch thick. The new float process glass is inexpensive. The lifetime of plastic covers is in

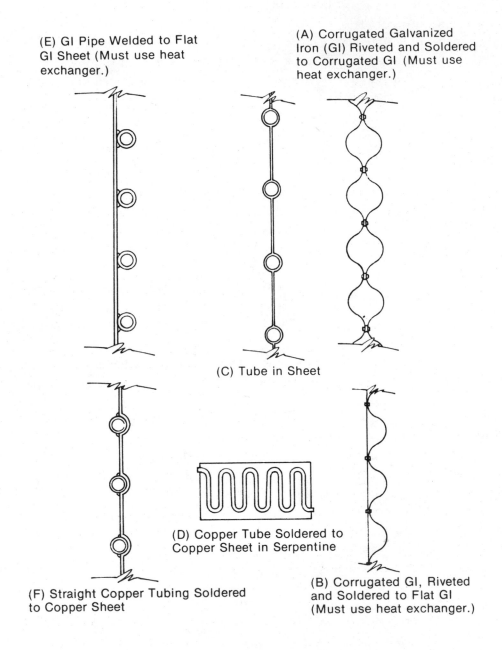

(E) GI Pipe Welded to Flat GI Sheet (Must use heat exchanger.)

(A) Corrugated Galvanized Iron (GI) Riveted and Soldered to Corrugated GI (Must use heat exchanger.)

(C) Tube in Sheet

(D) Copper Tube Soldered to Copper Sheet in Serpentine

(F) Straight Copper Tubing Soldered to Copper Sheet

(B) Corrugated GI, Riveted and Soldered to Flat GI (Must use heat exchanger.)

FIGURE 18-15: TYPES OF ABSORBERS

debate, though polycarbonates look promising. Make sure to use the MONO or silicone caulks mentioned in Chapter 11 to seal the cover glass in place. An airtight, watertight bond is essential here. Also, a felt plug at the bottom of the collector between the inner cover sheet and the plate is a good feature, as it allows this area to breathe.

● **The Tank**

A storage tank for potable (drinking) water should be glass, plastic, or stone lined to prevent corrosion. A standard hot water tank available from a plumbers' supply will work well in the Steve Baer type of heat exchanger. The fluid through the collector loop must have automobile antifreeze and rust inhibitor. (Propylene glycol, grain alcohol, and silicone oils are safer than ethylene glycol.) In the coil-type heat exchanger, no potable water tank is required; a barrel is sufficient for hot liquid storage, and copper tubing is code-acceptable for the coil. See Figure 18-16 for tank inlet and outlet locations.

In a forced-circulation system with an "Aquamate" storage tank, use inlets and outlets recommended by the manufacturer. Also, one-way valves and pressure vents

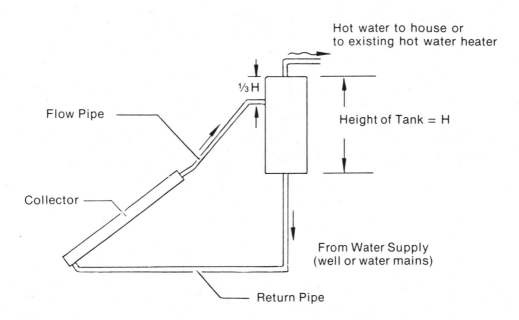

The storage tank must hold the heat harvested by the sun for use.
The last inlet to the storage tank is ⅓ from the top.

FIGURE 18-16: STORAGE TANK CONNECTIONS

are needed for safety and are required by plumbing codes in forced-circulation systems. See Figure 18-8 for details.

Costs

The cost of a solar hot water heating system will depend on the climate where you live and whether or not you build the unit yourself. However, even if you buy and have installed a ready-made commercial unit, it's still a very good investment. In the case of ready-made units, the costs will break down roughly as follows (based on 1975 prices): collector—$6 to $7 per square foot, storage tank—$1 per gallon, pump—$50, heat exchanger—$100, connections—$100, collector supports and rigging—$100, and installation—$100. The total could run up to $1,300 in some areas.

The payback period for such a system would be about 5 to 15 years over most of the U.S. (This was calculated using the prevailing interest rates on money in 1975 and assumed a fuel escalation clause equal to the rate of inflation.) The payback period would be closer to the lower end of the above scale in southern parts of the U.S. and Mexico, and closer to the high side of that scale in the northern U.S. and Canada. Another way to look at the savings is in terms of energy conservation. A solar hot water heating system can save the energy equivalent of about 1 gallon of oil per square foot of collector each year in the northern areas of the U.S. and southern Canada, and about 1½ gallons of oil in the southern parts of the U.S. and Mexico.

WHERE TO GET EQUIPMENT

Build-It-Yourself Units

Several people have put together plans on the owner-built solar water heater. Listed in order of the authors' knowledge about them:

1. *Solar Water Heater Plans:* $5.00; 1974
    Zomeworks Corp. (Steve Baer)
    Box 712
    Albuquerque, NM 87103
    Large, well-detailed shop prints. Two versions: one for all climates (including heat exchanger) and one for nonfreezing climates only.
2. *How To Build a Solar Water Heater:* $1.25; 12 pp, 1973 by the Brace staff
    Brace Research Institute
    MacDonald College of McGill University
    Ste. Anne de Bellevue 800
    Quebec, Canada

Originally developed for Third World countries. Low technology. Good detail on absorber construction.

3. *Solar Domestic Water Heating:* $1.00; 9 pp, 1974
   Sunworks, Inc.
   669 Boston Post Road
   Guilford, CT 06437
   A short paper on suggested systems using Sunworks collectors.

4. *Hot Water (Solar Water Heaters & Stack Coil Heating Systems):* $2.00; 24 pp, 1974 by Morgan & Taylor.
   Hot Water Co.
   350 East Mountain Drive
   Santa Barbara, CA 93108
   Good book outlining the basics of hot water from solar or wood power.

Suppliers of Solar Hot Water Heater Kits

1. Energex Corp.
   5115 S. Industrial Rd.
   Las Vegas, NV 89118
   Kits with an absorber that has a 10-year guarantee. Natural convection flow, freeze protection, all-copper absorber plate. Prices on request.

2. Garden Way Laboratories
   Charlotte, VT 05445
   Kits for forced-circulation units. Freeze protection, aluminum absorber plates sized for individual installations. Write for estimate.

3. Slan Research
   525 N. 5th Street
   Brighton, MI 48116
   Sells water heater kit. Freeze protection by heat exchanger. Write for prices.

4. Sol-Therm Corp.
   7 West 14th Street
   New York, NY 10011
   $700.00 per unit. Includes 32-gallon tank, 32 square feet of collectors, piping, and mounting stand for flat roof. Freeze prevention by manually draining collector. Unit has been marketed in Israel for years.

5. Daystar Corporation
   41 Second Street
   Burlington, MA 01803

   Complete system including freeze protection. Residential system has 2 collectors and 80 gal. storage tank with built-in auxiliary heater. Prices on request.

6. Fred Rice Productions
   6313 Peach Avenue
   Van Nuys, CA 91411

   This system is similar to photo 18-1. Not used in freezing climates. Prices on request.

7. Ray Pack, Inc.
   3111 Agoura Road
   Westlake Village, CA 91361

   Complete system. Non-freezing climates only. Prices on request.

8. Solaron Corporation
   4850 Olive Street
   Commerce City, CO 80222

   Complete system. Unique collector which heats air. Water is heated by heat exchanger. Freeze proof, 80 gal. storage. Write for nearest distributor.

9. Sunworks
   669 Boston Post Road
   Guilford, CT 06437

   Complete system using 2 "solectors" and a 65 gal. storage tank. Thermosyphon system available for nonfreezing climates. Write for nearest distributor.

10. *Spectrum: An Alternate Technology Equipment Directory:* $2.00
    Alternative Sources of Energy
    Rt. 2, Box 90-A
    Milaca, MN 56353

    Lists 500 items in the field, including collectors, plastic and glass cover materials, system parts like differential thermostats, thermal cements, etc.

Section V

PUTTING IT ALL TOGETHER

chapter 19

INTEGRATIONS

by ALEX WADE

Now that we have broken down into detailed elements all the various steps in acquiring efficient housing at low cost and have explained each at length, we proceed in this chapter to reassemble the elements into specific recommendations as applied to actual houses. We will begin with the most involved solutions and work our way downward to simpler housing solutions that require less effort on your part. Study all of the plans carefully as they are annotated to show how the various ideas presented earlier are applied.

A NEW HOUSE: DESIGNED BY YOURSELF—FOR YOURSELF

This kind of housing solution is obviously the most difficult but by far the most rewarding. Traditionally, people the world over have designed *and* built their own shelter. Even if you don't do the actual construction work, you can tailor a house much closer to your own needs if you design it yourself.

If you design a new house yourself, you will probably need a set of plans and specifications to show to the bank, the building inspector, and the contractor (if you have one build the house). Since it is obviously impossible for us to draw your plans for you, we've included a reproduction of working drawings and specifications for a very economically designed house as an example of the kind of detailing you'll need. This plan is one of the more conventional designs of the USDA series. Taken together, these drawings and specifications comprise a complete set of plans for a basic house.

We decided to include one of the USDA plans in this chapter because the full-scale set of drawings, with complete specifications and materials lists for this series of homes, is very inexpensive—in the $1.50 to $2.00 range. Hence, you can follow up on the ideas in this chapter with supplementary materials that are low in cost. However, the USDA plans were put together before the energy crisis and have not yet been updated. Thus, you will have to rework some of the designs to include the winterizing, insulating, ventilating, and solar heating designs found in this book before you actually build one of these houses.

To help you with these energy-saving design features and the low-cost building techniques described in Chapter 6, Alex Wade, a professional architect and the author of this chapter, has drawn up floor plans and construction sections for a salt-box house and a shed-type house, several floor plans for renovations, and a floor plan for a garage-studio. If the reader feels that he or she wants to build using the plans for the salt-box house or the shed house, a full set of working drawings can be obtained for $20 from Alex Wade, Station Road, Barrytown, N.Y. 12507. The author will also be happy to make modifications on these basic designs for a reasonable fee.

The most important single decision to be made is the siting of your new house. Study the sample site plans carefully to see that you are not making any obvious blunders. Also, study with particular care the foundation plan, floor plan, and cross sections. These drawings show all the vital information that is necessary for construction of your house.

When you draw up your own set of plans, make sure to have someone else double-check your work before you consider the plans complete. If you don't have the time or don't feel that you have the necessary skills to draw the plans, then by all means put an ad in your local newspaper for a draftsman to do the work for you. Let the draftsman read this book if he is inclined to, as he may be able to add some valuable ideas to your basic design.

In order to make absolutely sure that you correctly visualize the finished results of your house, make a model at the scale of ¼ or ½ inch to the foot to check yourself. Let the size of your house determine the scale of your model. Generally, the ½-inch scale is preferable—especially for a small house. The model will be very helpful in understanding just how roof slopes and window placements will appear on the final house. Making a model will also prove helpful if you are using stock plans or remodeling. The best material for a beginner to use is "styroboard" or "foam core board," both available at art supply stores. (See Photo 19-1 for various views of a single-house model.)

One final note of caution: make sure that you look up the references for post and beam structural design that are listed in the bibliography for Chapter 6. Follow them *carefully,* making adequate allowances for snow loads, heavy roofing materi-

Photo 19-1: Four views of an owner-builder's scale model of a future house.

als, and unusual code requirements. If you don't fully understand the procedure, get a contractor, architect, or engineer to check your work for you. If you would like to see a full set of plans at large scale (including a detailed materials list), send $1.50 to the Superintendent of Documents, U.S. Government Printing Office, Washington, D.C. 20402. Request plan #FS-FPL-4.

PLANS

What follows is a set of plans for USDA house #FS-FPL-4, plus a site plan for this house. *Note that these plans have been reduced to fit into the format of this book.* Therefore, pay attention to the dimensions marked in the drawings, *not* to the drawing size.

If you do your own drawings for a house, there are several basic procedures you should use. A good standard drawing size for a small house is 18 inches × 24 inches. It will pay to have your blueprint shop make about six sets of prints from your tracings because prints wear out rather quickly. One set of plans should be mounted on cardboard and covered with plastic to keep them dry and stop them from blowing away on the job. You will need other sets for the bank, the electrician, the plumber, etc. Also, the plans fade in sunlight, wear out, and get lost, so you will need extras. Next, the size of each drawing is important. On most drawings ¼ inch = 1 foot is a good size. However, on construction sections, either 3/8 or 1/2 inch to the foot will be a good size. On special details, it's best to use a size of at least 1 inch to the foot to make things clearer.

Two methods of signifying length have been used in the drawings that follow— the usual feet and inch symbols and another with feet marked in large print and inches marked as a superscript with a line under it as, for example, 3'10" or $3^{\underline{10}}$. Construction terms used in the plans and specifications have been explained in the Glossary in the Appendix.

SPECIFICATIONS

The specifications that follow are for a typical small house with standard wood framing for walls, floor, and roof. Slight modifications in the text will have to be made if you wish to use post and beam construction or the winterizing, insulating, ventilating, or solar heating ideas in this book. Two alternate foundation systems designed to save considerable amounts of money are shown in Chapter 6: concrete slab on grade and embedded, treated wood posts.

These specifications taken in conjunction with the working drawings comprise a full set of contract documents for a small house.

All work related to construction of this house shall be done in a first-class manner. Any detail not included in specifications or plans shall comply with accepted practices for wood-frame construction.

Excavation, Grading. Sod, growing plants, shrubs, stumps, and trees shall be removed and ground smoothed in the building area and 5 feet outside of building line. Excavate footings to required depth and size shown in plans. No forms are required if soil is stable. Locate excavated soil conveniently for backfilling around treated wood posts or concrete slab.

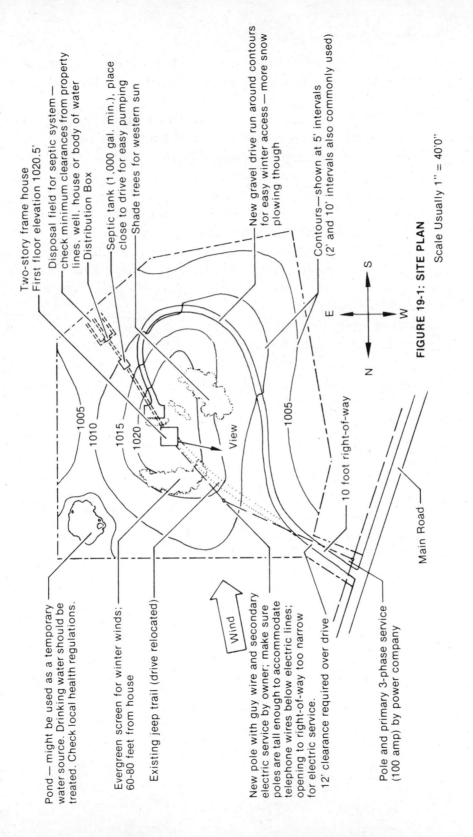

Two-story frame house
First floor elevation 1020.5'

Disposal field for septic system — check minimum clearances from property lines, well, house or body of water

Distribution Box

Septic tank (1,000 gal. min.), place close to drive for easy pumping

Shade trees for western sun

New gravel drive run around contours for easy winter access — more snow plowing though

Contours — shown at 5' intervals (2' and 10' intervals also commonly used)

1005
1010
1015
1020

View

1005

10 foot right-of-way

Main Road

Pond — might be used as a temporary water source. Drinking water should be treated. Check local health regulations.

Evergreen screen for winter winds; 60-80 feet from house

Existing jeep trail (drive relocated)

Wind

New pole with guy wire and secondary electric service by owner; make sure poles are tall enough to accommodate telephone wires below electric lines; opening to right-of-way too narrow for electric service.
12' clearance required over drive

Pole and primary 3-phase service (100 amp) by power company

N   S

E   W

FIGURE 19-1: SITE PLAN

Scale Usually 1" = 40'0"

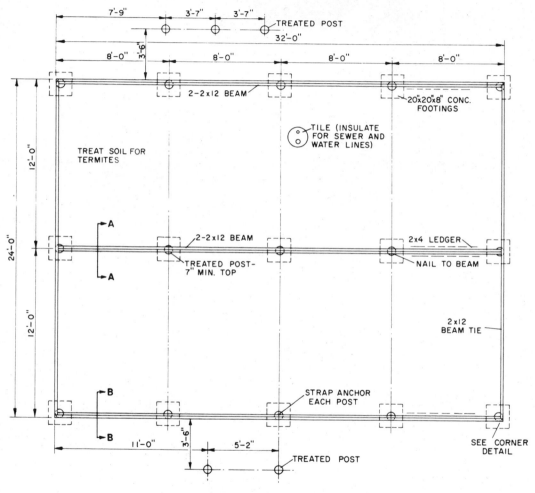

FOUNDATION PLAN

FIGURE 19-2

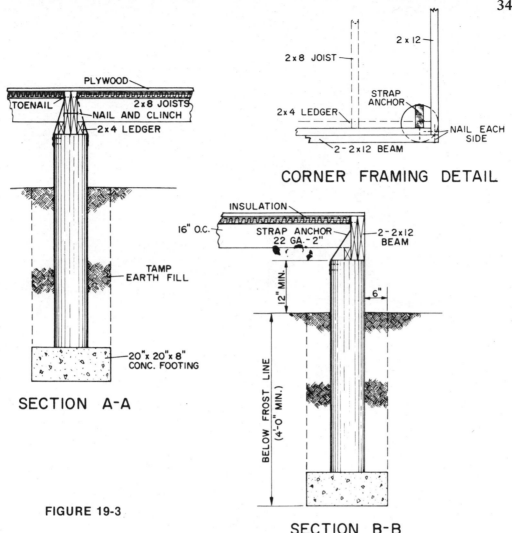

CORNER FRAMING DETAIL

SECTION A-A

FIGURE 19-3

SECTION B-B

Notes on Foundation Plan

Dimensions are the key item to be checked on your foundation plan. They are basic to the house; and, if they don't agree with those on your floor plans, it will be too bad, as the contractor will start the construction based upon the foundation plans. The foundation plan for a pole house is simpler than for conventional masonry construction. However, if any mistakes are made, there will either have to be expensive changes in the foundation or the upper floor plans must be altered, possibly disastrously.

An economical alternate foundation plan would be for a slab on grade. This plan is so simple that it may not even be necessary to draw one. If you have a square or rectangular house, all you need is a sketch of the slab with the length and width noted along the edges. If you plan a heavy masonry wall, chimney, or other special conditions, you will have to dimension their location, size, and reinforcements on the plan. We have described in the accompanying specifications the necessary reinforcements for a typical foundation slab. Any entrances for piping (water, sewage, or underground electric) should be shown on the foundation plan.

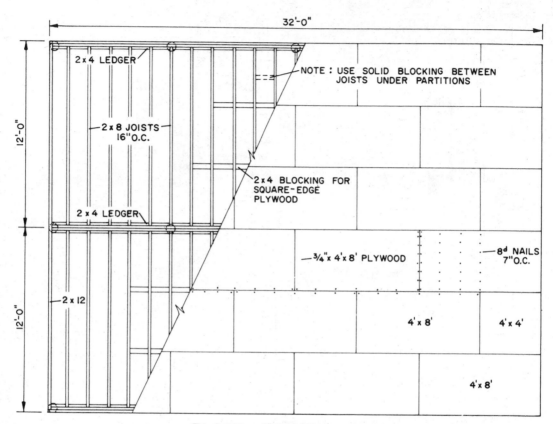

FLOOR FRAMING PLAN

FIGURE 19-4

Notes on Floor Framing Plan

For a slab on grade structure, this plan is not necessary. If you use post and beam framing, the structural members will be spaced further apart and be simpler to draw. It is not necessary to draw each joist as in the illustration, but it may help you to visualize the structure. Again, check dimensions very carefully. Make sure that the arrows go to exactly the same points and that the overall totals read the same on all plans. Check structural tables to see that your floor joists are strong enough to carry the load, particularly if you use widely spaced beams for post and beam construction.

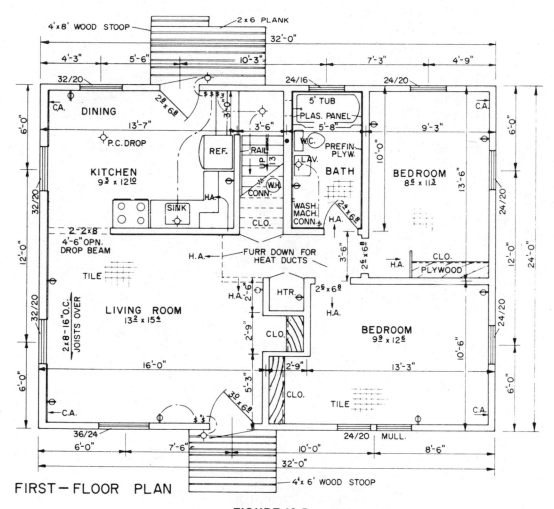

FIRST—FLOOR PLAN

FIGURE 19-5
Notes on First-Floor Plan or Floor Plan for One-Story Houses

This sheet will contain most of the important information for your house. After you draw the actual outlines on tracing paper, you may want to get your blueprint shop to make a reproducible copy for you so that you can split up some of the information. This will make the plan less confusing and save redrawing it.

Check to make sure that your plan contains the following information:

1. Complete dimensions locating all partitions, doors, windows, etc.
2. Names for each space for identification purposes (saves any confusion)
3. Sizes and spacing of second floor or roof beams or joists
4. Sizes of windows and doors
5. Direction of door swings (check carefully that they don't interfere with light switches, furniture placement, etc.)
6. Location of plumbing fixtures and vent stack
7. Location of chimney, heater, or furnace, and locations of supply and return diffusers
8. Location of electric lights, switches, and outlets
9. Locations of any furred down ceilings (dashed lines)
10. Check carefully with basement and second floor plan to make sure that vent stacks, partitions, stairs and any other necessary items line up from floor to floor

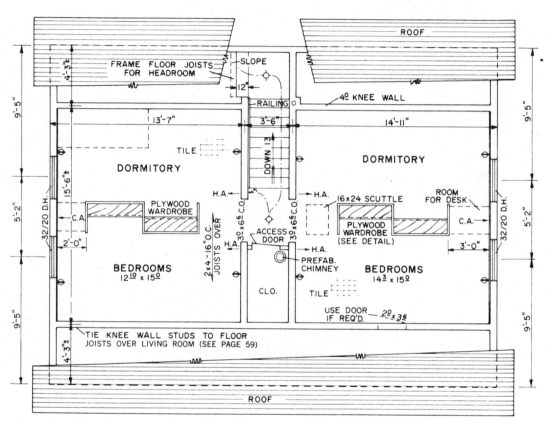

SECOND—FLOOR PLAN

Notes on Second-Floor Plan
Same comments apply as for First-Floor Plan

FIGURE 19-6

FIGURE 19-7

Notes on Construction Section

This drawing describes the overall shape of your house and shows ceiling heights, clearances and the like.

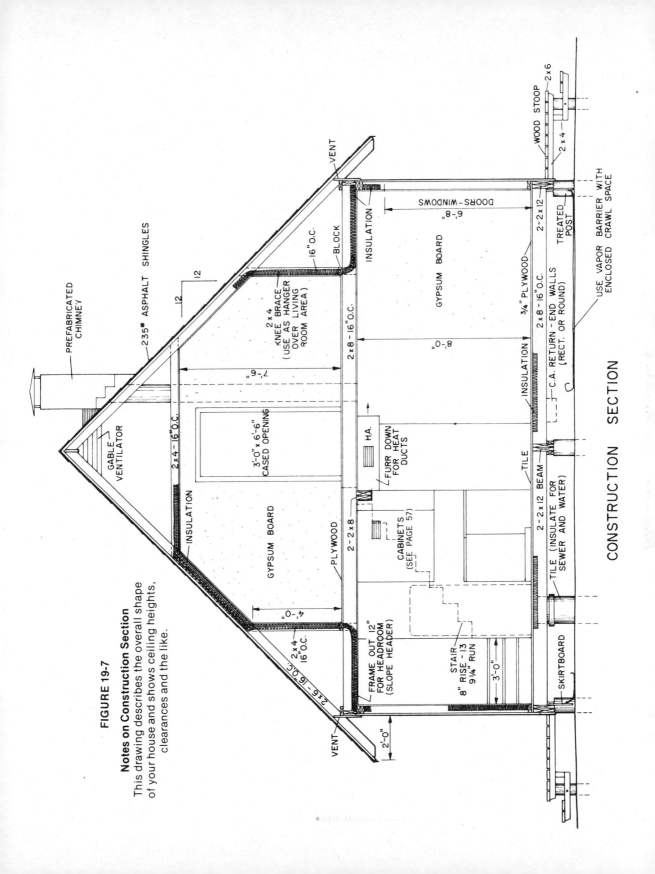

CONSTRUCTION SECTION

PREFABRICATED CHIMNEY

235# ASPHALT SHINGLES

GABLE VENTILATOR

12
12

VENT

BLOCK

16" O.C.

2 x 4
KNEE BRACE
(USE AS HANGER
OVER LIVING
ROOM AREA)

INSULATION

2 x 8 - 16" O.C.

DOORS - WINDOWS
6'-8"

GYPSUM BOARD

8'-0"

INSULATION

3/4" PLYWOOD

2 x 8 - 16" O.C.

C.A. RETURN - END WALLS
(RECT. OR ROUND)

2 - 2 x 12

TREATED POST

USE VAPOR BARRIER WITH
ENCLOSED CRAWL SPACE

WOOD STOOP

2 x 6

2 x 4

7'-6"

2 x 4 - 16" O.C.

3'-0" x 6'-6"
CASED OPENING

INSULATION

GYPSUM BOARD

PLYWOOD

2 - 2 x 8

H.A.

FURR DOWN
FOR HEAT
DUCTS

CABINETS
(SEE PAGE 57)

TILE

2 - 2 x 12 BEAM

TILE (INSULATE FOR
SEWER AND WATER)

SKIRTBOARD

2'-6" - 16" O.C.

4'-0"

2 x 4
16" O.C.

FRAME OUT 12"
FOR HEADROOM
(SLOPE HEADER)

STAIR
8" RISE – 13
9 1/4" RUN

3'-0"

2'-0"

VENT

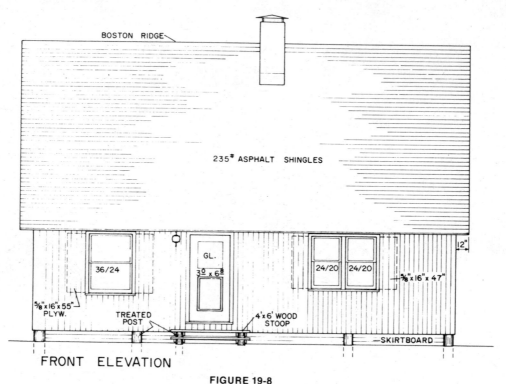

BOSTON RIDGE

235# ASPHALT SHINGLES

GL.
3⁰ x 6⁸

36/24

24/20   24/20

12"

⅝"x 16"x 55"
PLYW.

⅝"x 16"x 47"

TREATED
POST

4'x 6' WOOD
STOOP

SKIRTBOARD

FRONT ELEVATION

FIGURE 19-8

Notes on Elevations

The elevations primarily show what the exterior of your house looks like. Materials which are visible on the exterior are noted here. Window and door sizes are usually repeated here.

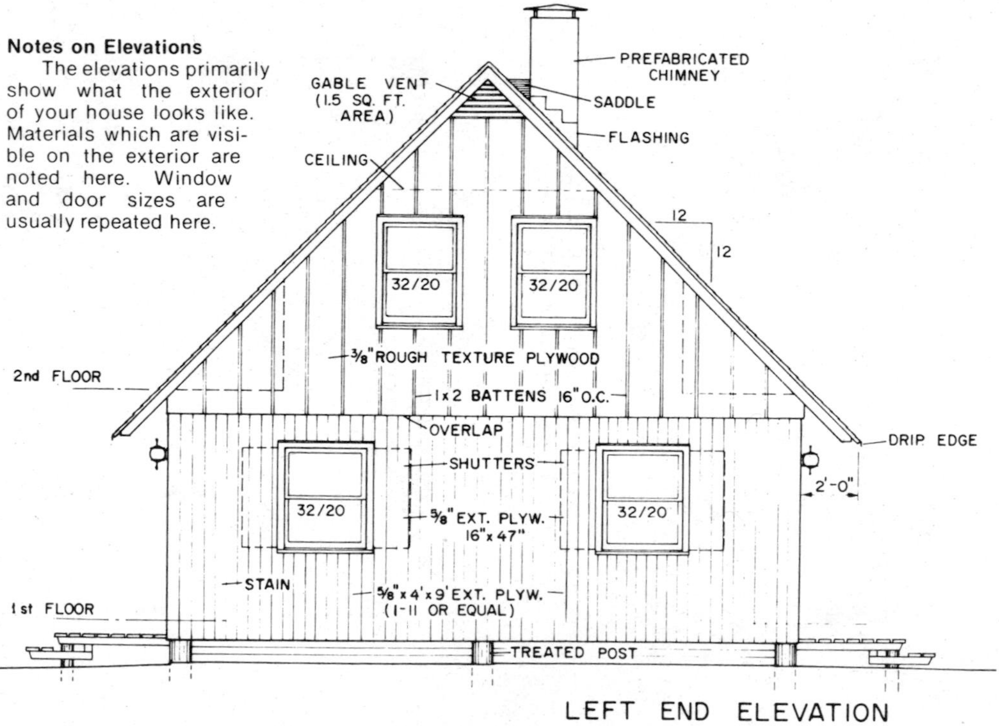

PREFABRICATED
CHIMNEY

GABLE VENT
(1.5 SQ. FT.
AREA)

SADDLE

FLASHING

CEILING

32/20   32/20

12

12

⅜"ROUGH TEXTURE PLYWOOD

2nd FLOOR

1 x 2 BATTENS 16"O.C.

OVERLAP

DRIP EDGE

SHUTTERS

2'-0"

32/20

32/20

⅝" EXT. PLYW.
16"x 47"

STAIN

⅝"x 4'x 9' EXT. PLYW.
(1-11 OR EQUAL)

1 st FLOOR

TREATED POST

LEFT END ELEVATION

FIGURE 19-9

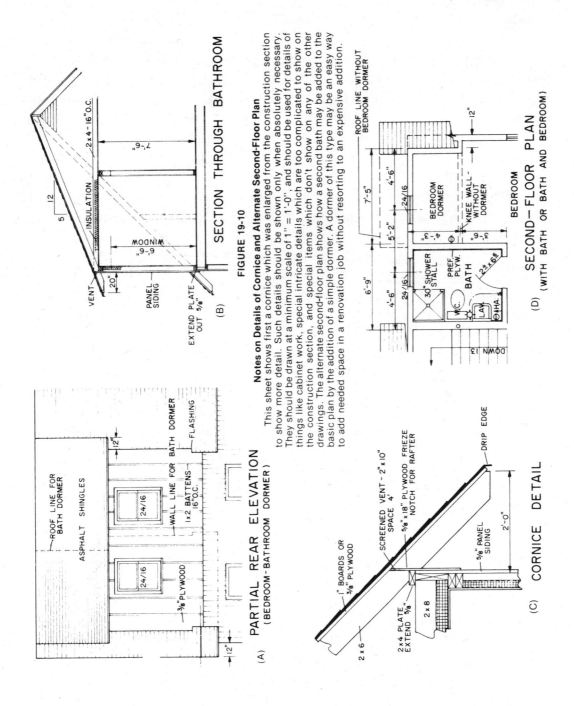

SECTION THROUGH BATHROOM

FIGURE 19-10

Notes on Details of Cornice and Alternate Second-Floor Plan

This sheet shows first a cornice which was enlarged from the construction section to show more detail. Such details should be shown only when absolutely necessary. They should be drawn at a minimum scale of 1" = 1'-0", and should be used for details of things like cabinet work, special intricate details which are too complicated to show on the construction section, and special items which don't show on any of the other drawings. The alternate second-floor plan shows how a second bath may be added to the basic plan by the addition of a simple dormer. A dormer of this type may be an easy way to add needed space in a renovation job without resorting to an expensive addition.

(A) **PARTIAL REAR ELEVATION**
(BEDROOM-BATHROOM DORMER)

(C) **CORNICE DETAIL**

(D) **SECOND-FLOOR PLAN**
(WITH BATH OR BATH AND BEDROOM)

Concrete Work. Footings shall be poured over undisturbed soil in excavations to indicated thickness and size as shown on plan. Top surface to be level.

Mix concrete to a $1:2\frac{1}{2}:3\frac{1}{2}$ mix. If premixed concrete is available, use 5-bag mix.

Concrete Floor Slab Foundation (Alternate Foundation #1). Concrete slab shall be poured of concrete as just specified. Edges of slab shall be turned down 3 to 4 feet (below frost line). Top surface of slab shall be machine-troweled smooth. Reinforce with $6 \times 6 \times 10/10$ mesh turned down at edges. In addition, reinforce bottom edge of turned down footing with one #4 bar continuous. Place slab over 4-inch bank and run gravel and 6-mil polyethylene vapor barrier. Test any buried piping and double-check location before pouring slab.

Treated Wood Posts (Alternate Foundation #2). Pressure-treated wood posts with 7-inch minimum top diameter shall be treated to conform to Federal Specification TT-571.

Carpentry

This branch of the work comprises all rough and finish carpentry necessary to complete the house, as shown on the plans and as specified. This includes layout, cutting and fitting, framing, and other carpentry items. Any phase that is necessary for the completion of the house and not specifically covered in the plans or specifications shall be included as required.

Wood Framing. Dimension material for studs to be standard (third grade); and for floor joists and framing, ceiling joists, rafters, beams, and trusses construction (second grade) in Douglas fir, southern pine, or equivalent,* unless otherwise noted on the plans. All floor and ceiling joists, studs, and rafters shall be spaced 16 inches on center. Moisture content of framing lumber is not to exceed 15 percent.

Subfloor. The subfloor shall be plywood to serve as a floor alone or as a base for finish floor material. It shall consist of ¾-inch, C-C plugged, exterior-grade, touch-sanded Douglas fir, southern pine, or equivalent. Use 2×4-inch blocking for all longitudinal joints. Toenail 2×4-inch blocks flatwise in each joist space.

Roof Sheathing. Roof sheathing shall be 3/8-inch Douglas fir or southern pine plywood or its equal in standard sheathing grade, or nominal 1×6 or 1×8-inch boards in No. 3 Douglas fir, southern pine, or equivalent. Boards shall be square edge, shiplap, or dressed and matched, and laid up tight at a moisture content of not more than 12 percent.

---

*See wall-framing grades and span tables for allowable spans of other species of wood (Federal Housing Administration tables) and American Lumber Standards for sizes and other information.

Insulation. Floors and walls of the first floor, walls and ceiling of the second floor, and the second-floor area outside of the knee walls to be insulated with standard batt- or blanket-type, flexible insulation with vapor barrier placed toward the inside of the building. Unless otherwise specified, minimum insulation thicknesses in the central and northern tiers of states shall be as follows: ceiling—10 inches; wall and floor—4 inches. Vapor barrier shall be placed toward the inside of the building. Vapor barrier to be 6-mil polyethylene.

Siding. Panel siding at gable ends shall be 3/8-inch thick and 4-foot wide exterior-grade plywood in roughsawn pattern with 1 X 2-inch batten at each stud. Remaining walls shall be panel siding of 5/8-inch X 4 X 9-foot exterior-grade plywood in texture 1-11 (4-inch grooves) or equivalent. Edges of the sheets shall be dipped or brush coated with a water-repellent preservative before installation. A pigmented stain finish is recommended. Plywood shall be nailed at each stud and at ends of panels with galvanized or other rust-resistant nails spaced 7 to 8 inches apart. Use sixpenny nails for the 3/8-inch plywood and eightpenny for the 5/8-inch plywood. Optional: 1-inch roughsawn pine, random-width board and batten pattern applied over 25/32-inch insulating sheathing (Celotex, Homosote, or equal).

Exterior Millwork

Exterior finish. Exterior trim and similar materials shall be No. 2 ponderosa pine or equivalent suitable for staining.

Window frames and sash. Complete double-hung windows shall be used with upper and lower sash cut to two horizontal lights, glazed with single-strength glass. Units to be treated with water-repellent preservative as outlined in Commercial Standard CS 190-64. Sash to be furnished fully balanced and fitted and with outside casing in place. Set in openings, plumb, and square. Screens shall be furnished, and storms when required.

Exterior door frames and doors. Exterior door frames shall have 1 3/8-inch rabbeted jambs and 1 5/8-inch oak sill or softwood sill with metal edge, all assembled. Set in openings, plumb, and square.

Exterior doors to be standard 1 3/4 inches with solid stiles and rails; panel type with glazed openings as shown on plans. Screen doors, or combinations when required, shall be furnished.

Screens. Galvanized fly screen shall be used for gable end-outlet ventilators and for inlet ventilators located in the plywood frieze board, as shown on plans.

Interior Millwork

Interior door frames and doors. Interior door frames and cased openings shall be nominal 1-inch ponderosa pine or equal in "D" select. Stops shall be installed only where doors are specified.

Interior doors shall be 1 3/8-inches thick, five-cross-panel style with solid stiles and rails.

Interior trim. Interior trim shall be ponderosa pine or equal in "D" select in ranch pattern in the following sizes:

Casings   – 11/16 × 2 1/4 inches
Stops   – 7/16 × 1 3/8 inches or wider
Base   – 7/16 × 2 1/4 inches
Base shoe – 1/2 × 3/4 inch (when required)
    Note: (1) Casing to be used at bottom of windows in place of stool and apron.
    (2) Base shoe used along freestanding plywood wardrobes.

Option: Use 1 × 2-inch roughsawn pine with butt joints for all above trims.

Walls and ceilings. All ceilings shall be finished with ½-inch gypsum board with recessed edges and with the length applied across the ceiling joists. End joints shall be staggered at least 16 inches. Walls shall be finished with ½-inch gypsum board with recessed edges and applied horizontally. Application and joint treatment shall follow accepted practices.

Walls of tub recess shall be covered with plastic-finished hardboard panels over gypsum board. Install with mastic in accordance with manufacturer's directions. Inside corners, edges, and tub edges shall be finished with plastic moldings.

Flooring. Finish flooring throughout shall be 1/8-inch-thick asphalt tile in 9 × 9-inch size, "B" quality. Combination plywood subfloor shall be cleaned, nails driven flush, and joints sanded smooth where required. Tile shall be applied in accordance with manufacturer's recommendations. Rubber baseboard shall be furnished and installed in the bathroom; wood base in the remainder of the house.

Hardware. Furnish and install all rough and finish hardware complete as needed for perfect operation. Locks shall be furnished for all outside doors. Outside doors to be hung with three 4 × 4-inch, loose-pin butt hinges and inside doors with two 3½ × 3½-inch loose-pin butt hinges. Bathroom door shall be furnished with standard bathroom lock set. Standard screen door latches, hinges, and door closers shall be furnished and installed. Furnish and install semiconcealed cabinet hinges, pulls, and catches where required for cabinet or closet doors. All finish hardware to be finished in dull brass.

Sheet-Metal Work and Roofing

Sheet Metal. Sheet-metal flashing, when required, shall be 28-gauge galvanized iron or painted terneplate.

Roofing.   Roofing shall be a minimum of 235-pound, square-tab, 12 X 36-inch asphalt shingles. A minimum of four 7/8-inch galvanized roofing nails shall be used for each 12 X 36-inch shingle strip. Any defects or leaks shall be corrected.

Electrical Work

The work shall include all materials and labor necessary to make the systems complete as shown on the plans. All work and materials shall comply with local requirements or those of the National Electrical Code. All wiring shall be concealed and carried in BX or other approved wire to each outlet, switch, fixture, and appliance or electrical equipment such as furnace, hot water heater, and range when required and as shown on the plan. Panel shall be 100-ampere capacity with overload cutout. Wall fixtures to consist of the following:

> Two outside wall fixtures with crystal glass
> Two (wall) fixtures, for kitchen and bath

Heating

Heater and prefabricated chimney shall be installed as shown on the plans with supply ducts located in furred-down ceiling of hall. Heater shall be for LP or natural gas with a 100,000 minimum BTU input or as required by design for each specific area (usually a *much* smaller unit in a well-insulated small house). Cold air return at furnace base and from each corner of the house. Cold air duct shall consist of 1/8-inch transite covered joist spaces along outside walls and connection to 10-inch-diameter or 6 X 12-inch rectangular galvanized or aluminum ducts to heater.

Plumbing

All plumbing shall be installed in accordance with local or national plumbing codes. Hot and cold water connections shall be furnished to all fixtures as required. Sewer and water and gas lines (when required) shall extend to building line with water shutoff valve. Framed and insulated box shall be used where required to protect water and sewer lines from freezing in crawl space. Cover with 1/8-inch transite or equal and insulate with 3 inches of fiberglass or styrofoam when required. Use a 16 X 16-inch vitrified tile or equal below the groundline.

Furnish and install the following fixtures:

> One kitchen sink—21 X 15 inches, self rim, steel, white, with fixtures.
> One bathtub—5 feet, cast iron; white; right-hand drain; complete with shower rod, shower head, and fixtures.
> One water closet—reverse trap; white; with seat, fixtures, and shutoff valve.
> One lavatory—19 X 17 inches, steel, white, with fixtures and shutoff valves.

One hot water heater—50 gallon (minimum), gas or electric.
Washing machine connection with hot and cold water and drain.

Painting and Finishing

Exterior.   Exterior plywood panel siding, facia, shutters, and soffit areas shall be stained with pigmented stain as required. Use light gray, yellow, or another light-color stain for the trim and shutters and a darker stain (brown, olive, gray, etc.) for the panel siding or a reverse color selection as desired by owner. (Optional shutters to be stained on faces and edges.)

Window and door frames, window sash, screen doors, and similar millwork shall be stained a light color. Types of paint and procedures to comply with recommended practices and materials.

Interior.   All woodwork shall be stained. Walls and ceilings finished in latex flat; bathroom woodwork and ceiling shall be finished with semigloss.

Termite Protection

Termite protection shall be provided by means of soil treatment or termite shields or both as required by local practices and regulations.

Summary

The *working drawings* and *specifications* taken together comprise the *contract documents* for your house if you hire a contractor to build it for you. Obviously, if you do the work yourself, you will need less in the way of drawings. The bank may require formal drawings, however. Since the drawings and specifications work together, you do not need to repeat information shown on one in the other. Check for conflicts of information between the two items. If you show something in either the plans or specifications, it is legally included. If it's not included, you will have to pay the contractor extra for it. The corollary of this is not to overdo and make the plans too complicated or the contractor will bid more money for the job just because it looks like a lot of work. Many architects trying to do a thorough job fall into this trap.

MAKING A COMPLETE SET OF PLANS FOR RENOVATIONS

Plans for renovating a house are at once simpler and much more difficult than for a new house. They are simpler because you can get by with fewer drawings (usually just floor plans showing partition changes and maybe some elevations will be enough). They are more difficult because much more information concerning finishes must be conveyed. What to take out and what to add? Also, when you open up walls, you can find rotted structural members, piping in the way, etc.; and this may necessitate a change in your plans. It can be very confusing, and the possibilities for misunderstandings with the contractor are great. A way of avoiding much of this trouble is to walk through the house, room-by-room, with the contractor after the basic scope of the project has been established. Discuss each item in each room and record your decision on tape. The transcription of the tape will serve to record for all parties what decisions were made.

Later in this chapter, you will find further information on renovations.

DESIGNING A NEW HOUSE BASED UPON OUR DRAWINGS OR USDA DESIGNS

Next, we present supplemental drawings for our basic two-story square house and our low-cost shed-roof four-pole house, both of which were shown in Chapter 6. The two-story house is excellent for a flat site slab on grade solution, while the four-pole house should be strongly considered for remote or steeply sloping rugged sites. The cross sections we show reflect these types of terrain. They also show you how to take advantage of otherwise wasted attic areas for sleeping lofts. These drawings are not intended to provide complete working drawing information for these houses, but rather to provide the basic information necessary so *you* can prepare working drawings to suit your own needs and your own particular site. Since these are self-supporting post and beam structures, it is possible to move the interior partitions around almost at will without disturbing the structure. The same applies to window design and placement. We show a variety of window designs on the elevations to act as suggestions. These are shown only as examples. You will be much better off if you carefully study the prevailing wind patterns and the desired views and privacy considerations that were discussed earlier in the book; size and place your windows and partitions accordingly. Extra appendages such as decks, greenhouses, and vestibules should be carefully located in conjunction with the windows. Again, build a model.

What follows now is a variety of economical house plans that you may wish to use as a source of ideas for your own house plans. The first plans are from the USDA series, which we have mentioned throughout the book. The first of these plans is #FS-FPL-2. It is designed *specifically* for very large families and hence is bigger than many of the other designs. It has the added advantage of having complete plans and specifications available for nominal cost—$1.50 as opposed to $25-$35 for typical plans. (This is one way to get back some of your tax dollars!)

Very few house plans on the commercial market other than the USDA series are

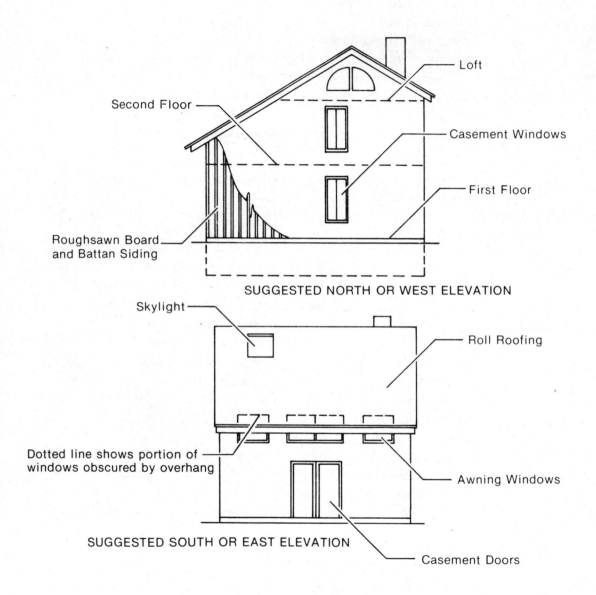

SUGGESTED NORTH OR WEST ELEVATION

Loft

Second Floor

Casement Windows

First Floor

Roughsawn Board and Battan Siding

Skylight

Roll Roofing

Dotted line shows portion of windows obscured by overhang

Awning Windows

SUGGESTED SOUTH OR EAST ELEVATION

Casement Doors

FIGURE 19-11: ELEVATIONS — TWO-STORY BASIC HOUSE

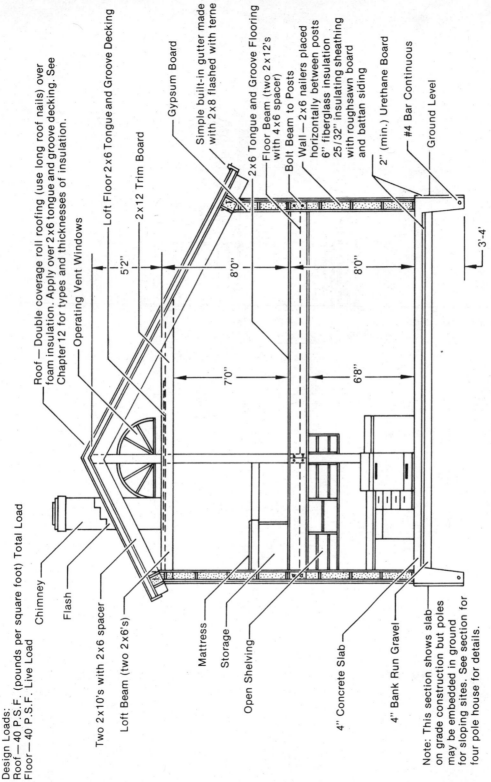

Design Loads:
Roof — 40 P.S.F. (pounds per square foot) Total Load
Floor — 40 P.S.F. Live Load

Chimney

Flash

Two 2x10's with 2x6 spacer

Loft Beam (two 2x6's)

Mattress

Storage

Open Shelving

4" Concrete Slab

4" Bank Run Gravel

Note: This section shows slab on grade construction but poles may be embedded in ground for sloping sites. See section for four pole house for details.

Roof — Double coverage roll roofing (use long roof nails) over foam insulation. Apply over 2x6 tongue and groove decking. See Chapter 12 for types and thicknesses of insulation.

Operating Vent Windows

Loft Floor 2x6 Tongue and Groove Decking

2x12 Trim Board

Gypsum Board

Simple built-in gutter made with 2x8 flashed with terne

2x6 Tongue and Groove Flooring

Floor Beam (two 2x12's with 4x6 spacer)

Bolt Beam to Posts

Wall — 2x6 nailers placed horizontally between posts 6" fiberglass insulation 25/32" insulating sheathing with roughsawn board and batten siding

2" (min.) Urethane Board

#4 Bar Continuous

Ground Level

5'2"

8'0"

7'0"

6'8"

8'0"

3'-4"

FIGURE 19-12: CONSTRUCTION SECTION — TWO-STORY BASIC HOUSE

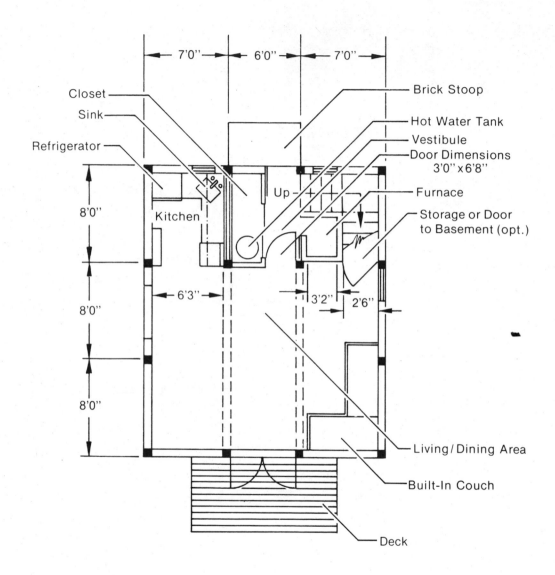

FIGURE 19-13: FIRST-FLOOR PLAN — TWO-STORY BASIC HOUSE

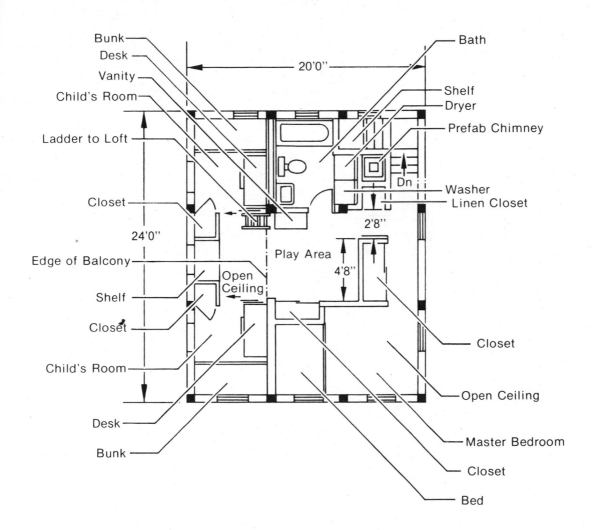

Bunk
Desk
Vanity
Child's Room
Ladder to Loft
Closet
Edge of Balcony
Shelf
Closet
Child's Room
Desk
Bunk

Bath
Shelf
Dryer
Prefab Chimney
Washer
Linen Closet
Closet
Open Ceiling
Master Bedroom
Closet
Bed

20'0"
24'0"
2'8"
4'8"
Dn
Play Area
Open Ceiling

FIGURE 19-14: SECOND-FLOOR PLAN — TWO-STORY BASIC HOUSE

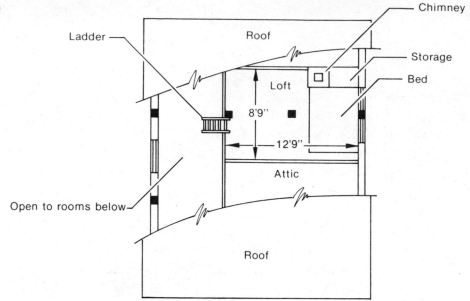

FIGURE 19-15: LOFT FLOOR PLAN — TWO-STORY BASIC HOUSE

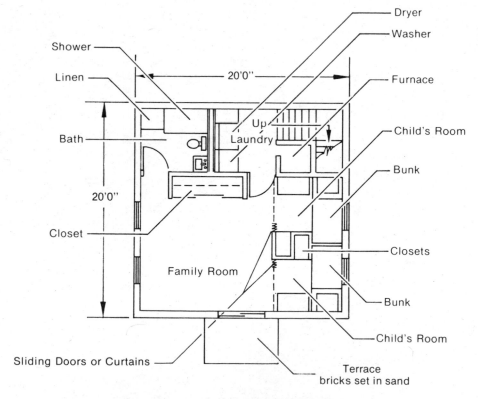

FIGURE 19-16: ALTERNATE BASEMENT PLAN
(for 20 foot square version of two story house shown in Chapter 6)

NOTE: Numerous variations are possible. This house can also be built as a slab on grade house. The roof line in that case could be changed to a saltbox in similar fashion to the two story house. Siding and window treatment could also be similar to the two story house.

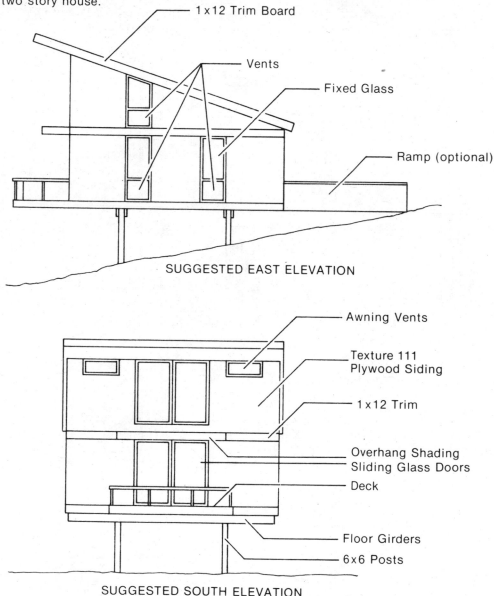

SUGGESTED EAST ELEVATION

SUGGESTED SOUTH ELEVATION

FIGURE 19-17: ELEVATIONS — FOUR-POLE HOUSE

(See Chapter 6 for a floor plan that would be useful in a shed house.)

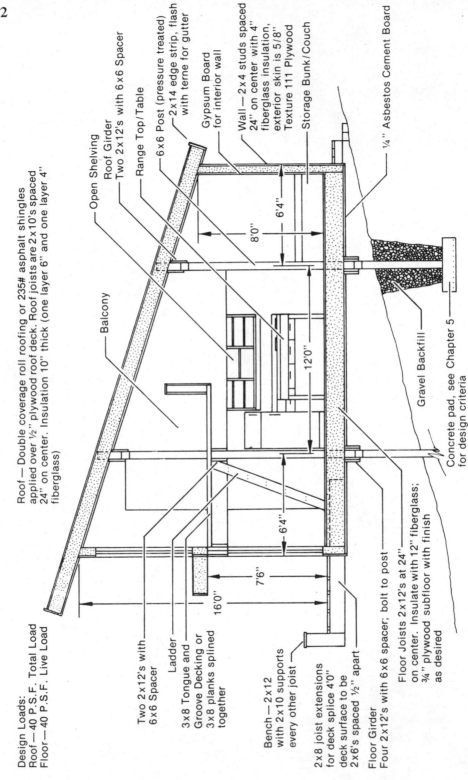

Design Loads:
Roof — 40 P.S.F. Total Load
Floor — 40 P.S.F. Live Load

Roof — Double coverage roll roofing or 235# asphalt shingles applied over 1/2" plywood roof deck. Roof joists are 2x10's spaced 24" on center. Insulation 10" thick (one layer 6" and one layer 4" fiberglass)

Open Shelving

Roof Girder
Two 2x12's with 6x6 Spacer

Range Top/Table

6x6 Post (pressure treated)

2x14 edge strip, flash with terne for gutter

Gypsum Board
for interior wall

Wall — 2x4 studs spaced 24" on center with 4" fiberglass insulation, exterior skin is 5/8" Texture 111 Plywood

Storage Bunk/Couch

1/4" Asbestos Cement Board

Balcony

Gravel Backfill

Concrete pad, see Chapter 5 for design criteria

Two 2x12's with 6x6 Spacer

Ladder

3x8 Tongue and Groove Decking or 3x8 planks splined together

Bench — 2x12 with 2x10 supports every other joist

2x8 joist extensions for deck splice 4'0" deck surface to be 2x6's spaced 1/2" apart

Floor Girder
Four 2x12's with 6x6 spacer; bolt to post

Floor Joists 2x12's at 24" on center. Insulate with 12" fiberglass; 3/4" plywood subfloor with finish as desired

6'4"

8'0"

12'0"

6'4"

7'6"

16'0"

FIGURE 19-18: CONSTRUCTION SECTION — FOUR-POLE HOUSE

This home (Plan FS-FPL-2) was developed for a large family of up to 12 children at a reasonable cost. It is 24 by 36 feet in size and is one and one-half stories. The first floor has 864 square feet, consisting of three bedrooms, a bath, and a living-dining-kitchen area. The second floor contains about 540 square feet and consists of two large dormitory-type bedrooms. Each is divided by a wardrobe-type closet which, in effect, contains space for two single beds on each side.

The plan was developed by the Forest Products Laboratory of Madison, Wis., and is one of a series of low-cost houses of wood being designed by the Forest Service, U.S. Department of Agriculture.

Area = 1404 sq. ft.

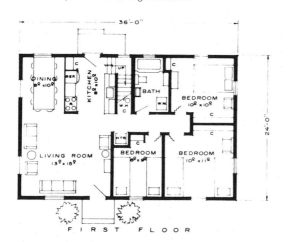

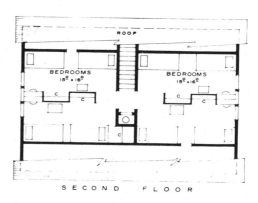

FIGURE 19-19

economical or efficient. An exception are some plans that are put out by *Popular Science* and available from Vacation Plans Service, P.O. Box 622, Princeton, New Jersey 08540. A book showing all the plans costs $2.50. Full working drawings and specifications are $35 per set. Many of their plans are "1950's modern" with oddly angled roofs, triangular rooms, and inefficient sloping walls. However, hidden among the many less-worthy designs are several excellent ones. The quality of suggested materials and the technical content of the plans is excellent. A two-story, 500-square-foot hillside house called the Olympic is a gem. Also included is a very fine plan for a four-unit, *two one-bedroom efficiencies,* and *two two-bedroom apartments.* This plan would be fine for a joint venture, and this particular layout is a model of efficiency.

RENOVATIONS OF EXISTING SMALL HOUSES

Large existing houses, as well as city townhouses and the like, are so variable that it's impossible to cover all the possibilities individually within the scope of this book. On these structures, you will have to apply the advice we give throughout the book on a job-by-job basis. We have, however, selected two designs typical of the small rural houses that are relatively common throughout the country. These have been selected primarily as exercises in what to look for and how to deal with problems that may arise. These two designs are typical of the sort of house that may be considered virtually worthless by the seller because of its small size and possibly poor condition. Such houses were usually constructed during the early 1800's through the early 1930's. If you are lucky, no one has messed them up by trying to install plumbing or by making clumsy additions. As you will see, both of these houses can easily be made into very handsome dwellings by utilizing efficient design principles. Yet, they would both be considered useless by customers looking for a typical ranch house. Try to use your imagination and study our examples carefully to see if you can come up with your own ways of transforming one of these abandoned houses into a new house.

Design 1

The first design that we present is for a typical small, wood-frame farm house. These generally consist of a small (approximately 18 X 24 feet), one and one-half-story, gable-roofed rectangle with various appendages, usually a front porch and a low-pitched kitchen wing. (See plan below.) By now, these houses frequently present a forbidding appearance. The front porch may be falling off and the kitchen-wing roof, because of its low pitch, has leaked badly (maybe the wing is even falling

This design, intended for a flat site, encloses 1,024 square feet. The square plan provides much more usable space, within the same exterior walls, than a typical rectangular plan.

While essentially conventional, the design features a novel floating-wood-floor system. The space under the floor serves as a return-air plenum from a centrally located forced-air furnace. Air will flow into each room through the opening above the door, then enter a space behind the baseboard for return under the floor to the furnace. Alternate designs are available for a conventional wood floor over a crawl space and for a concrete slab floor.

This rural home of wood was designed by the Forest Service, U.S. Department of Agriculture. It is one of a series of plans for low-cost houses developed by the Southeastern Forest Experiment Station, Forestry Sciences Laboratory, Athens, Ga., to illustrate effective utilization of wood products.

Area = 1024 sq. ft.

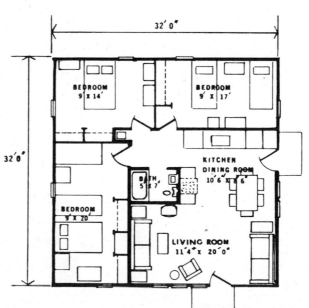

FIGURE 19-20

This attractive house was designed especially for large families wanting a low-cost three-bedroom home. The bedrooms and dining room will accommodate a large family without crowding. A utility room has been provided next to the bathroom, and the laundry tub will be useful as a second lavatory. The sizable open space of the kitchen-dining-living area will make this house seem much larger than its 1008 square feet.

When this house is constructed on a sloping site, a carport and storage room under the house are bonus features. The furnace and hot water heater will usually be located in the storage room under the house; however, if the site is level, the furnace could be in the linen closet location, and the under-counter type of hot water heater could be installed in the kitchen.

This rural home of wood was designed by researchers of the Forest Service, U.S. Department of Agriculture. It is one of a series of plans for low-cost homes developed at the Southeastern Forest Experiment Station, Forestry Sciences Laboratory, Athens, Ga., to illustrate effective utilization of wood products.

Strong, durable, pressure-treated wood poles support this house, which is particularly suitable for sloping sites.

Area = 1008 sq. ft.

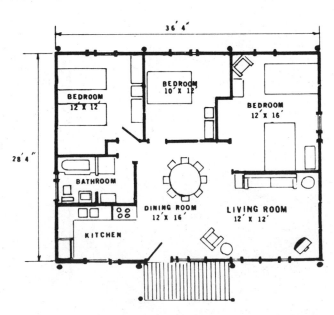

FIGURE 19-21

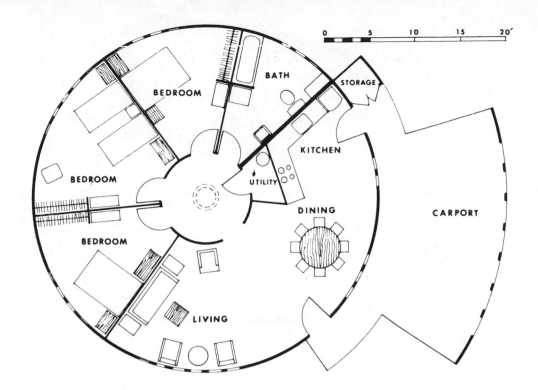

It is estimated the round house will cost about half as much as a conventional house with an equal amount of floor space. The house should be easy to build and suitable for self-help programs.

Floor System

A circular concrete slab is placed within a low brick foundation wall. The perimeter insulation and vapor barrier are conventional. Where needed, the soil is treated or poisoned to prevent attack by termites and other insects. A preservative-treated wood member is fixed around the edge of the slab. The ends of the exterior plank walls are nailed to this member.

Partition Walls

The interior partition walls are particleboard panels located under the roof beams and fitted into slots in 2 x 2 members at the vertical joints. The panels can be moved to alternate locations to provide two, three, or four bedrooms. Closet shelving and sidewalls serve to stiffen the particleboard walls. The walls may be finished with conventional wall paints or with natural finishes, either pigmented or clear.

Roof System

The roof system, which is essentially flat, consists of radially placed 4 x 6 beams, roughsawn or finished, inserted into slots cut in the tops of the perimeter planks and those around the atrium. The beams are covered with 1 x 6 tongue-and-grooved lumber decking in a herringbone pattern. The lower surface of the decking and the exposed beams are finished with natural stains. The roof surface of the decking is covered with a 1-inch layer of foamed-in-place rigid polyurethane insulation (2 pounds per cubic foot).

Area: Plans come for 804 ft.^{2} and 1134 ft.^{2} sizes

FIGURE 19-22

down). Despite all of this deterioration, it is likely that the basic center rectangle is still sound and in good shape. A primary reason for this is the better quality wood used in these old houses. You might wonder why this is the case. First, these houses were usually built of virgin timber, which is more dense than second growth. Second, until fairly recently, timbering was done in the winter months when logs could be hauled out by sled over frozen ground. There was no sap in the logs then, and the wood cured evenly, producing more-durable timbers. Therefore, if you are lucky, you may be able to get the shell of your new house virtually free along with a piece of property.

The primary basis of our remodeling is to replace the kitchen wing with a new kitchen-bath wing; however, if the existing wing is in good shape, it will be foolish to

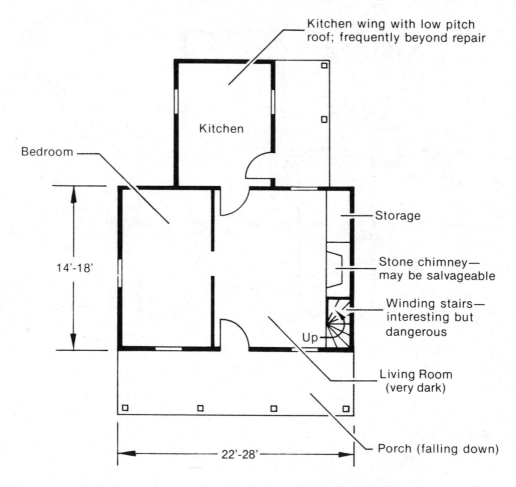

FIGURE 19-23: TYPICAL SMALL WOOD-FRAME FARMHOUSE (Before)

tear it off. Many of these houses had a large stone fireplace at one end with an incredible tiny winding stair crawling up the side to a second-floor sleeping loft. Be sure to check the fireplace carefully as it may be a fire hazard. You might want to rebuild it if it is handsome, or you may wish to remove it completely in order to gain more space and replace it with a small, centrally located chimney for a wood-stove or small furnace. The small stair is hazardous and poorly located. You may want to leave it as a second exit from upstairs or temporarily as the only stairs. If

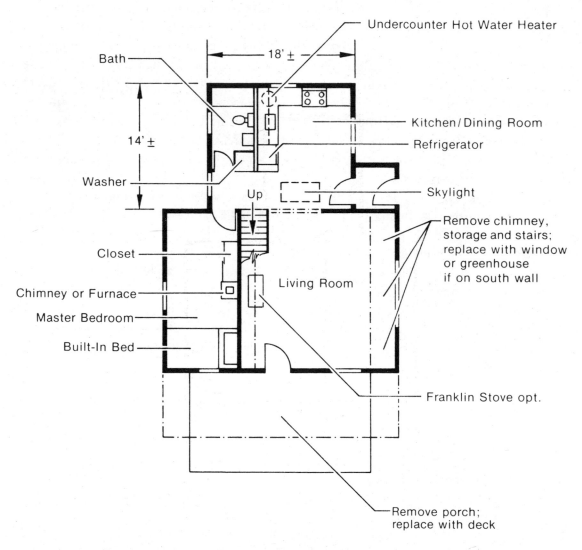

FIGURE 19-24: TYPICAL SMALL WOOD-FRAME FARMHOUSE (Revised)

you do leave it, make sure that you provide a good window exit from the second floor.

Our revised plan shows the stair relocated. We also show an alternate second-floor plan with a second bath above the kitchen. The dimensions of these old houses may vary considerably, so be prepared to make some revisions in our plans.

Design 2

The second solution is for a two-story square house 16 to 20 feet square. For purposes of illustration, we shall assume that it is constructed of brick, although the planning comments would just as easily apply to a frame house. Usually, these houses had rather steep roofs with a sizeable attic. In our example, we have opened up a portion of the second-floor ceiling right at the head of the staircase to provide sleeping space for children on a third level. Notice that this layout provides excellent ventilation through the house. Again, the arrangement of windows, decks, vestibules, greenhouses, and the like should be governed by natural site conditions. We have shown this plan with the typical straight-run staircase left in place. You should also note that our basic two-story square house plan can be readily adapted to this house if the staircase is removed.

Another possibility for finding additional space without making major and often difficult additions to a small old house is to build a garage-studio separate from the house. See Figure 19-26. In this way the older children could have their own apartment, which might be rented at a later date after the children are grown. This solution might also help with getting a bank loan if the bank felt that the house by itself was too small to be saleable.

When you renovate an older house, put the ideas in this book to work. While you have the structural members exposed on the house, it will pay to check very carefully for signs of termites or dry rot. Replace any structural members that are suspicious. A perpetually wet basement can cause "dry rot." Many of these old houses had basements built of stone. Since they were built before the days of modern waterproofing techniques, they tend to be wet. One expensive waterproofing technique is to dig up the fill outside, parge the outside of the wall smooth with mortar, and then apply waterproofing drains and gravel backfill as described in Chapters 5 and 6. This will be particularly important if you wish to use the basement as a living area. If you just want to get rid of the excess water, a sump pump may be the answer. With a fully buried basement, in some areas a sump pump will be essential in order to remove water even with a perimeter drain (as there may be no place for the drain to empty). If your basement is somewhat newer, it may be constructed of concrete or block walls. While it is easy to repair settlement cracks, it's very difficult to repair leakage in these old walls. Shop at a professional masonry

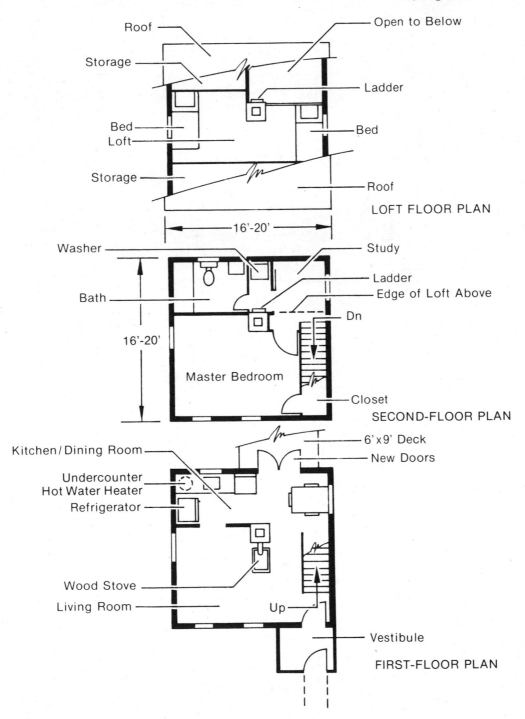

Roof

Open to Below

Storage

Ladder

Bed
Loft

Bed

Storage

Roof

LOFT FLOOR PLAN

16'-20'

Washer

Study

Bath

Ladder

Edge of Loft Above

Dn

16'-20'

Master Bedroom

Closet

SECOND-FLOOR PLAN

6'x9' Deck

Kitchen/Dining Room

New Doors

Undercounter
Hot Water Heater

Refrigerator

Wood Stove

Living Room

Up

Vestibule

FIRST-FLOOR PLAN

FIGURE 19-25: SMALL BRICK SHELL

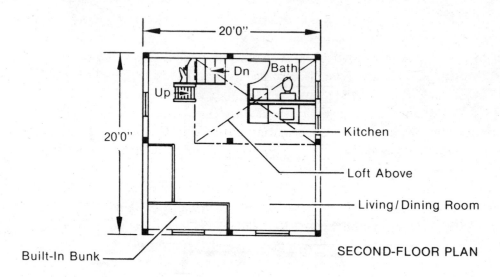

SECOND-FLOOR PLAN

20'0"

20'0"

Dn

Bath

Up

Kitchen

Loft Above

Living/Dining Room

Built-In Bunk

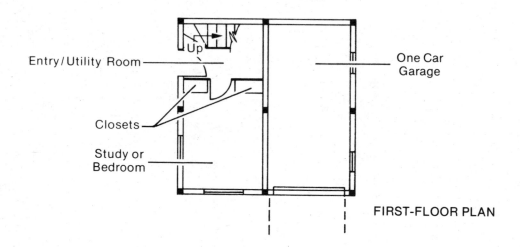

Entry/Utility Room

Up

One Car
Garage

Closets

Study or
Bedroom

FIRST-FLOOR PLAN

FIGURE 19-26: GARAGE APARTMENT
Supplemental space for an old house that is too small

supply store for a cement-based hydraulic compound. Sonneborne Chemical Company or Thoro Chemical Company are manufacturers of excellent products for repair of masonry work. Also, if you make any additions or major repairs to concrete, make sure that you use a commercial bonding material. Check with your building supplier for these materials.

If you have to repair windows, sheet vinyl, unlike the plastic stormwindow material normally sold at hardware stores, is water clear and has a life expectancy of 6 to 10 years. It can be applied in two layers inside and out, and you can still readily see through the window. The disadvantage is that it's available only from really large plastics dealers (check the Yellow Pages in the nearest large city). The price is about 30 percent of the cost of ordinary sheet glass. Also, make sure to put the techniques in ventilating, winterizing, and insulating outlined in Chapters 9 to 12 to work for you. Finally, make sure that you choose efficient appliances and heating systems. Oversize units of any type tend to be wasteful, so select your appliances and equipment with great care. Then think about some of the other options outlined in the book. Wood or coal stoves make an excellent primary or secondary source of heat for a small, well-insulated house. Remember to plan for chimneys when you are laying out your house. Also, when you renovate your home, think about adding the solar windows, a greenhouse, and the solar water heaters outlined in Chapters 16, 17, and 18.

We have endeavored throughout this book to lay out for you various alternatives to the almost prohibitively expensive housing options now on the market. We hope you will use our book as a shopping list, selecting various cost- and energy-saving options to help you tailor your home to your pocketbook and your spatial and esthetic needs. Some people may not have the time to put all these designs to work. For these instances, we have included a list of architects, engineers, contractors, and design groups who are familiar with low-cost post and beam houses as well as with the energy conservation techniques and use of solar options in this book. See the list in the Appendix if you find that you need help.

May you build your home, sweet home!

BIBLIOGRAPHY

CHAPTER 1

Kahn, Lloyd. *Shelter.* Bolinas, CA: Shelter Publications, 1973.

Rapoport, Amos. *House Form and Culture.* Englewood Cliffs, NJ: Prentice-Hall, 1969 (ref. ed. & pap.).

Rudofsky, Bernard. *Architecture Without Architects: A Short Introduction to Non-Pedigreed Architecture.* Garden City, NY: Doubleday and Co., 1964 (pap. 1969).

CHAPTER 2

Fitchter, R., and Turner, J. F. C. *Freedom To Build.* New York: Macmillan Co., 1972.

Hoffman, George C. *Don't Go Buy Appearances.* Westminster, MD: Ballantine Books, Inc., 1972.

CHAPTER 4

Anderson, L. O., and Zornig, Harold F. *Build Your Own Low Cost Home.* New York: Dover, 1972.

Daniels, George. *Home Guide to Plumbing and Air Conditioning.* New York: Harper & Row, 1967.

——*How To Be Your Own Home Electrician.* New York: Popular Science, 1965.

Pole House Construction. McClean, VA: American Wood Preservers Institute. Detailed instructions for installing pressure-treated poles (round or square) on sloping sites. How to handle difficult soil conditions. Description of various methods of preservative treatment for lumber. Several interesting house designs are included. The American Wood Preservers Institute offers this, *FHA Pole House Construction,* and many other useful publications. The Institute also provides lists of suppliers of pressure-treated lumber. Many useful pamphlets are published from time to time. Write for list.

Roberts, Rex. *Your Engineered House.* New York: M. Evans & Co., 1964.

U. S. Department of Agriculture. *House Plans No. FS-FPL-4 and 5.* Washington, DC: U. S. Government Printing Office, Division of Public Documents. Includes full plans, specifications, and detailed materials list.

——*Low Cost Homes for Rural America: Construction Manual, USDA Handbook #364.* Washington, DC: U. S. Government Printing Office, Division of Public Documents. This manual is filled with clear illustrations of step-by-step house construction. It illustrates post and beam framing and has many excellent and inexpensive cabinet and trim details. It is a must for the inexperienced owner-builder. It does not size structural members for you and it was written prior to the energy crisis, so insulation is not up to our standards. Otherwise, it's a very complete handbook. Price: $1.00. Comes with house plans for $1.00 more.

CHAPTER 5

Carpenters and Builders Guide. Indianapolis: Theodore Audel & Co. This is a practical guide, with clearly illustrated step-by-step methods.

Fields, Curtis P. *The Forgotten Art of Building a Stone Wall.* Dublin, NH: Yankee, Inc., 1971.

Griffith, John F. *Applied Climatology: An Introduction.* New York: Oxford University Press, 1966.

Hornbostel, Caleb. *Materials for Architecture: An Encyclopedic Guide.* New York: Van Nostrand and Reinhold, 1961. For the serious owner-builder, this book contains in-depth descriptions of materials and procedures.

Kern, Ken. *The Owner Built Home.* Oakhurst, CA (P. O. Box 550). 1961. An excellent book for site planning and low-cost techniques.

——*The Owner Built Homestead.* Oakhurst, CA (P. O. Box 550).

Olgyay, Victor. *Design with Climate.* Princeton, NJ: Princeton University Press, 1963.

Petrides, George A. *Field Guide to Trees and Shrubs,* 2d ed. Boston: Houghton Mifflin Co., 1972 (Peterson Field Guide Series).

Ramsey, Charles G., and Sleeper, Harold R. *Architectural Graphic Standards,* 6th ed. Somerset, NJ: John Wiley & Sons, Inc., 1970. The architects' bible. In the midst of its hundreds of pages of details is a small section showing excellent post and beam details; a comparison between standard framing, showing space savings; and, most important, simple and easy to understand structural design tables. Check your library because this is an expensive book.

Soil-Cement—Its Use in Building. United Nations Publication #E.64.W.6, Sales Section, New York.

Stoner, Carol H., ed. *Producing Your Own Power.* Emmaus, PA: Rodale Press, 1974. This book contains several essays on constructing methane digesters for family and farm use.

U. S. Department of Agriculture. *Wood Handbook.* Washington, DC: U. S. Government Printing Office, Division of Public Documents, 1955.

CHAPTER 6

Boericke, Art, and Shapiro, Barry. *Handmade Houses (A Guide to the Wood-butcher's Art).* San Francisco: The Scrimshaw Press, 1973. Owner-built houses photographed in color. Interesting items and loads of inspiration.

Downing, Andrew J. *The Architecture of Country Houses.* 1850 (reprint. New York: Da Capo Press, Inc., 1968; pap., Dover, 1969).

——*Hints to Persons About Building.* Even though written in the 1860's, this book contains a wealth of useful information on design siting, color, ventilation, and construction in general. Many of the exterior designs would be worthy of reproduction today. Check your library for a copy.

Plank and Beam Framing for Residential Buildings. Washington, DC: National Forest Products Association (1619 Massachusetts Avenue N.W.).

Pole House Construction. McClean, VA: American Wood Preservers Institute.

Pole Building Construction. Charlotte, VT: Garden Way Publishing Co.

Ramsey, Charles G., and Sleeper, Harold R. *Architectural Graphic Standards,* 6th ed. Somerset, NJ: John Wiley & Sons, Inc., 1970.

Small Homes Council—Building, Research Council, University of Illinois, Urbana, IL. Numerous publications that are a must for those who intend to do their own designing and building. Back issues of bulletins are available on various subjects, such as "Kitchen Design" and "Post and Beam Framing." Subscriptions are also a good source for new ideas, as several bulletins are published per year.

U. S. Department of Agriculture *Low Cost Homes for Rural America: Construction Manual, USDA Handbook #364.* Washington, DC: U. S. Government Printing Office (see listing in bibliography to Chapter 4).

CHAPTER 7

Burke, and Moger. *How to Buy a House.* New York: Lyle Stuart, Inc.

Cobb, Hubbard H. *Dream House Encyclopedia.* New York: Peter H. Wyden, Inc., 1970.

Price, Irving. *Buying Country Property: Pitfalls & Pleasures.* New York: Harper & Row, 1972.

Walton, Harry. *How To Build Your Cabin or Modern Vacation Home.* New York: Harper & Row, 1964.

Watkins, Arthur M. *The Complete Book of Home Remodeling, Improvement & Repair: A Handbook for the Owner Who Wants To Do It Right—But Not Do It Himself.* New York: Doubleday & Co., 1963.

Wren, Jack. *Home Buyer's Guide.* New York: Barnes & Nobles Books, Division of Harper & Row, 1970 (pap.).

CHAPTER 9

Callender, J. H. *Time-Saver Standards: A Handbook of Architectural Design,* 4th ed. New York: McGraw-Hill, Inc., 1966.

Diamont, R. M. E. *The Internal Environment of Dwellings.* London: Hutchinson Educational Limited, 1971.

Kern, Ken. *The Owner Built Home.* P. O. Box 550, Oakhurst, CA: 1961.

Olgyay, Aladar. *Solar Control and Shading Devices.* Princeton, NJ: Princeton University Press, 1957. Xerox University Microfilms, Ann Arbor, MI.

Reader's Digest Complete Do-It-Yourself Manual. Pleasantville, NY: Reader's Digest Association, 1973.

Roberts, Rex. *Your Engineered House.* New York: M. Evans & Co., 1964.

Small Homes Council. *F 11.0 Window Planning Principles.* Urbana, IL: University of Illinois.

——*F 11.1 Selecting Windows.* Urbana, IL: University of Illinois.

——*F 11.2 Insulating Windows and Screens.* Urbana, IL: University of Illinois.

Solar Effects on Building Design. Washington, DC: Building Research Institute, Inc. (1725 Desales St.).

Windows and Glass in the Exterior of Buildings. Washington, DC: National Research Council, Building Advisory Board, 1957.

CHAPTER 10

ASHRAE Handbook, Book of Fundamentals. New York: American Society of Heating, Refrigerating and Air-Conditioning Engineers (345 E. 47th St.), 1972.

Hutchinson, F. W. *Design of Heating and Ventilating Systems.* New York: The Industrial Press, 1955.

Kern, Ken. *The Owner Built Home.* P. O. Box 550, Oakhurst, CA: 1961.

Morrell, William. *The Energy Miser's Manual.* Eliot, ME: The Grist Mill Publication. Price: $2.25.

U. S. Department of Agriculture. *Trees for Shade and Beauty.* Bulletin 117. Washington, DC: U. S. Government Printing Office.

CHAPTER 12

Insulation Manual—Homes and Apartments. Rockville, MD: National Association of Home Builders (P.O. Box 1627), 1971. Price: $4.00.

Spies, Henry, et al. *350 Ways To Save Energy and Money in Your Home and Car.* New York: Crown, 1974.

CHAPTER 13

Directory of Certified Unitary Air Conditioners and Unitary Heat Pumps. Arlington, VA: Air Conditioning and Refrigeration Institute.

Electric Space Conditioning in Residential Structures. New York: Electric Energy Association (90 Park Avenue).

"Evaluate Burner Performance," *Fuel Oil & Oil Heat* 32 (1973): 36.

Exploring Energy Choices. Washington, DC: Energy Policy Project Ford Foundation (P. O. Box 23212). Price: $.75.

Gay, Larry. *The Complete Book of Heating with Wood.* Charlotte, VT: Garden Way Publishers, 1974.

Hamilton, L. S., and Winch, F. E. *Making and Using Wood Fuel.* Ithaca, NY: Cooperative Extension Service, Cornell University.

Havens, David. *Woodburners Handbook.* Portland, ME: Media Publications, 1973.

"Heat-Saving Ideas," *Popular Mechanics* 142 (1974): 142-56, 175.

The Home Comfort Handbook. New York, Popular Science Library, 1966 (out of print).

How To Choose the Room Air Conditioner Best Suited for You. Chicago, IL: Association of Home Appliance Manufacturers (20 N. Wacker Drive).

Insulation Manual: Homes—Apartments. Rockville, MD: National Association of Home Builders Research Foundation (P. O. Box 1627).

New York State Interdepartmental Fuel and Energy Committee. *Appliance and Apparatus Efficiency.* Albany, NY: New York State Public Service Commission, June 1973 (out of print).

Powell, Evan. "Tune Your Heating System to Peak Efficiency," *Popular Science* 203 (1973): 124-28, 172.

Save Energy: Save Money, (OEO Pamphlet 6143-5). Washington, DC: Community Services Agency. Free.

Spectrum: An Alternate Technology Equipment Directory. Milaca, MN: Alternative Sources of Energy, Inc. (Rt 2, Box 90A). This booklet contains listings of manufacturers who make greenhouse kits, glazing materials, and other equipment used in greenhouses. Price: $2.00.

Spies, Henry, et al. *350 Ways To Save Energy and Money in Your Home and Car.* New York: Crown, 1974.

Standard for Application of Year-Round Residential Air Conditioning, (#230-62). Arlington, VA: Air Conditioning and Refrigeration Institute (1815 N. Ft. Meyer Drive).

Tips for Energy Savers. Washington, DC: Federal Energy Administration.

CHAPTER 14

Bailey, J. R., et al. *A Study of Flow Reduction and Treatment of Waste Water From Households.* Washington, DC: Superintendent of Documents, U. S. Government Printing Office, December 1969. Price: $1.25.

Design Criteria for Lighting Interior Living Spaces. New York: Illuminating Engineering Society (345 E. 47 St.).

Facts on Energy Use and Conservation for Refrigerator-Freezers. Chicago, IL: Association of Home Appliance Manufacturers (20 N. Wacker Drive), 1974.

Major Appliances and Energy, Pub. No. 1-9574-1. Louisville, KY: General Electric Co. (P. O. Box 1661).

Philco Consumer Information Report—Cold Guard Refrigerator-Freezers. Blue Bell, PA: Philco-Ford Corporation, May 1974.

Residential Energy Consumption, Phase I Report. Columbia, MD: Hittman Associates, Inc. (9190 Red Branch Rd.), March 1972.

"Sewage Treatment in Small Towns and Rural Areas." Hanover, NH: Dartmouth College (proceedings of a conference, March 3, 1971).

A Statement of Policy. Hyattsville, MD: Washington Suburban Sanitary Commission (4017 Hamilton St.), October 29, 1971.

Stop the 5 Gallon Flush. Montreal, Canada: Minimum Cost Housing, School of Architecture, McGill University, July 1973.

CHAPTER 15

Appliance and Apparatus Efficiency. New York: New York State Interdepartmental Fuel and Energy Committee, June 1973. Out of print.

Exploring Energy Choices. Washington, DC: Energy Policy Project, Ford Foundation (P. O. Box 23212). Price: $.75.

Growing Ground Covers. Washington, DC: Superintendent of Documents, U. S. Government Printing Office. Price: $.15.

Manual of Septic Tank Practice. Washington, DC: Superintendent of Documents, U. S. Government Printing Office. Price: $.50.

Spies, Henry, et al. *350 Ways To Save Energy and Money in Your Home and Car.* New York: Crown, 1974.

Tips for Energy Savers. Washington, DC: Federal Energy Administration.

CHAPTER 16

Brinkworth, B. J. *Solar Energy for Man.* Somerset, NJ: John Wiley & Sons, Inc., 1973. An excellent introduction to the subject, though this one stresses more of an overview of basic physical principles.

Daniels, Farrington. *Direct Use of the Sun's Energy.* New Haven, CT: Yale University Press, 1964. Another excellent introduction to the whole field of solar energy, with good explanations of the design principles involved as well as a lot of practical information.

Olgyay, Aladar. *Solar Control and Shading Devices.* Ann Arbor, MI: Xerox University Microfilms. This book, and Victor Olgyay's *Design with Climate,* are the best overall sources of information available on designing with the sun in mind. They contain many examples and are profusely illustrated.

Olgyay, Victor. *Design with Climate.* Princeton, NJ: Princeton University Press, 1963.

Thomason, Harry. *Solar House Plans.* Barrington, NJ: Edmund Scientific Co. The fine detailing and the integrated design presented in these plans make them worthy of study. Price: $10.00.

Zomeworks Corporation. C/O Steve Baer, P. O. Box 712 Albuquerque, NM 87103. If you want to study a set of plans for a solar house, the really elegant and efficient designs this group offers are worth getting. Steve Baer has come up with a heat storage wall composed of 55-gallon drums filled with water, as well as several other designs for insulating over solar windows at night. Really first-rate stuff. Prices for different plans vary, so write for their brochure.

CHAPTER 17

Abraham, George (Doc), and Abraham, Katy. *Organic Gardening Under Glass.* Emmaus, PA: Rodale Press, 1975. There's much information here about buying and building all types of greenhouses, maintaining the right under-glass environment, and growing ornamentals, especially edibles. It is written by two well-known gardeners/writers who have been in the commercial greenhouse growing business for over 25 years.

Neal, Charles D. *Build Your Own Greenhouse.* Radnor, PA: Chilton, 1975. This book covers the large and the small of building greenhouses, from coldframes to lean-tos, to full-sized structures.

Spectrum: An Alternate Technology Equipment Directory. Milaca, MN: Alternative Sources of Energy, Inc. (Rt. 2, Box 90A) (see listing in bibliography to Chapter 13).

U. S. Department of Agriculture. *List of Sources of Information on Greenhouses, Correspondence Aid 34-134.* Washington, DC: U. S. Government Printing Office, 1970.

——*Electric Heating of Hotbeds, Leaflet 445.* Washington, DC: U. S. Government Printing Office, 1969.

——*Plastic Covered Greenhouse Coldframe, Miscellaneous Publication 1111.* Washington, DC: U. S. Government Printing Office, 1969.

CHAPTER 18

Chinnery, D. N. W. *Solar Water Heating in South Africa,* National Buildings Research Council Bulletin #44. Pretoria, South Africa: Council for Scientific and Industrial Research, 1967.

Close, D. J. "The Performance of Solar Water Heaters with Natural Circulation," *Solar Energy* 6 (1962): 33.

Daniels, Farrington. *Direct Use of the Sun's Energy.* New Haven, CT: Yale University Press, 1964.

Paige, Steve. *Solar Water Heating.* Barrington, NJ: Edmund Scientific Co. $4, 30 pp.

Portola Institute. *Energy Primer.* Fremont, CA: Fricke-Parks Press, Inc., 1974. This book can be obtained for $4.50 from the Whole Earth Truck Store, Santa Cruz, CA.

Solar Water Heaters—Principles of Design, Construction, and Installation, Circular #2. Melbourne, Australia: Division of Mechanical Engineering, CSIRO, 1964.

Technical Committee on Solar Energy Utilization. *Low Temperature Engineering Application of Solar Energy.* New York: American Society of Heating, Refrigerating and Air-Conditioning Engineers (345 E. 47th St.), 1967.

LIST OF DESIGN GROUPS

The following individuals and groups have experience in the areas of low-cost housing, energy conservation, or the use of solar energy. If you need help with design work or need a contractor-builder, contact the person in your area.

Ron Alward
  c/o Brace Research Institute
  MacDonald College of McGill University
  Ste. Anne de Bellevue 800
  Quebec, Canada HOA ICO
  engineer
Bruce Anderson
  Total Environment Action
  Churchhill
  Harrisville, NH 03450
William Bettridge
  East Mountain Construction Company
  Napanoch, NY 12458
  contractor/builder
Richard Blazej
  RFD #1
  Newfane, VT 05345
  contractor/builder
John Browne
  628 Maxwell
  Boulder, CO 80302
  architect/engineer
Peter Clark
  P.O. Box 797
  Berkeley, CA 94701
  builder

Steve Coffel
  Rt. 4, Box 90
  Golden, CO 80401
  builder
Richard Crowther
  2830 East 3rd Avenue
  Denver, CO 80206
  architect
Eugene Eccli
  Design Alternatives Inc.
  1448 N. Lancaster St.
  Arlington, VA 22205
Fred S. Dubin
  Dubin Bloom Assoc.
  42 W. 39th St.
  N.Y., NY 10018
  engineer
Steve Gibson
  Box 209
  Port Royal, VA 22535
  contractor/builder
Robert Godwin
  129 Troop St.
  Rochester, NY 14603
  architect
David L. Hartman
  1337 Wilmot
  Ann Arbor, MI 48104

Burt Hill & Assoc.
  610 Mellon Bank Bldg.
  Butler, PA 16001
  architects
Homestead Housing Group, Inc.
  Box 52
  Woodbury, VT 05681
  contractors for post and
  beam construction
Jerome Kerner
  c/o D.A.W.N. Assoc.
  Box 66
  Phoenicia, NY 12464
  architecture
Richard Lamar AIA
  201 Woodrow St.
  Columbia, SC 29205
  architect
Bill Langdon
  P.O. Box 7163
  Asheville, NC 28807
Charles Liston
  Kingswood, WV 25312
  contractor/builder
Chris Logan
  Grant Rd.
  Newmarket, NH 03857
  engineer
Steve Ridenour
  c/o Approtech
  1700 Meadow Rd.
  Southampton, PA 18966
  engineer
Steven Schatz
  5714 Indian Trail
  Madison, WI 53716
  architect

Robert Schubert
  Box 499
  Blacksburg, VA 24060
  designer
Charles Simms
  160 Duene St.
  N.Y., NY 10013
  contractor/builder
Soltec
  P.O. Box 6844
  Denver, CO 80206
  designers
Alex Wade
  Station Road
  Barrytown, NY 12507
  architect
John Wymore
  82 Hancock St.
  Lexington, MA 02173
  architect
Zomeworks
  c/o Steve Baer
  P.O. Box 712
  Albuquerque, NM 87103
  architectural and engineering
  design work

GLOSSARY

air change—the movement of a volume of air in a given period of time; if a house has one air change per hour, it means that all the air in the house will be replaced in a 1-hour period. Air changes are expressed in cubic feet per minute.

ambient temperature—the temperature out of doors.

attic ventilators—in houses, screened openings provided to ventilate an attic space. They are located in the soffit area as inlet ventilators and in the gable end or along the ridge as outlet ventilators. They can also consist of power-driven fans used as an exhaust system.

auxiliary heater—a conventional hot water heater that acts as a back-up to bring the water up to temperature on cloudy days.

axial fan—a fan with blades that rotate within a duct or large pipe to move air along the direction of the duct or pipe.

backfill—the replacement of excavated earth into a trench or pier excavation around and against a basement foundation.

bank run gravel—a mixture of sand and gravel, with the gravel size no larger than 2½ inches in diameter, used as backfill.

baseboard raceway—a series of wires and outlets encased in metal containers so that wiring can be run along exposed areas, like baseboards, without danger of damage.

batch solar water heater—a unit that is filled once in the morning. The water sits to be heated during the day and drained off at night.

batten—narrow strips of wood used to cover joints or as decorative vertical members over plywood or wide boards.

beam—a structural member transversely supporting a load.

bearing partition—a partition that supports any vertical load in addition to its own weight.

blind stop—a rectangular molding, usually 3/4 × 1 3/8 inches or more in width, used in the assembly of a window frame. It serves as a stop for storm and screen or combination windows and helps to resist air infiltration.

boiler—a device used to heat water or produce steam. It includes burner, heat exchanger, flue, and container and controls; may be fueled by oil, gas, or electricity.

bolts, anchor—bolts to secure a wooden sill plate to concrete or masonry floor or wall or pier.

Boston ridge—a method of applying asphalt or wood shingles at the ridge or at the hips of a roof as a finish.

brace—an inclined piece of framing lumber applied to wall or floor to stiffen the structure. Often used on walls as temporary bracing until framing has been completed.

BTU (British thermal unit)—a unit of heat roughly equal to the amount of heat given off by burning a kitchen match.

built-up roof—a roofing composed of three to five layers of asphalt felt laminated with coal tar, pitch, or asphalt. The top is finished with crushed slag or gravel. Generally used on flat or low-pitched roofs.

c.a.—an abbreviation for *conditioned air.*

cantilever—any structural member extending from a support at one end but unsupported at the other end.

casement frames and sash—frames of wood or metal enclosing part or all of the sash, which may be opened by means of hinges affixed to the vertical edges.

casing—molding of various widths and thicknesses used to trim door and window openings at the jambs.

c.f.m.—an abbreviation for *cubic feet per minute.*

clo—an abbreviation for *closet.*

c.o.—an abbreviation for *cased opening;* e.g., an opening with no door but with framing all around it.

comfort region—the range of temperatures and humidity of room air in which 50 percent of the adults "feel" comfortable.

condensation—beads or drops of water, and frequently frost in extremely cold weather, that accumulate on the inside of the exterior covering of a building when warm, moisture-laden air from the interior reaches a point where the temperature no longer permits the air to sustain the moisture it holds. Use of louvers or attic ventilators will reduce moisture condensation in attics. A vapor barrier will reduce condensation in walls.

conduction—a process of heat transfer whereby heat moves directly through a material.

conduit, electrical—a pipe, usually metal, in which wire is installed.

construction, dry-wall—a type of construction in which the interior wall finish is applied in a dry condition, generally in the form of sheet materials or wood paneling, as contrasted to plaster.

construction, frame—a type of construction in which the structural parts are of wood or depend upon a wood frame for support. In building codes, if masonry veneer is applied to the exterior walls, the classification of this type of construction is usually unchanged.

continuous-flow water heating systems—those systems, either thermosyphon or forced circulation, in which the water circulates continuously through the solar collector as long as there is heat available.

convection—a transfer of heat created by the motion of air resulting from a difference in temperature (density) and the action of gravity.

corner braces—diagonal braces at the corners of frame structure to stiffen and strengthen the wall.

cornice—overhang of a pitched roof at the eave line, usually consisting of a facia board, a soffit for a closed cornice, and appropriate moldings.

counterflashing—a flashing usually used on chimneys at the roofline to cover shingle flashing and to prevent moisture entry.

crawl space—a shallow space below the living quarters of a basementless house, sometimes enclosed.

daylighting—the use of natural light from windows, clerestories, and skylights to illuminate a part of a house.

deck paint—an enamel with a high degree of resistance to mechanical wear; designed for use on such surfaces as porch floors.

degree-day—a measure of temperature difference between the inside and outside of a building. The usual base temperature is 65°F, so that if the average ambient (e.g., outdoor) temperature were 60°F over an entire day, there would be five degree-days over that 24-hour period. The total number of degree-days over the heating season indicates the relative severity of the winter in that area.

density—the mass of substance in a unit volume. When expressed in the metric system (in g. per cc.), it is numerically equal to the specific gravity of the same substance.

d.h.—an abbreviation for *double-hung windows.*

differential thermostat—a special thermostat used on forced-circulation systems. It actuates the circulation pump *only* when the collector is warmer than the fluid in the storage tank.

dormer—a projection in a sloping roof, the framing of which forms a vertical wall suitable for windows or other openings.

downspout—a pipe, usually metal, for carrying rainwater from roof gutters.

drip cap—a molding placed on the exterior top side of a door or window frame to cause water to drip beyond the outside of the frame.

drop beam—a horizontal structural support member that goes below other framing to provide clearance.

eaves—the overhang of a roof projecting over the walls.

electric resistance heat—electric baseboard or electric boiler used for space heating. Heat is generated by current flowing in a resistance wire similar to those used in a hair dryer.

energy efficiency ratio (eer)—the cooling energy output from an air conditioner divided by the watt input; about 6 for most air conditioners.

facia or **fascia**—a flat board, band, or face, used sometimes by itself but usually in combination with moldings; often located at the outer face of the cornice.

fenestration—a term used to signify an area of glass.

flashing—sheet metal or other material used in roof and wall construction to protect a building from seepage of water.

flat paint—an interior paint that contains a high proportion of pigment and dries to a flat or lusterless finish.

flow pipe—the pipe that leaves the top of the solar collector, carrying the heated water to the upper part of the storage tank.

flow rate—amount of water (or air) moving through a pipe or duct per unit of time. Normal units are gallons per minute for water and cubic feet per minute for air.

flue—the space or passage in a chimney through which smoke, gas, or fumes ascend. Each passage is called a flue, which, together with any others and the surrounding masonry, make up the chimney.

footing—a masonry section, usually concrete in a rectangular form wider than the bottom of the foundation wall or pier it supports.

foundation—the supporting portion of a structure below the first-floor construction, or below grade, including the footings.

frieze—a horizontal member connecting the tip of the siding with the soffit of the cornice.

frostline—the depth of frost penetration in soil. This depth varies in different parts of the country. Footings should be placed below this depth to prevent movement.

furnace—similar to a boiler except that it is used to heat air, and wood may be used as a fuel.

furred-down ceiling—the addition of nonstructural framing, in addition to roof structure, in order to create a lower ceiling.

furring—strips of wood or metal applied to a wall or other surface to even it and usually to serve as a fastening base for finish material.

gable—the triangular vertical end of a building formed by the eaves and ridge of a sloped roof.

girder—a large or principal beam of wood or steel used to support concentrated loads at isolated points along its length.

glass types—The following are a few types of glass mentioned in Chapters 9 and 11, along with a few others that you will probably encounter. The starred ones (*) are recommended.

*double-strength glass (1/8 inch) is the most useful to you. Also made in 3/16 and 7/32-inch strengths. Single-strength (3/32 inch) breaks very easily.

heat-absorbing glass—a glass similar to tinted glass but which screens out only the infrared or invisible heat rays while allowing the visible light to penetrate. This is an expensive way to aid in summer cooling. An inside shade is more effective and can be raised on winter days to allow this radiant solar heat to enter.

*insulating glass—a factory-sealed, double-paned glass with an air space in the middle. Edges are fused either with additional glass or a metal edge. A common overall thickness is 5/8 inch, though it can range from a ½ inch to 1 inch.

*plate and float glass—a glass of clearer quality than sheet glass and one which therefore can be made thicker and stronger. Plate glass can often be picked up secondhand at bargain prices. Thicknesses range from 1/8 inch up to 1 inch but 1/4 inch will be most useful to you.

Plexiglas—not a glass at all but a clear plastic. Not as durable as glass, but there is less danger of breakage.

*sheet glass—an economy grade of glass that may have slightly detectable distortions in larger sheets.

*tempered glass—all thicknesses of sheet glass (except single-strength). Float and plate glass can be tempered. The glass is heat treated to become more than three times stronger. Tempered glass cannot be drilled or cut, for it will disintegrate into small pieces. However, breakage is not a serious safety hazard here, as broken pieces do not have the jagged edges that regular broken glass has.

tinted glass—a darkened glass that screens visible light and infrared or radiant heat. Each square foot of glass area in your home has initial costs and on-going heating costs in exchange for the benefits of transparency. Tinted glass not only costs more but lowers the benefits of having a glass area.

wired glass—a glass with a wire mesh embedded within it for safety and as an effective fire shield.

glazing bar—a structural support used to hold glass or other transparent materials in place on a greenhouse.

gloss (paint or enamel)—a paint or enamel that contains a relatively low proportion of pigment and dries to a sheen or luster.

grade—the natural slope of the land.

grading—adjusting the natural slope of the land to new contours.

grain—the direction, size, arrangement, appearance, or quality of the fibers in wood.

gusset—a flat wood, plywood, or similar member used to provide a connection at the intersection of wood members. Most commonly used at joints of wood trusses. They are fastened by nails, screws, bolts, or adhesives.

gutter or **eave trough**—a shallow channel or conduit of metal or wood set below and along the eaves of a house to catch and carry off rainwater from the roof.

h.a.—an abbreviation for *heated air*.

header—(a) a beam placed perpendicular to joists and to which joists are nailed in framing for chimney, stairway, window, or other opening; (b) wood lintel.

heartwood—the wood extending from the pith to the sapwood, the cells of which no longer participate in the life processes of the tree.

heat capacity—sometimes called the thermal capacity; a measure of how much heat is required to raise the temperature of a given amount of material by a certain amount.

heat exchanger—an extra circulation loop hooked into solar systems to prevent freezing of the water at night. *Propylene glycol* or *ethylene glycol* is used in the collector circuit as an antifreeze.

heat pump—combination heating and cooling unit; operates like a normal air conditioner in summer, and in winter operates in reverse, ejecting warm air indoors and cool air (or water) outdoors.

humidity relative—the ratio of the amount of water vapor at a given temperature to the maximum amount of water vapor that could be held as vapor.

infiltrated air—outside air that gets into the house at an uncontrolled rate (through walls, cracks, around doors, etc.) because of the naturally occurring forces (wind, temperature differences, etc.).

infrared radiation—the radiant heat given off by objects at low temperatures. Generally, the range includes radiation given off by objects between room temperature and about 500°F.

insulation, thermal—any material high in resistance to heat transmission that, when placed in the walls, ceilings, or floors of a structure, will reduce the rate of heat flow.

insulation board, rigid—a structural building board made of wood or cane fiber in 1/2 and 23/32-inch thicknesses. It can be obtained in various size sheets, in various densities, and with several treatments.

jamb—the side and head lining of a doorway, window, or other opening.

joint—the space between the adjacent surfaces of two members or components joined and held together by nails, glue, cement, mortar, or other means.

joist—one of a series of parallel beams, usually 2 inches thick, used to support floor and ceiling loads and supported in turn by larger beams, girders, or bearing walls.

kilowatt-hour (KWH)—unit of electrical energy consumption; equals about 3,400 BTU's.

knee wall—a short wall used to close off space that has inadequate head room.

lag screw—a fastener with screw-type threads and a square head. Large sizes may be used to attach post and beam framing members.

lath—a building material of wood, metal, gypsum, or insulating board that is fastened to the frame of a building to act as a plaster base.

ledger—a strip of lumber nailed onto the bottom of the side of a girder on which joists rest.

lintel—a horizontal structural member that supports the load over an opening such as a door or window.

live load—a superimposed load such as the weight of people, furniture, etc.

load-bearing wall—a wall that supports any vertical load in addition to its own weight.

lumber—lumber is the product of the sawmill and planing mill not further manufactured other than by sawing, resawing, passing lengthwise through a standard planing machine, crosscutting to length, and matching.

lumber, board—yard lumber less than 2 inches thick and 2 or more inches wide.

lumber, matched—lumber that is dressed and shaped on one edge in a grooved pattern and on the other in a tongued pattern.

lumber, pressure-treated—lumber impregnated with preservatives or decay and fire retardants via a pressurized process.

lumber, shiplap—lumber that is edge dressed to make a close rabbeted or lapped joint.

lumber, yard—lumber of those grades, sizes, and patterns that are generally intended for ordinary construction, such as framework and rough coverage of houses.

lumber dimension—yard lumber from 2 inches up to, but not including, 5 inches thick, and 2 or more inches wide. Includes joists, rafters, studs, planks, and small timbers. The actual size dimension of such lumber after shrinking from green dimension and after machining to size or pattern is called the *dress size*.

masonry—stone, brick, concrete, hollow tile, concrete block, gypsum block, or other similar building units or materials, or a combination of the same, bonded together with mortar to form a wall, pier, buttress, or similar mass.

millwork—generally all building materials made of finished wood and manufactured in millwork plants and planing mills are included under this term. It includes such items as inside and outside doors, window and door frames, blinds, porchwork, mantels, panelwork, stairways, moldings, and interior trim. It normally does not include flooring, ceiling, or siding.

moisture content of wood—weight of the water contained in the wood, usually expressed as a percentage of the weight of the oven-dry wood.

molding—a wood strip having a curved or projecting surface used for decorative purposes.

mortise—a slot cut into a board, plank, or timber, usually edgewise, to receive the tenon of another board, plank, or timber to form a joint.

mull—an abbreviation for *mullion,* a vertical bar or divider in the frame between windows, doors, or other openings.

natural finish—a transparent finish that does not seriously alter the original color or grain of the natural wood. Natural finishes are usually provided by sealers, oils, varnishes, water-repellent preservatives, and other similar materials.

non-load-bearing wall—a wall supporting no load other than its own weight.

o.c., on center—the measurement of spacing for studs, rafters, joists, and the like in a building from the center of one member to the center of the next.

paint—a combination of pigments with suitable thinners or oils to provide decorative and protective coatings.

panel—in house construction, a thin, flat piece of wood, plywood, or similar material, framed by stiles and rails, as in a door, or fitted into grooves of thicker material with molded edges for decorative wall treatment.

parapet—that portion of a wall that projects up past the roof.

parging—the process of applying a cement-plaster mixture over masonry elements for waterproofing purposes.

partition—a wall that subdivides spaces within any story of a building.

penny—as applied to nails, it originally indicated the price per hundred. The term now serves as a measure of nail length and is abbreviated by the letter *d.*

pier—a column of masonry, usually rectangular in horizontal cross section, used to support other structural members.

pitch—the incline slope of a roof, or the ratio of the total rise to the total width of a house; i.e., an 8-foot rise and a 24-foot width are a 1/3 pitch roof. *Roof slope* is expressed in inches of rise per 12 inches of run.

plate—*Sill plate:* a horizontal member anchored to a masonry wall. *Sole plate:* bottom horizontal member of a frame wall. *Top plate:* top horizontal member of a frame wall supporting ceiling joists, rafters, or other members.

plenum—in forced hot air heating systems, a large duct used as a distributor for the hot air from the furnace going to different parts of the house via smaller ducts leaving the plenum.

plumb—exactly perpendicular; vertical.

plywood—a piece of wood made of three or more layers of veneer joined with glue and usually laid with the grain of adjoining plies at right angles. Almost always an odd number of plies is used to provide balanced construction.

potable water—the drinking water for the house. In solar water heating systems where antifreeze is used, this *must* be kept separate by use of a heat exchanger.

preservative—any substance that, for a reasonable length of time, will prevent the action of wood-destroying fungi, borers of various kinds, and similar destructive life when the wood has been properly coated or impregnated with it.

pressure drop—the reduction in the pressure of air in a ducting system due to the friction between the moving air and the duct walls.

p.s.f.—an abbreviation for *pounds per square foot.*

rabbet—a continuous slot in a wood member into which another wood member will be placed to create a mechanical joint.

rafter—one of a series of structural members of a roof designed to support roof loads. The rafters of a flat roof are sometimes called *roof joists.*

raw source energy—amount of nuclear fuel, oil, gas, or coal input at a power plant to produce electrical energy output; approximately three times the electrical output.

rebar—steel tensile reinforcements placed in concrete footings and walls.

reflectance—a property of a material indicating the percentage of light that is reflected when a certain amount of light strikes the surface of the material. The remainder that is not reflected is either absorbed by the material or transmitted through it.

repointing—the process of chipping out old, dry mortar between masonry units and then reapplying cement grout.

return pipe—the pipe going to the bottom of the solar collector, carrying cold water from the bottom of the storage tank to be heated.

ridge—the horizontal line at the junction of the top edges of two sloping roof surfaces.

ridge board—the board placed on edge at the ridge of the roof into which the upper ends of the rafters are fastened.

roll roofing—roofing material, composed of fiber and saturated with asphalt, that is supplied in rolls containing 108 square feet in 36-inch widths. It is generally furnished in weights of 45 to 90 pounds per roll.

roof sheathing—the boards or sheet material fastened to the roof rafters on which the shingle or other roof covering is laid.

saddle—two sloping surfaces meeting in a horizontal ridge; used between the back side of a chimney and other vertical surfaces.

salt-box house—a style generally associated with early New England Cape Cod houses, where the ridge line was offset towards the front of the house.

sash—a single light frame containing one or more lights of glass.

scuttle—a hatch or lid that can be raised or lowered for access from above.

sealer—a finishing material, either clear or pigmented, that is usually applied directly over uncoated wood for the purpose of sealing the surface.

shake—a thick, handsplit wood shingle, usually edge grained.

sheathing—the structural covering, usually wood boards or plywood, used over studs or rafters of a structure. Structural building board is normally used only as wall sheathing.

shingles—roof covering of asphalt, asbestos, wood, tile, slate, or other material cut to stock lengths, widths, and thicknesses.

shiplap—see *lumber, shiplap.*

short-wave radiation—the radiation given off by a high-temperature object like the sun.

siding—the finish covering of the outside wall of a frame building, whether made of horizontal weatherboards, vertical boards with battens, shingles, or other material.

sill—the lowest member of the frame of a structure, resting on the foundation and supporting the floor joists or the uprights of the wall. The member forming the lower side of an opening, as a doorsill, windowsill, etc.

soffit—usually the underside covering of an overhanging cornice.

soil cover (groundcover)—(a) a light covering of plastic film, roll roofing, or similar material used over the soil in crawl spaces of buildings to minimize moisture permeation of the area; (b) vegetation substitute for grass.

soil stack—a general term for the vertical main of a system of soil, waste, or vent piping.

solar tempering—a design process of using windows and masonry elements within a house to partially solar heat the structure.

spackle—a powder that is usually mixed with water and used for joint treatment in gypsum-wallboard finish.

span—the distance between structural supports such as walls, columns, piers, beams, girders, and trusses.

square—a unit of measure (e.g., 100 square feet) usually applied to roofing material. Sidewall coverings are sometimes packed to cover 100 square feet and are sold on that basis.

stile—an upright framing member in a panel door.

storm sash or **storm window**—an extra window usually placed on the outside of an existing window as additional protection against cold weather.

story—a term used to indicate one level of a building or house.

stud—one of a series of slender, wood or metal, vertical structural members placed as supporting elements in walls and partitions. (Plural: studs or studding.)

subfloor—boards or plywood laid on joists over which a finish floor is to be laid.

sump pump—a pump used to remove water from basement areas.

swale—a trench or ditch that is sloped to one side as well as along its length.

t and g—an abbreviation for *tongue and groove*.

termite—an insect that superficially resembles an ant in size, general appearance, and habit of living in colonies; hence, frequently called a "white ant." Subterranean termites *do not* establish themselves in buildings by being carried in with the lumber but, instead, by entering from ground nests after the building has been constructed. If unmolested, they eat out the woodwork, leaving a shell of sound wood to conceal their activities; damage may proceed so far as to cause collapse of parts of a structure before discovery. There are about 56 species of termites known in the United States; but the two main species, classified from the manner in which they attack wood, are subterranean (ground-inhabiting) termites, the most common, and dry-wood termites, found almost exclusively along the extreme southern border of the U.S. and along the Gulf of Mexico.

termite shield—a shield, usually of noncorrodible metal, placed in or on a foundation wall or other mass of masonry or around pipes to prevent passage of termites.

thermal wall—a masonry wall used to store heat from the sun. This heat is stored at a time when there is usually more than enough (such as at noon), and then is given off again by the wall when it cools down after sunset.

toenailing—to drive a nail at a slant with the initial surface in order to permit it to penetrate into a second member.

total load—the combined load, including the live load and the weight of any structural materials.

transite—a cement-asbestos mixture that is formed into sheets, pipes, etc.

trim—the finish materials in a building, such as moldings, applied around openings (window trims, door trim) or at the floor and ceiling of rooms (baseboard, cornice, picture molding).

truss—a frame or jointed structure designed to act as a beam of long span, while each member is usually subjected to longitudinal stress only, either tension or compression.

vapor barrier—material used to retard the movement of water vapor into walls and prevent condensation in them. Applied separately over the warm side of exposed walls or as a part of batt or blanket insulation.

variance—a procedure that allows for a change in the building codes, under exceptional circumstances, whereby an individual can build something not normally permitted.

vent—anything that allows the flow of air, as an inlet or outlet.

ventilated air—outside air that is introduced into the house by mechanical means (e.g., a fan) or by opening a window.

water gauge—a term used to indicate pressure differences; or the ability of a fan to overcome resistance to air flow due to friction.

water-repellent preservative—a liquid designed to penetrate wood and impart water repellency and a moderate preservative protection. It is used for millwork, such as sash and frames, and is usually applied by dripping.

w.c.—an abbreviation for *water closet* or *toilet.*

weatherstrip—narrow or jamb-width sections of thin metal or other material to prevent infiltration of air and moisture around windows and doors.

wetted area—that area on the solar collector in contact with the circulating fluid on continuous-flow systems. The greater the number of flow pipes or channels in the absorber, the greater the wetted area.

window types—*Clerestory:* a window, usually fixed, that is placed vertically in a wall above one's line of vision to provide natural light. *Fixed Sash:* a window that cannot be opened. *Operating Sash:* a window that slides or pivots to open. *Skylight:* a clear or translucent panel set into a roof to provide natural light.

w.s.—an abbreviation for *weatherstripping.*

CONTRIBUTORS

CHRIS AHRENS

Chris Ahrens received a degree in Administrative Engineering from New York University in 1949 and, in 1972, an MA in Rural Community Planning and Development from Goddard College, Plainfield, Vermont.

Over the years, Chris, a professional engineer, has worked with many international and rural development agencies, including CARE, the Peace Corps, and the Foundation for Cooperative Housing. Housing for the less affluent is his specialty, and he has served as a rural housing specialist in Kentucky, West Virginia, and Virginia, as well as elsewhere. For some years now, he has been a housing specialist for the Community Services Administration (formerly Office of Economic Opportunity), New York City region, where he advises on energy, self-help housing, and economics.

He is, like many of the other contributors to this book, a member of the International Solar Energy Society.

RON ALWARD

On receiving his Bachelor's degree in Engineering Science from the University of Western Ontario, London, Ontario, in 1963, Ron Alward taught physics and mathematics in Ghana before returning to the same university for an MS in Engineering in 1968.

From that time on, Ron has been a research associate with the Brace Research Institute, MacDonald College of McGill University, Montreal, where he works on the development of solar and wind power systems for small-scale community operations. In this capacity, he has written over 25 technical pamphlets, papers, and articles on alternate energy development and was chosen by the Canadian International Development Agency and the Canadian Hunger Foundation in 1973 to edit *The Appropriate Technology Handbook*—a manual of ideas and designs on the development of small-scale, alternate energy resources for underdeveloped communities. Another one of his specialties is greenhouse design.

He is a member of the International Solar Energy Society.

EUGENE ECCLI

Eugene Eccli graduated from Stevens Institute of Technology in 1968 with a BS in Physics and has taken subsequent graduate work in that field. Working for a while as a maintenance engineer, in 1972 he joined the faculty of the State University of New York at New Paltz, where he served as coordinator of the Environmental Studies Program until June 1974. While at New Paltz, he and his students built many alternate energy systems, including a small solar house. He is currently a design engineer and private consultant for energy-conserving housing, as well as a lecturer and consultant in the Social Ecology Program, Goddard College, Vermont. He has served as guest lecturer at many colleges and at EXPO '74.

From 1971 to 1975, Eugene was an editor of *Alternative Sources of Energy* magazine; and also the technical editor of the books *Alternative Sources of Energy* (Seabury Press, 1974) and *Producing Your Own Power* (Rodale Press, 1974). He contributed an article to the latter on "Energy Conservation in Existing Housing," a subject that has occupied more and more of his time during the past 2 years. Early in 1975, his booklet *Save Energy: Save Money* was published by the Community Services Agency (formerly OEO), Washington, D.C.

Eugene is on the Board of Advisors, Energy Information Systems, Inc., New York City, and the Board of Directors, Ecology House Associates, New York. He is a member of the International Solar Energy Society and President of Integrative Design Associates, Arlington, Virginia.

DAVID L. HARTMAN

After receiving a BSE in Mechanical Engineering from the University of Michigan in 1972, David Hartman came soon thereafter to Dubin-Mindell-Bloome Associates, New York City, a firm specializing in solar energy design and consulting work. As a project manager, he has been responsible for the research and design on several projects, including the U.S. Home Corporation's "Resource Saving House," Grassy Brook Village Condominiums, and he has done feasibility studies on solar heating and energy conservation for NASA, Argonne National Laboratories, and the New York Botanical Gardens, among others.

He is a member of the Energy Subcommittee of the New York Scientists' Committee for Public Information, and has given a number of talks on energy conservation and solar energy potentials for buildings.

JEROME KERNER

Jerome Kerner, a 1958 graduate of Pratt Institute, New York City, with a Bachelor's degree in Architecture, has been a registered architect since 1964. For the last several years, he has addressed himself to the fundamentals of new kinds of building design for today's realities, while at the same time he has worked as a professional architect on many projects in the northeastern United States.

As a lecturer and teacher, Jerome has stressed the owner-builder tradition and the skills needed to create low-cost, low-energy shelter. For the past 2 years, he has been on the faculty of the State University of New York at New Paltz; and for a year, he has been coordinator of the Alternate Community Technology (formerly Environmental Studies) program there. Presently, he has served as a lecturer on social ecology at Goddard College in Vermont, where he shares with others his knowledge of how housing needs for tomorrow can be met by a renewed understanding of the building techniques of the past.

He is a member of the American Institute of Architects and the National Council of Architectural Registrants and is active in the Audubon Society.

WILLIAM K. LANGDON

During interludes of Bill Langdon's education, his experience ranged from urban planning work to rural homesteading. After graduating from the University of Michigan in 1973 with a master's degree in Architecture, he worked for a while as a carpenter and general building construction worker. While doing so, he studied rural community design and solar energy.

Now settled in Asheville, North Carolina, Bill is presently working for the city of Asheville with several professional architects on a project to improve the quality of the downtown area. He also teaches solar heating and energy-conserving home design at Asheville-Buncombe Technical Institute.

Bill is a contributor to *Alternative Sources of Energy* magazine for which he wrote an article on his research into energy-saving designs for windows.

CHRIS LOGAN

Chris Logan, a registered professional engineer, graduated from Lowell Technological Institute, Lowell, Massachusetts, in 1965 with a BS in Nuclear Engineering. He subsequently attended the University of New Hampshire for graduate studies in Mechanical Engineering (fluid dynamics). As a design engineer for the Portsmouth (N. H.) Naval Shipyard, he worked for several years on heat transfer and ventilation systems.

About 5 years ago, Chris began to study energy alternatives other than nuclear.

He later became a private engineering consultant to New Hampshire communities, as well as a teacher of solar and wind systems at the University of New Hampshire. In 1975, he and others founded the Energy Information Center, a nonprofit corporation of which he is Executive Director. The Center offers a broad range of energy-related services to industry, private and public agencies, and the consumer.

Chris is a contributor to *Alternative Sources of Energy* and is presently designing a solar heating system for the retrofitting of existing structures as well as for new construction.

STEVE RIDENOUR

After graduating with a Master of Science in Engineering degree from Purdue University in 1972, Steve Ridenour taught ecological engineering techniques and, for a year, established student programs as staff engineer, Student Ecological Community Project, State University of California at Santa Cruz. For the next two years, Steve worked as a research associate at the Institute of Energy Conversion, University of Delaware, Newark, Delaware, where he evaluated the performance of solar collectors and taught courses on alternate energy sources. He has also been a solar project engineer, analyzing solar collector performance, with Dubin-Mindell-Bloome Associates, New York City. Presently employed as a development engineer with AMETEK, Inc., he is working in the research and development of flat-plate solar collectors and solar heating and cooling systems. He has recently accepted a teaching position in the School of Mechanical Engineering Technology at Temple University in Philadelphia.

Steve is especially fascinated by solar water heating and has made it a specialty. He authored the chapter on this subject for *Producing Your Own Power* (Rodale Press, 1974) and is a contributor to *Alternative Sources of Energy* magazine.

ALEX WADE

Alex Wade lives in New York State. A registered professional architect, he graduated in 1958 from the University of Illinois School of Architecture. He has worked as a field engineer on low-cost, precast concrete building systems and as an architect with Baker and Blake Associates of New York City, where he has worked on innovative modular designs.

Alex has also done extensive research on low-cost post and beam framing for houses, including designs for contractors and the actual construction of several post and beam homes. He has now expanded his services to include consultation to owner-builders and the development of sets of plans on low-cost post and beam construction.

INDEX